Lecture Notes in Networks and Systems

Volume 253

The series "Lecture Notes in Networks and Systems" publishes the latest developments in Networks and Systems—quickly, informally and with high quality. Original research reported in proceedings and post-proceedings represents the core of LNNS.

Volumes published in LNNS embrace all aspects and subfields of, as well as new challenges in, Networks and Systems.

The series contains proceedings and edited volumes in systems and networks, spanning the areas of Cyber-Physical Systems, Autonomous Systems, Sensor Networks, Control Systems, Energy Systems, Automotive Systems, Biological Systems, Vehicular Networking and Connected Vehicles, Aerospace Systems, Automation, Manufacturing, Smart Grids, Nonlinear Systems, Power Systems, Robotics, Social Systems, Economic Systems and other. Of particular value to both the contributors and the readership are the short publication timeframe and the world-wide distribution and exposure which enable both a wide and rapid dissemination of research output.

The series covers the theory, applications, and perspectives on the state of the art and future developments relevant to systems and networks, decision making, control, complex processes and related areas, as embedded in the fields of interdisciplinary and applied sciences, engineering, computer science, physics, economics, social, and life sciences, as well as the paradigms and methodologies behind them.

Indexed by SCOPUS, INSPEC, WTI Frankfurt eG, zbMATH, SCImago.

All books published in the series are submitted for consideration in Web of Science.

More information about this series at http://www.springer.com/series/15179

Juan M. Corchado · Saber Trabelsi
Editors

Sustainable Smart Cities and Territories

Springer

Editors
Juan M. Corchado
Department of Computing Science
Universidad Salamanca
Salamanca, Spain

Saber Trabelsi
Texas A&M University at Qatar
Doha, Qatar

ISSN 2367-3370 ISSN 2367-3389 (electronic)
Lecture Notes in Networks and Systems
ISBN 978-3-030-78900-8 ISBN 978-3-030-78901-5 (eBook)
https://doi.org/10.1007/978-3-030-78901-5

This Springer imprint is published by the registered company Springer Nature Switzerland AG
The registered company address is: Gewerbestrasse 11, 6330 Cham, Switzerland

Preface

It is expected that the world's population will reach 9.7 billion in 2050. By then, two-thirds of the population will live in urban environments. Critical social and ecological challenges that cities will face include urban violence, inequality, discrimination, unemployment, poverty, unsustainable energy and water use, epidemics, pollution, environmental degradation, and increased risks of natural disasters. The concept of smart cities, which emerged in the early 2000s, attempts to provide solutions to these challenges by implementing information and communication technologies. The initial concept of smart cities focused on the modernization of megacities. However, most of the so-called smart cities are just cities with several 'smart' projects. The most promising trend is the creation of smart territories, defined as small hi-tech towns, districts or satellite towns near megacities. With the current availability of an enormous amount of data, the challenge is to identify intelligent and adaptive ways of combining the information to create valuable knowledge. Sensorization plays a fundamental role in the collection of data, which, once analyzed on IoT and smart city platforms, can be used to make multiple decisions regarding governance and resource consumption optimization.

The Sustainable Smart Cities and Territories International Conference (SSCt2021) is an open symposium that brings together researchers and developers from academia and industry to present and discuss the latest scientific and technical advances in the fields of smart cities and smart territories. It promotes an environment for discussion on how techniques, methods, and tools help system designers accomplish the transition from the current cities toward those we need in a changing world. The SSCt2021 program includes keynote abstracts, a main technical track, two workshops and a doctoral consortium. This year's technical program presents both high quality and diversity. Ninety-eight papers were submitted to the conference. After careful review, twenty-nine articles were selected containing the latest advances on IoT, blockchain, edge and fog computing, distributed programming, artificial intelligence, and many more. Submissions were received from thirteen countries (Qatar, Spain, India, Portugal, Colombia, Malaysia, Morocco, Costa Rica, Pakistan, France, Egypt, Iran, and Denmark).

The symposium is organized by the Texas A&M University at Qatar. We would like to thank all the contributing authors, the members of the local committee, scientific committee, organizing committee, and the sponsors (Texas A&M University of Qatar, AIR Institute, and the IoT Digital Innovation Hub). This year's symposium counts with the collaboration of more than a hundred people from twenty-five countries (Qatar, Spain, Japan, Portugal, Morocco, Malaysia, Turkey, Colombia, Ireland, China, Germany, Uruguay, UK, Indonesia, USA, France, India, Costa Rica, Sweden, México, Brasil, New Zealand, United Arab Emirates, Australia, and Iran). Thank you all for your hard work and dedication, which has been essential for the success of the SSCt2021.

April 2021

Juan Manuel Corchado
Saber Trabelsi

Organization

Local Committee Chair

Saber Trabelsi	Science Program, Texas A&M University, Qatar

Local Committee

Otman Aghmou	Msheireb Downtown Doha, Qatar
Roberto Di Pietro	Hamad Bin Khalifa University, Qatar
Nora Fetais	Qatar University, Qatar

Scientific Committee Chair

Juan Manuel Corchado	University of Salamanca, Spain

Workshop Chairs

Fernando de la Prieta	University of Salamanca, Spain
Kenji Matsui	Osaka Institute of Technology, Japan

Doctoral Consortium Chairs

Paulo Novais	Universidade do Minho, Portugal
Sara Rodríguez	University of Salamanca, Spain

Scientific Committee

Ben Ahmed Mohamed	Faculty of Sciences and Technologies Tangier, Morocco
Zulaiha Ali Othman	Universiti Kebangsaan, Malaysia
Cesar Analide	Universidade do Minho, Portugal

Reyhan Aydogan	Ozyegin University, Istanbul, Turkey
Carlos Jaime Barrios	Universidad Industrial del Santander, Colombia
Holger Billhardt	University Rey Juan Carlos, Spain
José Luis Calvo-Rolle	University of A Coruña, Spain
José Carpio Pinedo	Universidad Politécnica de Madrid, Spain
Carlos Carrascosa	Universidad Politecnica de Valencia, Spain
Eduardo Carrillo Zambrano	Universidad Autónoma de Bucaramanga, Colombia
Roberto Casado Vara	University of Salamanca, Spain
José-Luis Casteleiro-Roca	University of A Coruña, Spain
Luis Castillo	Autonomous University of Manizales, Colombia
Luis Fernando Castillo Ossa	Facultad de Ingenierías, Universidad de Caldas, Colombia
Roberto Centeno	Universidad Nacional de Educación a Distancia, UNED
Pablo Chamoso	University of Salamanca, Spain
Cesar Alberto Collazos	Universidad del Cauca, Colombia
Angelo Costa	University of Minho, Portugal
Faber Danilo	Universidad del Quindío, Colombia
Fernando de la Prieta	University of Salamanca, Spain
Kapal Dev	Cork Institute of Technology, Ireland
Yucheng Dong	University of Sichuan, China
Jürgen Dunkel	Hannover University of Applied Sciences and Arts, Germany
Nestor Duque	Universidad Nacional de Colombia, Colombia
Liliana Durón Figueroa	University of Salamanca, Spain
Ponciano Jorge Escamilla-Ambrosio	Facultad de Ingeniería, UdelaR, Uruguay
Bruno Fernandes	Universidade do Minho, Portugal
Luis-Alfonso Fernández-Serantes	University of A Coruña, Spain
Alberto Fernández	Universidad Rey Juan Carlos, Spain
Ângela Ferreira	Instituto Politécnico de Bragança, Portugal
Hamido Fujita	Iwate Prefectural University, Japan
David García Retuerta	University of Salamanca, Spain
Ana Belén Gil González	University of Salamanca, Spain
Adriana Giret	Universitat Politècnica de València, Spain
Alfonso González-Briones	Universidad Complutense de Madrid, Spain
Jian Guo Zhang	London South Bank University, UK
Yusuf Hendrawan	Universitas Brawijaya, Indonesia
Ramon Hermoso	University of Zaragoza, Spain
Guillermo Hernández	AIR Institute, Spain
Javier Hernández Fernández	Iberdrola - Iberdrola Innovation Middle East, Qatar
Elena Hernández Nieves	University of Salamanca, Spain

Francisco Herrera Viedma	University of Granada, Spain
Carlos A. Iglesias	Universidad Politécnica de Madrid, Spain
Gustavo Adolfo Isaza	Universidad de Caldas, Colombia
Jaume Jordan	Universitat Politècnica de València, Spain
Esteban Jove	University of A Coruña, Spain
Vicente Julian Inglada	Universitat Politècnica de València, Spain
Omar Khan	University of Maine, USA
Arun Kumar Sharma	Sagam University, India
Sandeep Langar	University of Texas at San Antonio, USA
Tianchen Li	Northwestern Polytechnical University, China
Marin Lujak	IMT Lille Douai, France
José Machado	Universidade do Minho, Portugal
Shalva Marjanishvili	Hinman Consulting Engineers, Inc, USA
Abdelhamid Mellouk	Université Paris-Est Créteil, France
Carlos Meza	Tecnológico de Costa Rica, Costa Rica
Yeray Mezquita Martín	University of Salamanca, Spain
Radu-Casian Mihailescu	Malmö University, Sweden
Pablo Monzon	Centro de Investigación en Computación-IPN, México
Leandro Nunes de Castro	Universidade Presbiteriana Mackenzie, Brazil
Sigeru Omatu	Hiroshima University, Japan
Sascha Ossowski	University Rey Juan Carlos, Spain
Sarita Pais	Auckland University of Technology, New Zealand
Javier Palanca	Universitat Politècnica de València, Spain
Dave Parry	Auckland University of Technology, New Zealand
María Eugenia Pérez Pons	University of Salamanca, Spain
Marta Plaza-Hernández	University of Salamanca, Spain
Javier Prieto	University of Salamanca, Spain
Héctor Quintián	University of A Coruña, Spain
Alberto Rivas	University of Salamanca, Spain
Milagros Rey	Naturgy, Spain
Mohd Saberi	United Arab Emirates University, United Arab Emirates
Cristobal Salas Alfredo	Universidad Veracruzana, México
Ali Selamat	UTM, Malaysia
Carlos Fernando da Silva Ramos	Polytechnic of Porto, Portugal
Richard Sinnott	University of Melbourne, Australia
Layth Sliman	Efrei-Paris, Villejuif, France
Andrei Tchernykh	Centro de Investigación CICESE, México
Mirwan Ushada	Universitas Gadjah Mada, Indonesia
Zita Vale	Polytechnic of Porto, Portugal
G. Kumar Venayagamoorthy	Clemson University, USA

SSCt2021 Sponsors

TEXAS A&M UNIVERSITY at QATAR

AIR INSTITUTE

IOT DIGITAL INNOVATION HUB

Abstracts of Keynotes

Communication Protocols and Infrastructure: From the Smart City to the Smart Territory

Eduardo Martínez-Gill

Smart Cities & Smart Territories, INDRA
emgil@minsait.com

Keynote Abstract

The concept of smart cities is nowadays widespread along the globe. We are all aware of the many smart city projects deployed in the last 5 or 10 years in the five continents. However, the original scope of the smart city projects has been shifting lately.

Current trends in the field of smart cities are diverging from the original focus on efficiency and safety through cameras and video analytics by focusing on two new large topics: the transfer of the smart city concept to a smart territory and the change of the service provision model toward a more sustainable management, alienated among others, to the SDGs.

Nowadays, it is becoming more and more complicated to be able to manage a city as a single entity that does not infer on neighboring cities both in the management of infrastructures, spaces and even talent, as in the provision of resources and the decision taking. That is why the smart territory approach is gaining weight on the classic smart city concept.

On the other hand, we have a second trend that is globally transforming the definition of smart city or smart territory projects, and this is sustainability. Sustainability can be understood as a broad concept, which refers to the environmental sustainability of projects, the focus on neutrality in emissions, in compliance with the SDGs, in the fight against climate change, in the efficient use of resources, in the circular economy. But we must also focus on the sustainability of the project itself, on making it maintainable and manageable in the medium and long term to build the transforming strategy of our city or territory on this cornerstone.

Finally, and every year with greater relevance, we have the use of data for decision making. It is not a new area, but it is perhaps the one with the greatest impact on the global management of cities and territories. A proper management of data through master platforms will allow a more efficient operation, better monitoring of results, and better optimization of the provision of services, helping cities and territories to enhance their performance and to do it following the path, mark by the SDGs.

To do so, this transformation must be aligned in a three phase's strategy:

- Change and transform: From within, based on the transformation of the territory current assets, with a focus on the digitization of operational models and processes and the interaction with the citizens.
- Revalue: Creating future growth by reinventing and dynamically adjusting the value proposition of the territory, its services and their provision and the improvement of citizens' well-being.
- Create the future: Through anticipation and intelligent planning actions that allow adapting products, channels of communication and interaction, resources, services, and infrastructures, offering a territory with all kinds of guarantees against any future challenge.

The Role of Responsible Artificial Intelligence (AI) in Achieving Desired Smart City Outcomes

Tan Yigitcanlar

Queensland University of Technology, Brisbane, Australia
tan.yigitcanlar@qut.edu.au

Keynote Abstract

Utilization of smart and innovative digital technologies has become a mainstream in the efforts of tackling urban crises—whether they are climatic, pandemic, natural disaster, or socioeconomic in nature. In recent years, particularly the advancements at the 'artificial intelligence' (AI) front—as one of the most prominent technologies of our time with significant impacts on our economy, society, environment, and governance—have resulted in invaluable opportunities for cities increasing their infrastructural efficiencies and predictive analytic capabilities, and hence to a degree improving the quality of living and sustainability in cities. Today, AI has become an integral part of the smart city structure that provides the required efficiencies and automation ability in the delivery of local infrastructures, services, and amenities.

Especially when coupled with other smart urban technologies, AI applications—e.g., expert systems, knowledge management systems, process automation systems, chatbots and virtual agents, predictive analytics and data visualization, identity analytics, cognitive robotics and autonomous systems, recommendation systems, intelligent digital assistants, speech analytics, cognitive security analytics and threat intelligence, and so on—provide new capabilities and directions for our cities, such as building the next-generation smart cities—i.e., 'artificially intelligent cities' of the future.

Nonetheless, as much as creating useful benefits—for instance generating operational infrastructure or service efficiencies—AI's impacts also pose substantial risks and disruptions for cities and citizens through the opaque decision-making processes and the privacy violations it causes—e.g., automating inequality, generating algorithmic bias due to bad or limited data and training, removing, or limiting human responsibility, and lacking adequate level of transparency and accountability.

The increasing concerns on the negative AI externalities, particularly in smart cities, proved the need for the development of more ethical AI systems. Subsequently, in recent years, the concept of 'responsible AI'—that ensures the ethical, transparent, and accountable use of AI applications in a manner consistent with user expectations, organizational values, and societal laws and norms—is coined. It is widely argued that such responsible approach—AI and urban technology in general—will help in maximizing the desired smart city outcomes and positive impacts for all, while minimizing the negative consequences.

This keynote presentation focuses on and further elaborates the role of responsible AI in achieving desired smart city outcomes. The structure of the presentation is as follows. First, the narrow and broad meanings of the popular smart city concept are introduced. Second, AI in the context of cities, particularly those are branding and promoting themselves as smart cities, is placed under the microscope of sustainable urban development and futures. Then, prospects and constraints of AI-driven smart city agendas are identified. This is followed by advocating the need for responsible urban innovation—hence responsible AI—to achieve desired smart city outcomes. Next, the examples on the best practice urban innovation cases, cases with negative consequences of urban innovation, success cases of AI adoption in the public sector, and failure cases of AI adoption in local governments are presented. Last, the key characteristics of responsible AI—i.e., explainable, ethical, trustworthy, and frugal—in the context of cities are discussed with application examples of responsible local government AI systems. The presentation is, then, concluded with some essential remarks on the critical importance of responsible AI systems and the directions for prospective research and development agendas.

Digital Transformation and Technology Trends For Smart Cities

Enrique Díaz-Plaza Sanz

Business Development of IBM, Energy Sector and Smart Cities
enrique.diaz-plaza@es.ibm.com

Keynote Abstract

The digital transformation is having a full impact on all areas of industry and society. Today, we find ourselves surrounded and imbued by terms and approaches that pivot around this concept. Although there are many definitions of digital transformation, we can say in a generic way that the process of transferring physical processes and activities to cyberspace is the result of digital transformation. Considering that, if we had to identify the three main pillars on which the digital transformation is based, without a doubt, we would identify these three fundamental facilitators: technological, social, and economic.

Technological is the catalysts capable of carrying out this revolution on a large scale. The emergence of new technologies and the improvement of many others have made society have a wide range of previously unimaginable possibilities and certainly not imaginable on a medium-term horizon. Within the social, it is necessary to emphasize that the digital transformation has introduced significant impacts on society, both positive and negative, since new products and services are causing a drastic change in the way of life. Finally, although technology is what enables the development of digital transformation, this one happens because there are strong economic and commercial motivations; motivations that can be grouped into areas of operational improvement and growth drivers, usually in this order because existing industries are primarily driven by operational efficiency, with the growth being a second priority over the first.

In addition, and although it may be transversal vectors to the previous areas, we could include two additional vectors: environmental and ethical. From an environmental point of view, no progress is sustained without considering its environmental component, thinking not only in the short term but in the long term. On the other hand, the introduction of disruptive technologies and the change that these are assuming on a day-to-day basis imply the need to consider ethical aspects that can significantly impact the development of the rest of the vectors.

As a system of systems and their complexity, cities and territories are not external to the revolution that digital transformation implies. Thus, cities and regions are not alien to these trends and are also taking advantage of new technologies and approaches. The main driver is developing new services and optimizing existing ones under the implementation of smart services. Undoubtedly, the different technological drivers are used in cities considering their overall holistic and integrative approach. But in addition, the adoption of technology is having a direct impact on both the social aspects and the economic well-being of society and its relationship with the different stakeholders, since the digital transformation, as well as collaboration and implementation techniques, is empowering citizens and relating them in an even greater way with the rest of the citizen stakeholders; this will not be a specific moment to be analyzed and managed, but with continuity over time.

Therefore, it is not an easy task for cities to combine the different drivers in an integrated and coordinated way, considering aspects such as the characteristics and roadmap of the different technologies, the socioeconomic impact on the city, and the difficulty cities impose by its intrinsic complexity.

Contents

Edge and Fog Computing

Hybrid and Machine Learning Algorithms

Distributed Programming and Applications

Internet of Things

Wind Energy Assessment for Small Wind Turbines in Different Roof Shapes Based on CFD Simulations

Carlos Oliveira[1], Adelaide Cerveira[2], and José Baptista[1,3](✉)

[1] Department of Engineering, University of Trás-os-Montes and Alto Douro, Vila Real, Portugal
baptista@utad.pt
[2] Department of Mathematics, University of Trás-os-Montes and Alto Douro, LIAAD-INESCTEC UTAD Pole, Vila Real, Portugal
[3] CPES-INESCTEC UTAD Pole, Vila Real, Portugal

Abstract. With a still high rate of use of energy from non-renewable sources, it is crucial that new energy generation solutions are adopted to reach greenhouse gas reduction targets. The integration of renewable energy sources in buildings is an interesting solution that allows reducing the need for energy from the power grid, contributing to a significant increase in the energy efficiency of buildings. The main aim of this paper is to evaluate the impact that the aerodynamics of the buildings in particular the roof shape has considering the integration of wind energy systems. The results of Computational Fluid Dynamics (CFD) simulations are presented in order to identify the effect of the two roof shapes on energy production by wind turbines (WT). For this purpose, the factor matrices (FM) that gives information about the wind profile around the building taking into account the building's roof profile were calculated. Comparing the results for the wind flow obtained by the FM and the CFD simulations for the flat and gabled roofs, similarities are observed for them, allowing to conclude that the CFD analysis results in a methodology with great accuracy for the aerodynamic study of buildings roof shape.

Keywords: Renewable energies · Wind turbine · Roof shape · Wind power utilization · Building aerodynamics · Factor matrices

1 Introduction

Population density growth in urban areas leads to the search for sustainability in energy consumption in buildings [1–3]. Therefore, it is necessary to give priority to energy efficiency, as well as to consider the integration of renewable energy sources in buildings to achieve the EU goals [4]. Local energy generation has several advantages in terms of not only environmental, but also technological and economic, such as reducing energy losses in the public electricity grid, increasing the autonomy and decision-making power of individual consumers, and contributing to the decrease in the energy import balance for the country in question [5, 6]. The strategy for integrating wind systems in an urban environment must include maximizing renewable energy generation in cities, together

J. M. Corchado and S. Trabelsi (Eds.): SSCTIC 2021, LNNS 253, pp. 3–13, 2022.
https://doi.org/10.1007/978-3-030-78901-5_1

with the aiming to minimize negative impacts on health and the local environment to attend to the essential energy needs of all those who live and work in these areas [7]. This distributed power production in urbanizations is a notable contribution to the sustainable design of new buildings relatively of energy consumption, produced directly where the demand for power is sought [8]. In addition to this growing interest in installing local generation within cities, small WT also gains a focus of interest in understanding their potential in the urban environment [9]. The local electricity generation in buildings by WT involves many different challenges compared to stand-alone wind energy systems and wind farms. It is known that the wind profile in an urban environment is very complex and the adaptability of WT to that environment is not known well enough, both in terms of productivity and compatibility with the structure of buildings. This is due to the high terrain roughness length and the presence of obstacles with different shapes strongly influences the airstream for wind power use. However, disturbed flows around buildings can locally increase wind speed and energy yield can be increased compared to open locations. As the energy efficiency of WT has a cubic relationship with the speed of the wind, the increase in wind speed because of the surrounding buildings can make the turbines favorable to the wind [7, 9].

Small wind turbines integrated into buildings are one of the low-cost renewable energy sources [10]. Despite its potential, Ledo et al. [11] pointed out that the reasons for the limited installation of wind microturbines in urban areas are the low average wind speed, high levels of turbulence and relatively high aerodynamic noise levels generated by the turbines. Blackmore [12, 13] stated that if a turbine is installed in the wrong place on a building roof, it is possible that the power will decrease to zero for significant periods of time, even when the wind is at a speed relatively favorable to energy production wind power. The effort put into optimizing the flow simulation models for different species should be valued. The potential of a CFD-based project practice that has wide application in the development of the most varied processes and equipment results in a huge reduction in the time and costs spent, in comparison with conventional practices [14].

This paper analyses and evaluates the energetic suitability of wind turbines, installed in different rooftops, flat and gable roof, in buildings with the same configuration and dimensions. It is important to evaluate numerically the influence on the airstream from buildings and rooftops and to model wind flows over buildings and their roofs to better analyze, locate and design WT based on local wind meteorological data and local urban terrain characteristics for assessing and improving local urban wind power utilization. The paper aims to present FM, based on local meteorological data and characteristics, where is possible to investigate the wind aerodynamics and wind flows over the rooftops, concluding the optimal roof shape and its orientation, as well as the results of CFD simulations to identify the effect of the two roof shapes on the wind flow, in order to compare the two methods and evaluate the results.

This paper is organized as follow: Sect. 2 addresses and characterizes the studied scenarios and refers to the data processing, Sect. 3 describes the scenarios for the CFD simulations and briefly how they were built, Sect. 4 presents the results of the case studies about two roof formats, and finally, Sect. 5 draws the main conclusions.

2 Material and Methods

This section briefly explains the methods used to carry out the proposed experiment, identifying within two different roof shapes the potential that they may have in the wind resource.

The place of study for this work was in the Energy Research Park Lichtenegg located geographically in southern Austria, at an altitude of 800 m above sea level. It is known that the buildings for the two scenarios are not oriented to the north, in this particular case, they are 18° offset from north to west.

2.1 Test Site

The test site consists of 6 masts, each with a sonic anemometer installed to perform various tasks, which are the measurement of wind speed, wind direction, pressure, temperature, and air humidity. The measurements were collected according to the criteria and recommendations of the International Energy Agency (IEA) in accordance with the standards of IEC 61400-12-1 [15].

Throughout the work, real data on wind speed as well as the power of several small turbines were registered and analyzed to obtain and optimize their electrical characteristics [16].

2.2 Record and Data Normalization

To perform an accurate energetical assessment and evaluate the location and type of building for the installation of WT, meteorological data must be collected over a long period of time, with respect to wind speed, wind direction, temperature, humidity, and air pressure. For this work, the time interval analyzed was between April 9, 2019, from 01:00 am to August 9, 2019, at 00:59 am, and the information provided by the sensors was captured in 60-s time intervals.

With the data collection of atmospheric pressure and the local temperature, it is possible to calculate the air density at the sampling frequency of the data, which is 60 Hz, representing one sample per minute. The selected data sets should be normalized to the reference air densities. The air density at sea level, referring to the ISO standard atmosphere is 1,225 kg/m^3. After data collection and processing, if the average air density for the study site differs more than 0,05 kg/m^3 from the reference, it must be normalized. Air density can be determined from the measured air temperature and the air pressure according

$$\rho = \frac{B}{R_0 \times T} \tag{1}$$

where ρ is the air density given in kg/m^3, B is the air pressure, in Pa, R_0 is the gas constant of dry air and T is the air temperature in degrees Celsius.

In the present case study, the average air density of the site was below the reference values by more than 0,05 kg/m^3. Thus, there was a need to normalize the recorded wind speed, for a WT with active power control, according to Eq. (2).

$$u_n = u \times \left(\frac{\rho}{\rho_0}\right)^{\frac{1}{3}} \tag{2}$$

where u_n is the normalized wind speed, u is the measured wind speed and ρ_0 is the reference air density at sea level. After normalizing all wind speeds for each sensor, the Bins method is implemented.

3 CFD Simulations

In this section the scenarios for the CFD simulations are described. ANSYS Fluent is an engineering software designed to simulate the flow of fluids and, by combining advanced mathematical models with the ability to configure complex geometries, it has capabilities that allow it to simulate a wide range of physical phenomena [17, 18].

3.1 Geometry Scenarios

To investigate the wind conditions between two different roof shapes of two buildings in terms of the effect of the roof shape in the wind flows over the building roof, two scenarios are chosen.

Scenario A and Scenario B are to find out the effect in the wind flow over a building with a flat roof and a gable roof, respectively, for an initial wind speed of 5 m/s at the reference height, 7 m. In each scenario, two cases are simulated: the first case is for the north as the orientation of the wind and the second case is for the east. In this way, we can analyze the effect of the wind on the length and width of the building roof. The buildings are both the same size, 4 m long and 2,5 m wide. The building with the flat roof is 3,5 m high and the one with the gable roof is 4,8 m high.

3.2 Tridimensional Model

When it is necessary to know in detail the configuration and the point-to-point properties of a given flow, small scale analysis is used. This technique is supported by the application of conservation laws to an infinitesimal volume of the fluid or to an elementary system. Together with the necessary thermodynamic relationships and the adoption of appropriate boundary conditions, this set of equations potentially solves any problem in the field of fluid flow.

4 Results and Discussion

This section presents the results measured and simulated using Matlab software for the wind speed and direction as a function of the roof shape as well as CFD simulations performed in Ansys Fluent. As result, it is possible to analyze the influence of the roof shape on the wind flow profile.

4.1 Wind Roses

According to the wind roses obtained, Fig. 1, the predominant wind direction in the place of implementation of the WT is from north to south, obtaining records of peak wind speeds from 10 m/s to 25 m/s. Bearing in mind the positioning of the building, it is known that both the northern and southern parts of the building are more exposed than its sides, at higher recorded wind speeds.

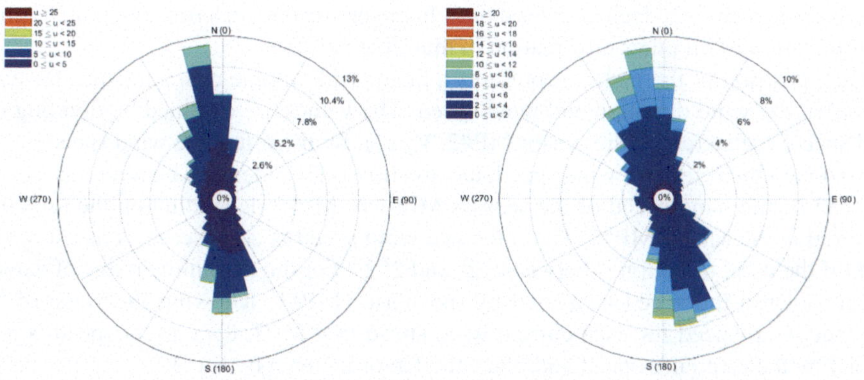

(a) (b)

Fig. 1. Wind roses for (a) flat roof, and (b) gable roof [16].

4.2 Factor Matrix

The FM gives information about the wind profile around the building considering the building's roof profile.

In a first stage, without any type of built structure, 4 masts were implemented with a sensor each, MP#1, MP#2, MP#3, MP#4, according to the layout shown in Fig. 2, to measure various characteristics of the site such as wind direction, wind speed, pressure, temperature and air humidity. The sensor MP#2 is strategically positioned exactly in the center of the building, which will later be replaced by a WT, allowing to obtain data to build the FM. At this stage, it is necessary to calculate the reference speed as function of the wind direction at the WT location.

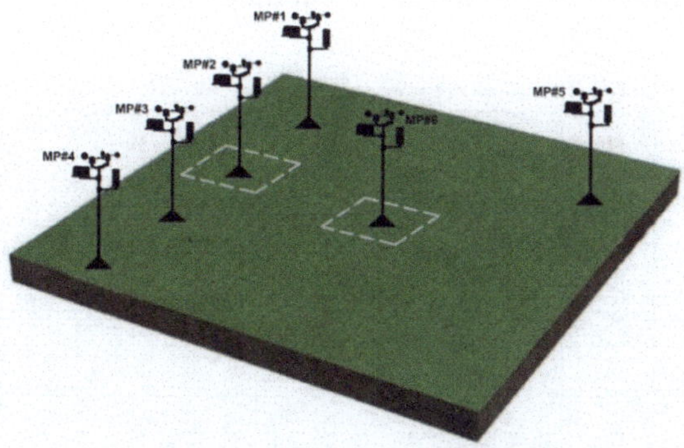

Fig. 2. Layout of the test site [16].

Due to the lack of uniformity of the data obtained by the different types of sensors, it was necessary to resort and filtering some data. Depending on the type of sensor, the data filtering was mainly due to the lack of information on time intervals of wind speed and wind direction. The lack of at least one characteristic data in a time interval, implied the elimination of all other information in that time period.

Each position of the matrix represents a factor with unique characteristics for each interval of the wind direction and wind speed. The values are obtained by dividing the wind speed, registered by the sensor MP#2, V_{actual}, by the reference wind speed.

To obtain the reference wind speed, its direction is considered. The reference speed, V_{ref}, can be the wind speed of the sensors MP#1 or MP#3 depending on the direction measured in the sensor MP#2. The reference wind speed is considered as measured by MP#3 if the wind direction is between 72° and 252°. For the remaining range of values, it is considered the speed measured by the sensor MP#1. Knowing the value of the reference wind speed for each current wind speed in # MP2, the current speed is then divided by the reference speed and, thus, the factor is given by Eq. 3.

$$Factor = \frac{V_{actual}}{V_{ref}} \tag{3}$$

Each factor is inserted in its correct position in the matrix and as the recursion of the cycle is completed, more factors are added to the matrix and increased when the intervals are the same. Later, to finalize the matrix, each matrix position is averaged through a counter matrix created as the factors are inserted in the FM.

To complete the factor matrix, a filter was applied to factor values where the registration rate was less than 10 samples, with the percentage of the associated factor replaced by 100%, that is, a neutral value that would have no influence on the final calculation of the power curve.

Figure 3 and Fig. 4 shows the FM for the two roof formats, flat and gable, respectively. These matrices have a color gradient, from blue to red, associated with the percentages obtained per cell. Cold colors are associated with low percentages and warm colors with high percentages, so the higher the value of the factor, the hotter the background color of the cell will be, which means an increase in the wind flow speed for that speed and direction to the location of the turbine in the building.

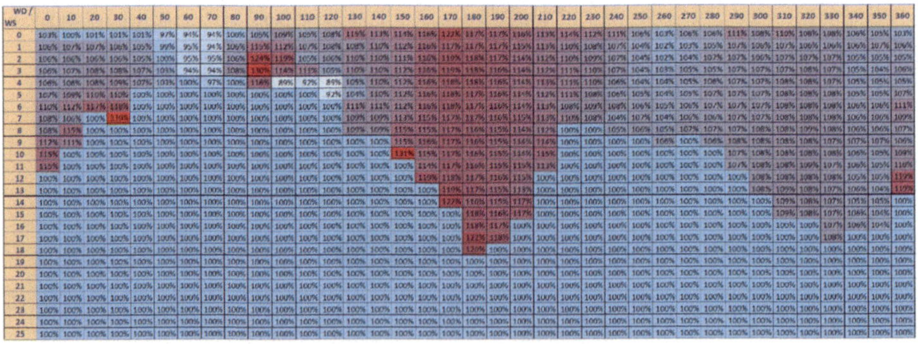

Fig. 3. Factor matrix for flat roof.

According to the wind rose, in Fig. 1, the predominant wind direction at the place where the turbine is installed is from north to south, with peak velocity records between

WD / WS	0	10	20	30	40	50	60	70	80	90	100	110	120	130	140	150	160	170	180	190	200	210	220	230	240	250	260	270	280	290	300	310	320	330	340	350	360
0	100%	106%	99%	98%	98%	94%	97%	93%	100%	107%	107%	105%	104%	109%	109%	105%	110%	123%	117%	119%	121%	121%	120%	107%	108%	108%	110%	114%	114%	114%	111%	108%	102%	105%	99%	101%	103%
1	101%	102%	102%	103%	103%	101%	92%	92%	104%	128%	118%	105%	107%	108%	107%	106%	108%	115%	115%	116%	119%	117%	109%	111%	109%	105%	105%	105%	108%	108%	104%	106%	104%	101%	98%	100%	101%
2	100%	102%	101%	101%	102%	100%	94%	93%	104%	127%	115%	108%	107%	106%	106%	106%	112%	114%	116%	116%	115%	114%	112%	109%	107%	105%	104%	106%	106%	106%	104%	103%	101%	100%	97%	99%	100%
3	101%	101%	102%	102%	102%	100%	94%	95%	107%	128%	116%	105%	105%	105%	106%	108%	112%	115%	117%	116%	115%	114%	115%	110%	109%	105%	105%	106%	108%	107%	105%	103%	102%	100%	97%	99%	100%
4	101%	102%	103%	103%	103%	100%	94%	95%	109%	131%	119%	107%	107%	108%	107%	108%	111%	115%	116%	117%	116%	114%	115%	110%	109%	106%	107%	106%	107%	109%	105%	103%	102%	100%	98%	99%	100%
5	102%	102%	103%	103%	103%	102%	94%	101%	116%	[illegible]	125%	107%	108%	108%	109%	109%	112%	114%	116%	117%	117%	114%	116%	111%	109%	107%	106%	106%	109%	109%	106%	104%	102%	100%	98%	99%	100%
6	102%	103%	104%	104%	105%	100%	100%	100%	100%	136%	128%	112%	110%	110%	110%	109%	112%	119%	117%	117%	117%	116%	100%	100%	108%	100%	104%	110%	109%	108%	107%	105%	103%	101%	98%	99%	101%
7	102%	104%	104%	105%	105%	100%	100%	100%	100%	100%	100%	100%	112%	110%	111%	110%	113%	116%	117%	117%	115%	116%	100%	100%	100%	100%	100%	100%	100%	109%	108%	108%	103%	101%	98%	100%	101%
8	103%	103%	105%	100%	100%	100%	100%	100%	100%	100%	100%	100%	100%	110%	112%	113%	115%	116%	117%	117%	120%	100%	100%	100%	100%	100%	100%	100%	100%	100%	107%	106%	103%	100%	98%	100%	101%
9	103%	100%	100%	100%	100%	100%	100%	100%	100%	100%	100%	100%	100%	100%	100%	115%	115%	117%	117%	117%	119%	114%	100%	100%	100%	100%	100%	100%	100%	100%	108%	106%	103%	100%	99%	100%	102%
10	103%	100%	100%	100%	100%	100%	100%	100%	100%	100%	100%	100%	100%	100%	100%	100%	117%	117%	117%	119%	122%	100%	100%	100%	100%	100%	100%	100%	100%	100%	100%	100%	103%	101%	98%	99%	102%
11	100%	100%	100%	100%	100%	100%	100%	100%	100%	100%	100%	100%	100%	100%	100%	100%	100%	117%	119%	118%	120%	100%	100%	100%	100%	100%	100%	100%	100%	100%	100%	100%	100%	101%	98%	99%	100%
12	100%	100%	100%	100%	100%	100%	100%	100%	100%	100%	100%	100%	100%	100%	100%	100%	100%	119%	118%	124%	100%	100%	100%	100%	100%	100%	100%	100%	100%	100%	100%	100%	100%	101%	98%	99%	100%
13	100%	100%	100%	100%	100%	100%	100%	100%	100%	100%	100%	100%	100%	100%	100%	100%	100%	120%	120%	118%	100%	100%	100%	100%	100%	100%	100%	100%	100%	100%	100%	100%	100%	100%	100%	100%	100%
14	100%	100%	100%	100%	100%	100%	100%	100%	100%	100%	100%	100%	100%	100%	100%	100%	100%	100%	100%	100%	100%	100%	100%	100%	100%	100%	100%	100%	100%	100%	100%	100%	100%	100%	100%	100%	100%
15	100%	100%	100%	100%	100%	100%	100%	100%	100%	100%	100%	100%	100%	100%	100%	100%	100%	100%	100%	100%	100%	100%	100%	100%	100%	100%	100%	100%	100%	100%	100%	100%	100%	100%	100%	100%	100%
16	100%	100%	100%	100%	100%	100%	100%	100%	100%	100%	100%	100%	100%	100%	100%	100%	100%	100%	100%	100%	100%	100%	100%	100%	100%	100%	100%	100%	100%	100%	100%	100%	100%	100%	100%	100%	100%
17	100%	100%	100%	100%	100%	100%	100%	100%	100%	100%	100%	100%	100%	100%	100%	100%	100%	100%	100%	100%	100%	100%	100%	100%	100%	100%	100%	100%	100%	100%	100%	100%	100%	100%	100%	100%	100%
18	100%	100%	100%	100%	100%	100%	100%	100%	100%	100%	100%	100%	100%	100%	100%	100%	100%	100%	100%	100%	100%	100%	100%	100%	100%	100%	100%	100%	100%	100%	100%	100%	100%	100%	100%	100%	100%
19	100%	100%	100%	100%	100%	100%	100%	100%	100%	100%	100%	100%	100%	100%	100%	100%	100%	100%	100%	100%	100%	100%	100%	100%	100%	100%	100%	100%	100%	100%	100%	100%	100%	100%	100%	100%	100%
20	100%	100%	100%	100%	100%	100%	100%	100%	100%	100%	100%	100%	100%	100%	100%	100%	100%	100%	100%	100%	100%	100%	100%	100%	100%	100%	100%	100%	100%	100%	100%	100%	100%	100%	100%	100%	100%
21	100%	100%	100%	100%	100%	100%	100%	100%	100%	100%	100%	100%	100%	100%	100%	100%	100%	100%	100%	100%	100%	100%	100%	100%	100%	100%	100%	100%	100%	100%	100%	100%	100%	100%	100%	100%	100%
22	100%	100%	100%	100%	100%	100%	100%	100%	100%	100%	100%	100%	100%	100%	100%	100%	100%	100%	100%	100%	100%	100%	100%	100%	100%	100%	100%	100%	100%	100%	100%	100%	100%	100%	100%	100%	100%
23	100%	100%	100%	100%	100%	100%	100%	100%	100%	100%	100%	100%	100%	100%	100%	100%	100%	100%	100%	100%	100%	100%	100%	100%	100%	100%	100%	100%	100%	100%	100%	100%	100%	100%	100%	100%	100%
24	100%	100%	100%	100%	100%	100%	100%	100%	100%	100%	100%	100%	100%	100%	100%	100%	100%	100%	100%	100%	100%	100%	100%	100%	100%	100%	100%	100%	100%	100%	100%	100%	100%	100%	100%	100%	100%
25	100%	100%	100%	100%	100%	100%	100%	100%	100%	100%	100%	100%	100%	100%	100%	100%	100%	100%	100%	100%	100%	100%	100%	100%	100%	100%	100%	100%	100%	100%	100%	100%	100%	100%	100%	100%	100%

Fig. 4. Factor matrix for gable roof.

10 m/s and 20 m/s being obtained. Bearing in mind the positioning of the building, it is known that both the northern and southern parts of the building are more exposed than its sides, at higher recorded wind speeds. This statement can then be corroborated by analyzing the FM of Fig. 4, the first column represents the wind speed in m/s and the first row the wind direction with a range of 360°. The range of directions with the highest associated percentages is contained between 140° and 220°, thus confirming a beneficial influence on the wind flow in the south wall of the building.

Regarding the results obtained in the FM of Fig. 4 for the gable roof, the principle of analysis is the same adopted for the flat roof. Through Fig. 1, it is observed that the predominant wind direction is practically similar to the wind rose of Fig. 1, with the exception of a small difference in the speed values, due to the building structure itself, since the gable roof favors the flow of wind in a different way from the flat roof. Analyzing the FM of the gable roof, in Fig. 4, this statement can be verified analyzing the range of directions from 170° to 220°, thus confirming itself due to the high percentage values that the south zone of the building benefits the wind flow.

4.3 CFD Results

This topic presents the computational results of fluids performed for the scenarios already described. Each scenario is presented in two cases, and for each case, two orientations from the wind to the building are considered, North and East. Through the analysis of these two orientations of the wind, it is possible to analyze the flow behavior for the front and later walls of the building with the flat roof and for the gable roof.

Scenario A, flat roof, is shown in Fig. 5 and Fig. 6. In these figures the planes of the wind flow to the flat roof are presented. For the Fig. 5.a, 6.a, 7a and 8a the horizontal planes of the wind flow are presented, 7 m above the ground. For the Fig. 5.b, 6.b, 7.b and 8.b the vertical planes of the wind flow are shown, centered on the building.

For the Fig. 5.a, considering that the wind orientation is taken as North, an increase of about 36% is observed in the South zone of the building for the red zones and in the central zone of the increase between 7% and 22%.

Through the Fig. 5.b it is possible to clearly see that as the wind flow moves away from the roof, the wind speed increases, so for areas less than 2 m away from the roof, the wind speed is low and for areas more than 3 m away from the roof the wind speed increases considerably and as it moves further it decreases.

Figure 6 shows the horizontal and vertical planes related to the second case of scenario A, with the objective of observing the flow of the wind from the east side to the west side of the building. There is a very similar behavior of the wind profile for the outflow of the wind on the North wall, however, there is an even greater increase in the wind speed as it passes through the roof, mainly towards the Northwest corner.

The behavior of the wind flow, for both cases 1 and 2, is quite identical as can be seen by comparing the vertical planes of Fig. 5.b and Fig. 6.b. The differences in flow on the roof are visible by comparing Fig. 5.a and Fig. 6.a, where it can be seen that in the

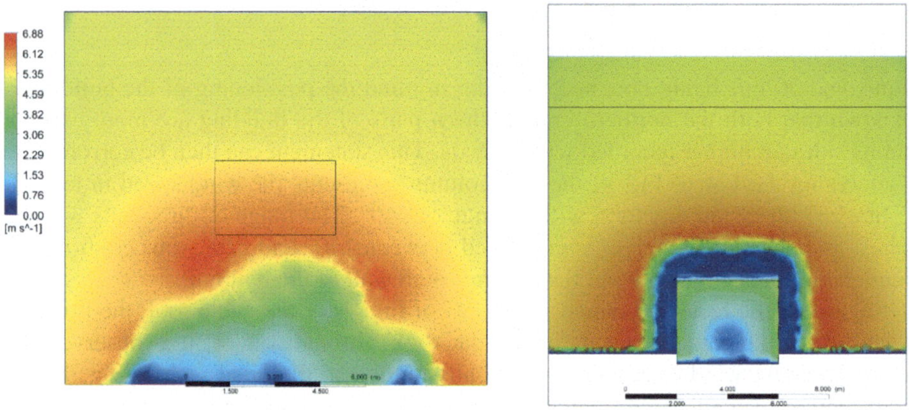

(a) (b)

Fig. 5. Planes of wind flows for Scenario A and Case 1: (a) top view, and (b) frontal view.

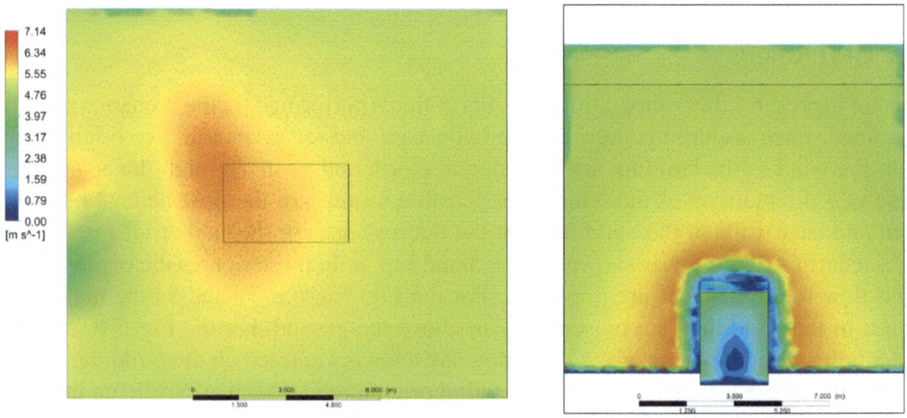

(a) (b)

Fig. 6 Planes of wind flows for scenario A: Case 2: (a) top view, and (b) side view.

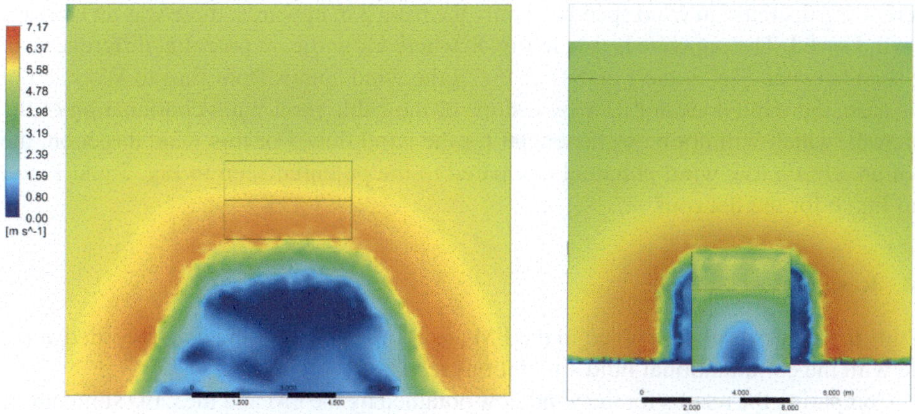

(a) (b)

Fig. 7. Planes of wind flows for Scenario B: Case 2: (a) top view, and (b) frontal view.

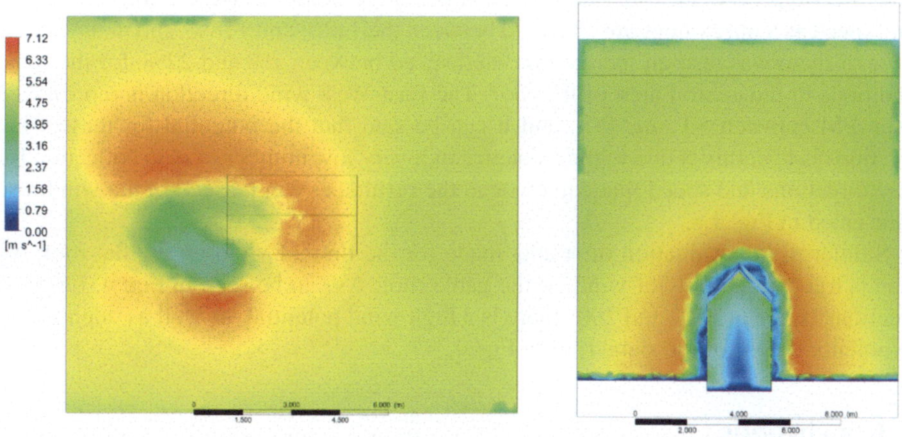

(a) (b)

Fig. 8. Planes of wind flows for Scenario B: Case 2: (a) top view, and (b) frontal view.

central area of the roof, in Fig. 6.a, there is an even greater increase than that observed in Fig. 5.a, an increase between 11% and 27%.

In Fig. 7 and Fig. 8 are presented the computational results related to the wind flow for the gable roof. The plane in Fig. 7.a shows a wind flow profile similar to that presented in Fig. 5.a. Where the greatest potential is observed, in relation to the increase in wind speed, is in the south slope of the roof, with an increase between 12% and 43%. The main difference in wind flow for the two roofs, for the wind direction as North, can be seen in Fig. 7.b, where it can be said that for this roof shape, after the wind has passed through the north slope of the same, the wind speed increases to a height of about 4 m above the peak of the roof, while on the flat roof up to a height of 2 m above the roof there was a decrease in wind speed and only the from 3 m upwards, there was an increase in wind speed. The opposite is true in Fig. 8, where a significant potential difference can be seen between Fig. 7 and Fig. 8. For Fig. 8 the wind comes from East to West and in this case, the wind does not flow by a slope of the gable roof, but a continuation of the sidewall, which will not be so beneficial for the wind flow. For this wind direction, the roof area has a low wind potential compared to the potential seen in Fig. 7 and Fig. 5 and Fig. 6 for the flat roof.

4.4 Results Assessment

In this topic, the results obtained in the FM are compared using the data recorded at the site with the computational fluid simulations results.

Comparing the results for the wind flow obtained by the FM and the CFD simulations for the flat and gabled roofs, similarities are observed for them. Ascertaining the results for the flat roof, it is known that both the North and South parts of the roof and the building are the same, so it is possible to predict an identical wind flow for either of these two wind directions. In this way, comparing the values present in Fig. 3, for the wind directions between 150° and 200°, that is, for the South wind direction, and for the wind speed as 5 m/s, values are observed between the 112% and 118%, and in the graphs in Fig. 5 there was also an increase in wind speed between 7% and 22%, for the same conditions in the central area of the roof. The East-West wind direction is represented in the FM between 50° and 90°, and it can be said that the potential for these wind directions is low, this is due to the existence of a very low number of data collection for these directions and speed that can mislead the results, as can be seen in the wind rose represented in Fig. 1.a.

Similar to the evaluation of results made for the flow of wind on the flat roof, the same is true for the flow of wind on the gable roof. As can be seen in Fig. 4, for wind directions between 150° and 200° there is a high wind potential, as well as, identical to the potential seen for the same roof in Fig. 7.

5 Conclusions

The research carried out in this paper shows the high relevance of the FM and CFD simulations to evaluate the wind profile in a site. The evaluation of the wind potential through the analysis of the flow of any roof of a building has advantages for the future since a database is being created as investigations of this kind are carried out. In this way, the bet on wind energy for urbanization can increase.

It was verified that through the results obtained by the two methods used, FM and CFD simulations, with the objective of evaluating the wind flow in a building with a flat roof and in another with a gable roof, both the flat roof and the gable roof for the predominant wind directions, North-South and South-North in the central area of the building's roof showed a potential increase in wind speed between 11% to 27%, and 12% to 43%, respectively.

It was concluded by comparing the results obtained by the two methods a coherence between them, and, thus, it can be said that CFD simulations provide reliable results.

Allowing to conclude that the CFD analysis results in a methodology with great accuracy for the aerodynamic study of the building's roof shape, having the advantage of not needing a complex sensors network and instrumentation in the field.

References

1. York, R., et al.: STIRPAT, IPAT and ImPACT: analytic tools for unpacking the driving forces of environmental impacts. Ecol. Econ. **46**(3), 351–365 (2003)
2. Jones, D.W.: How urbanization affects energy-use in developing countries. Energy Policy **19**(7), 621–630 (1991)
3. Al-Mulali, U., et al.: Exploring the relationship between urbanization, energy consumption, and CO_2 emission in MENA countries. Renew. Sustain. Energy Rev. **23**, 107–112 (2013)
4. European Parliament and of the Council: Directive (EU) 2018/844 of 30 May 2018 (2018)
5. Walker, S.L.: Building mounted wind turbines and their suitability for the urban scale-a review of methods of estimating urban wind resource. Energy Build. **43**(8), 1852–1862 (2011)
6. Chicco, G., Mancarella, P.: Distributed multi-generation: a comprehensive view. Renew. Sustain. Energy Rev. **13**(3), 535–551 (2009)
7. Lu, L., Ip, K.Y.: Investigation on the feasibility and enhancement methods of wind power utilization in high-rise buildings of Hong Kong. Renew. Sustain. Energy Rev. **13**(2), 450–461 (2009)
8. El-Khattam, W., Salama, M.M.A.: Distributed generation technologies, definitions and benefits. Electr. Power Syst. Res. **71**(2), 119–128 (2004)
9. Balduzzi, F., et al.: Feasibility analysis of a darrieus vertical-axis wind turbine installation in the rooftop of a building. Appl. Energy **97**, 921–929 (2012)
10. Stankovic, S., et al.: Urban Wind Energy. Routledge, Earthscan (2009)
11. Ledo, L., et al.: Roof mounting site analysis for micro-wind turbines. Renew. Energy. **36**(5), 1379–1391 (2011)
12. Blackmore, P.: Siting Micro-Wind Turbines on House Roofs. IHS BRE Press (2008)
13. Blackmore, P.: Building-Mounted Micro-Wind Turbines on High-Rise and Commercial Buildings. IHS BRE Press (2010)
14. Padoin, N., et al.: Heat and mass transfer modeling for multicomponent multiphase flow with CFD. Int. J. Heat Mass Transf. **73**, 239–249 (2014)
15. International Electrotechnical Commission: Wind energy generation systems - Part 12-1. In: Power Performance Measurements of Electricity Producing Wind Turbines (2017)
16. Oliveira, C., et al.: Wind energy exploitation in urban environment: the effect of roof shape. In: Second Edition of the International Conference on Innovative Applied Energy (2020)
17. Dutta, U., et al.: Ascent modes of jets and plumes in a stationary fluid of contrasting viscosity. Int. J. Multiph. Flow. **63**, 1–10 (2014)
18. Ansys, I.: Fluent Theory Guide version 14.5 (2012)

IoT-Based Human Fall Detection Solution Using Morlet Wavelet

Osvaldo Ribeiro[1,2], Luis Gomes[1,2], and Zita Vale[2](✉)

[1] GECAD—Research Group on Intelligent Engineering and Computing for Advanced Innovation and Development, Rua Dr. Antonio Bernardino de Almeida, 431, 4200-072 Porto, Portugal
{1180437,lufog}@isep.ipp.pt
[2] Polytechnic of Porto, Rua Dr. Antonio Bernardino de Almeida, 431, 4200-072 Porto, Portugal
zav@isep.ipp.pt

Abstract. Human fall detection is a problem that needs to be addressed to decrease the significant number of elderly people being affected, disabled, or even killed by falls. While the prevention of falls is a goal harder, or impossible, to be achieved, the fast detection and aid of people are two aspects that technological solutions can help with. With the support of internet of things devices, a fall detection solution for building deployment is proposed in this paper. The classification of fall is done using the Morlet wavelet in the fog-computing layer, enabling the detection of falls near the person and near the people who can provide first aid. The proposed solution of this paper was tested using a new dataset created using a human-body model. The results are promising, proving the efficiency of the proposed solution.

Keywords: Internet of Things · Fall detection · Wavelet transform

1 Introduction

"*A fall is an unexpected event in which the participant comes to rest on the ground, floor or lower without know loss of consciousness.*" (Archcare 2010). According to the World Health Organization (WHO) and the Prevention of Falls Network Europe (ProFaNE), a fall is an event that inadvertently takes a person to the ground or to a lower level, excluding intentional situations [1]. Falls occur in all age groups due to loss of balance or the inability to recover. However, it is in the older population that the prevalence of the risk of falling and the resulting damage is greater. Falls are the cause of significant morbidity or mortality, being one of the main causes of hospitalization. As a result, direct treatment costs and indirect costs, loss of productivity, and absenteeism increase, with an economic impact on families and society. Falls can also lead to states of dependence, loss of autonomy, confusion, immobilization, and depression, which lead to several restrictions on day-to-day activities [2].

Population aging is a reality worldwide, especially in cases such as the European continent where an older population prevails. People aged 65 or over, living in Europe, in 1991 represented 13.9% of the population and in 2011 they represented 17.5%. Eurostat

J. M. Corchado and S. Trabelsi (Eds.): SSCTIC 2021, LNNS 253, pp. 14–25, 2022.
https://doi.org/10.1007/978-3-030-78901-5_2

predicts that this age group will increase from a percentage of 19.2% in 2016, to around 30% in 2050 [3].

Extrapolating the statistics to the current context, and according to INE data [4], with a total of 2,244,225 elderly people in Portugal, during the current year 67,327 will be hospitalized due to a fall and of these, 4,040 will eventually die in the hospital. Falls have different causes and can be classified as [5]:

- Accidental falls: they happen due to external factors in people with no risk of falling, they are not predictable. For its prevention, measures include minimizing risks;
- Unpredictable physiological falls: they also occur in people with no risk of falling, but they happen for the first time due to physiological factors, such as seizures or pathological fractures. They represent about 8% of falls;
- Predictable physiological falls: this cause appears in individuals who have a potential risk of falling due to the existence of physiological changes. Although they constitute almost 80% of falls, they can be avoided with the application of the Morse Falls Scale (NDE) [6].

Fall detection systems must be studied, conceived, developed, and tested to achieve the market and start to minimize the impact of falls. These systems can efficiently detect falls and warn emergency contacts and people who can provide the needed support [7]. Near fall detection is also possible, but harder to achieve [8]. Although desirable, the complete prevention of falls, especially in elderly people, is a goal difficult to conquer.

Although there are currently products that detect falls and communicate them to information centers or emergency services, they all have vulnerabilities that lead to them being little used or even ignored. The options that depend on the user, such as wearable devices, suffer from limitations such as the person inadvertently actuating the device with an abrupt movement causing the detection of a false fall. The detection of falls through independent means of the user, in most cases, uses infrared cameras to monitor the behavior of users and detect fall events with computer vision techniques. However, the use of cameras causes constraints, since residents have the feeling of "being observed".

In the present years, we are seeing a massification of internet of things (IoT) devices in our daily life activities. Because these devices allow the sensing of environments and contexts, they can be used in fall detection systems [9]. One of the main IoT devices used in fall detection systems is wearable devices that accompany the person in their daily lives [10, 11].

This paper proposed an IoT-based solution for fall detection. The solution is to be deployed in the buildings and does not require any use or maintenance from the elderly. The proposed solution uses edge-, fog-, and cloud-computing layers for data acquisition, for fall classification using the continuous wavelet transform (CWT), and for the notification module. The CWT allowed a precise classification of falls, but it required a more capable processing hardware installed in the fog layer. Future work can study the application of soft sensing models to estimate the classification in the edge layer, reducing the communication flow and avoiding the use of fog layer.

The use of IoT devices allowed the continuous monitoring of the environment while providing the needed communication means to create a data stream from the IoT device to the cloud. The use of IoT devices and architecture brings benefits to detect human falls by being able to digitalize real contexts and provide multiple levels of processing, promoting distributed systems.

After this first introductory section, a review of fall detection systems is presented in Sect. 2. Section 3 presents the proposed solution and its architecture and components. The case study and results are shown in Sect. 4, and Sect. 5 presents the main conclusions of this work.

2 Fall Detection Systems

In the review published by Vallabh and Malekian (2018), the authors proposed a fall detection classifier model, comprising five steps, that is commonly used in fall detection monitoring systems [12]. Figure 1 shows the five steps commonly used in fall detection systems: data collection, feature selection, feature extraction, classifier, and evaluation.

Fig. 1. Classifier model for fall detection system [12].

According to Chaccour et al. (2017), a fall detection system can be defined in three groups: wearable system, non-wearable system, and fusion system [13]. One of the main used configurations is the wearable solution because it allows a continuity use of the device [14]. However, these solutions have some problems when built for elderly people, namely battery and usage issues. A fall detection system requires a significant battery usage that demands the user to periodically charge it. In elderly people, the charging, and the daily use of a device, can be an issue regarding its acceptance. Non-wearable systems can be deployed in inside and outside environments but are always restricted in their monitoring are. However, these systems are easier to maintain increases the freedom of elderly people. Some of these solutions use computer vision [15] and ambient sensors [16]. Fusion systems are described as being solutions that can fuse several data sources.

A wearable with a 3D accelerometer was used in the solution proposed in Yacchirema et al. (2018), where data were communicated to a big data server using the IPv6 over low power wireless personal area network (6LowPAN) [17]. Aziz et al. (2017) also used a 3D accelerometer but to feed a support vector machine (SVM) classifier [18]. The proposed solution was tested, and it was able to achieve an 80% accuracy. In Ajerla et al. (2019), a wearable device is used to detect falls in the edge-computing layer [19]. This work analysis the position of the sensor, at several heights in the user body, and the reading frequencies. The results are classified by a long short-term memory (LSTM) neural network, achieving the highest accuracy of 95.8%, using two wearable devices, one in the waist and one in the wrist.

In Miguel et al. (2017), a computer vision solution is proposed to feed a k-Nearest Neighbors (KNN) algorithm for classification [20]. The solution was tested indoors and tried to solve the current issue of elderly people often forget to wear wearable devices.

The results showed a 97.6% accuracy. A deep learning approach is proposed in Lu et al. (2019) where it was combined a convolutional neural network (CNN) with an LSTM neural to detect falls using kinematic data recorded in a video [21]. The proposed solution was tested with two datasets and it was able to achieve accuracies slightly above 90%.

A non-intrusive solution is proposed in Wang et al. (2017), by monitoring the channel state information of IEEE 802.11 signals the system is able to detect distortions in the environment [22]. The proposed model was used in a controllable environment and it was able to achieve a 90% detection precision.

Because of its relevance in society, it is common to find open data datasets created to allow the testing and validation of systems. A review of publicly available datasets can be seen in Casilari, Santoyo-Ramón, and Cano-García (2017) [23]. Two examples of datasets are UMAFall [24] and SisFall [25]. The UMAFall dataset was recorded using a smartphone in the right thigh pocket and four wearable devices located in the ankle, waist, right wrist, and chest. The SisFall dataset used two accelerometers and one gyroscope located in the user's waist. Part of the data was recorded by applying the wearable device in elderly people – below 75 years. Both datasets have records of different activities of daily living (ADL) and falls.

A fall detection solution is composed of hardware, to read the context, and by software algorithms, to analyze and identify/classify the data. The identification of falls can be done using several methods. In this paper, we will focus on the wavelet transform to enable the detection of falls. The wavelet transform allows the analyze of time series with nonstationary signals [26].

In Palmerino et al. (2015), a wavelet transform approach is used to analyze the impact moment [27]. The proposed model was tested using a dataset with 1170 acceleration signals and achieved an accuracy of 90%. In Yazar et al. (2012), it is proposed a solution where the wavelet transform is used to analyze vibration data [28]. This work was able to prove the benefits of wavelet transform against the Fourier transform and the mel-frequency cepstrum. SVM classifiers and Euclidean and Mahalanobis distances are used to identify the falls. To prevent false positives, the authors proposed the use of a passive infrared (PIR) sensor that can detect the radiant heat of human bodies.

3 Proposed Solution

Based on the fall detection classifier model identified by Vallabh and Malekian (2018), the proposed solution will use Arduino-based hardware modules for data collection. The Arduino microcontroller will be responsible for feature extraction and selection according to previously defined triggers that identify a distortion of monitoring values, detecting anomalies (i.e., values that go beyond the normality). The wavelet transform model, developed in a Raspberry Pi 4 Model B, will be used as a classifier. The evaluation will be made manually considering the accuracy of the model.

The use of computer vision can be an accurate solution for fall detection, but it implies an invasion of the person's life, opening the door to possible privacy attacks. For this reason, the proposed solution will avoid the use of sensitive private data. The solution proposes an IoT device with ambient sensors that can identify the context of a home division.

Figure 2 shows the proposed solution composed of three IoT layers: edge-, fog-, and cloud-computing. At the edge-computing layer, the proposed device will closely monitor the four sensors that were used: a sound senor, a Doppler sensor, and two accelerometers. The edge-computing proposed device is a microcontroller-based solution with trigger-based functionalities. In the proposed solution, an Arduino UNO is used in the edge layer. The current implementation uses the sound sensor to establish a trigger value. Readings above that value will trigger the microcontroller and reading data will be sent using two reading buffers: a buffer for past readings that will have the values read before the trigger, and a buffer for future readings that will be populated after the appearance of the triggering value. This way, it is possible to send to the fog-computing layer the sensor data read prior, during, and after of the trigger detection.

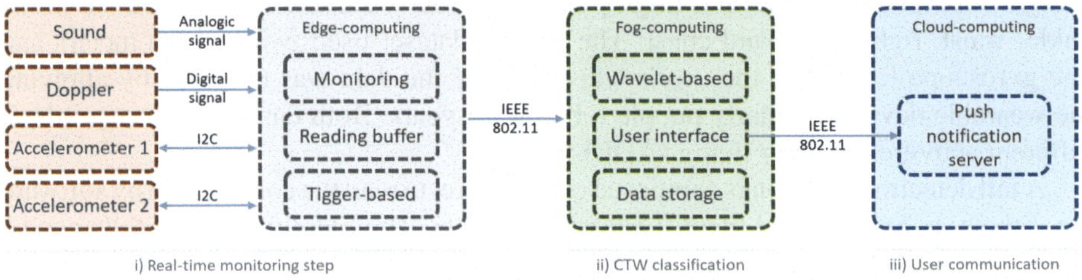

Fig. 2. Proposed solution overview

The proposed fog-computing layer used a single-board computer (SBC) to evaluate the wavelet transform and to identify if a fall had occurred. If so, the SBC will send a push notification request into the cloud to alert the responsible persons. In the proposed solution, a Raspberry Pi 4 Model B is used in the fog layer. The use of push notifications required a cloud-based server of a third-party, such as Firebase Cloud Messaging.

The interaction among edge, fog, and cloud layers defines the steps needed in the proposed fall detection system, these steps are sequential and are shown in Fig. 2. The edge layer is a continuous operating monitoring system that measures sensor data looking for triggering values. The edge layer is the only layer that has a continuous running routine, other layers are activated by the previous layer. The fog layer is activated if edge layer measures triggering values, at this moment the CTW classification is executed in the Raspberry Pi. If a positive fall classification is detected, the cloud layer is activated, as final step, to warn the users.

The sensors used in the proposed solution were:

- Analog sound sensor: they are normally used to detect ambient noise, it was used in this solution to evaluate the noise, especially noise spikes, the sensor used was adjusted using an inboard potentiometer;
- MPU-9250/6500: The MPU-9250 module from InvenSense, is an inertial 9-axis sensor (9 DoF) combining two chips in the same housing (SiP): an MPU-6500 responsible for the 3 axes of the accelerometer and the 3 axes of the gyroscope; and an AK8963 containing a 3-axis magnetometer. In addition to inertial information, the MPU-9250

has an internal thermometer, a 16-bit analog-to-digital converter, and SPI (Serial Peripheral Interface) and I2C (Inter-Integrated Circuit) communication ports.
- GY-521: GY-521 is a module based on the MPU-6050 sensor, contains an accelerometer and a MEM technology gyroscope. A 16-bit digital-analog converter for each analog channel allows capturing the *xx*, *yy*, and *zz* axes simultaneously. The module uses the I2C protocol to communicate with the microcontroller (Robots 2016). For its configuration and calibration, the sensitivity of ± 2 g were selected for the accelerometer and the range of 250° for the gyroscope. The data for each axis is stored in two bytes or registers (Dejan 2019);
- HB-100 Doppler: Unlike ordinary infrared detectors, they detect the movement of an object by detecting microwaves reflected from the object, thus, it is not affected by the ambient temperature, it has a long detection range and high sensitivity (Spark Fruit, nd). In combination with the PIR sensor, it can effectively determine if someone has passed, without being affected by other sources of heat, color of clothing, or other objects. The tested HB-100 Module consists of a transmitter, which transmits impulses at a frequency of 10.525 GHz, and a receiver, which captures the impulses reflected by the moving object (Spark Fruit, n.d.). A fall is a short-lived event, characterized by an increase in the speed of the human body, which translates into high-frequency Doppler in the Time / Frequency domain.

Figure 3 shows the proposed IoT device that was developed under this work. The Arduino-based device integrated the above-mentioned sensors and perform a first step data processing routine that reads and analyses the sensor data. The device was programmed to be triggered when a high-level sound is detected. The sensor data read before the trigger, during the trigger, and after the trigger is sent to the fog layer where the wavelet will be used to classify the data.

The continuous wavelet transform (CWT) is a tool for time series analyses that will be applied to the proposed solution as a classifier model to identify human falls. The CWT decomposes a signal into components of different scales, comparing the signal with the wavelet of different sizes, this is done by

$$c(a, b) = \int f(t)\psi(at + b)dt \tag{1}$$

where $\psi(at + b)$ is a set of smaller waves, a is a real value used for dilatation, b the displacement [29].

The CWT is suitable for analyzing signals that are characterized by transient behaviors or discontinuities, such as transient events typical of human movement [30]. The use of CWT allowed the simultaneously decomposing of the time series into a time/frequency space obtaining information about the amplitude of any "periodic" signals within the series and how that amplitude varies with time.

Consider the average fall pattern in the real world, such as the "fall of the human model" generating the reference time series "mother wave" to detect falls, each recorded acceleration signal can then be compared to this series, through the analysis of the waves. The similarity of the recorded signal with the fall of the "model", is the characteristic that can be used to discriminate between falls and other activities.

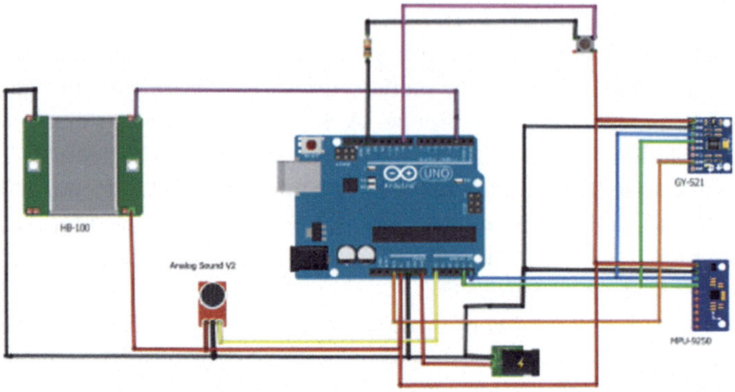

Fig. 3. Proposed IoT device

The average values of the acceleration signals were used to obtain the standard average fall signal, which was converted into an adapted mother wavelet, defined in the interval [0,1], and satisfying the definition of the wavelet (Eq. (1)).

After the creation of the mother wave, the degree of similarity between it and the signals to be tested is validated. Then, we compare the signal in the current window with the mother wavelet calculating the coefficients of CWT [26]

$$CWT_{coeff\,(a,b)} = \frac{1}{\sqrt{x}} \int_{-\infty}^{+\infty} SV_{candidate}(t)\,\psi_{fall}\left(\frac{t-b}{a}\right)dt \tag{2}$$

where a and b are the scale and translation parameters, $CWT_{coeff\,(a,b)}$ describes the similarity (the higher the coefficient, the greater the similarity) between the candidate and the mother wavelet, in different scales, a, and translations, b. Then, the maximum value of $CWT_{coeff\,(a,b)}$ is chosen [31].

To calculate the wavelet transform of the time series D, we chose the Morlet wavelet. This selection was made because this type of wavelet is recommended for the analysis of geophysical signals [32], similar to the non-stationary signal provided by the accelerometer when reading the floor vibrations. Additionally, Morlet wavelet is similar to the impact wave, whose mother wavelet is,

$$\psi_0(\eta) = \pi^{-\frac{1}{4}} e^{iw_0\eta} e^{-\eta^2/2} \tag{3}$$

The choice of w_0 defines the number of oscillations within the wavelet itself. Errors due to the average, other than zero, will be minor. The following steps will also be performed:

- Determine the Fourier Transform of the mother wavelet;
- Determine the Fourier transform of the time series;
- Selection of the minimum scale S_0 and all other scales.

The spectral analysis of the fall time series and ADL was performed, using the pycwt package [32] of Python using a Raspberry Pi 4 Model B in the fog-computing layer, of

Fig. 2. The data is received and classified inside Python. If a fall is detected, then a push notification is sent to the users.

The proposed solution uses non-wearable IoT devices to detect human falls. This approach allows the system to work autonomously without the need of user actions. However, it brings some limitations regarding the positioning of sensors. In our study, several falls were performed within a 3-m radius to validate the reading ability of the sensors. The proposed solution was design to have a reading device (i.e., a edge layer IoT device) per room. This also enable the identification of the room where the person fell.

4 Case Study

To test the proposed solution, a dataset was created using a human-body model built using a thick board card role, of 20 cm diameter and 120 cm height, filled with water bottles, with a total of 15 L, plus a 3 kg iron on top. The human-body model was overturned in several directions in a brick pavement and achieved a 4.1 m/s fall velocity. The data recorded in this dataset was: acceleration, the angular velocity (gyroscope) on the *xx*, *yy*, and *zz* axes, the sound, and the Doppler count. The data was recorded using a frequency of 100 Hz (T = 10 ms). The recorded drop patterns resulted in a non-stationary signal.

The spectral analysis was performed in the accelerometer data using the three, *xx*, *yy*, and *zz*, axis combined as

$$D = \sqrt[2]{a_x^2 + a_y^2 + a_z^2} \tag{4}$$

where a_x represents the acceleration in *xx* axis, a_y represents the acceleration in *yy* axis, and a_z represents the acceleration in *zz* axis.

The analysis was performed in the profile of falling, and the profile of activities of daily living (ADL) times series. The profiling of time series was done using an arithmetic average of D values.

In Fig. 4a, the time series of the vector sum of the acceleration of a fall is represented. Figure 4b shows the power spectrum of the wavelet, using Morlet's wavelet. The *xx* axis is the location of the wavelet in time, for a time series of 120 values. The *yy* axis is the wavelet period in milliseconds. The black outlines are the 10% significance regions, using a background spectrum of yellow noise. The orange areas indicate that the drop occurred between the samples 40 and 50, while between 0–20 and 60–120 there was no event.

The application of CWT to the acceleration patterns allows the localization of the transient state of the signal during impact. Thus, an incidence of fall can be detected, if a maximum surface (orange zone) is observed. On the contrary, an ADL will not generate a significant maximum, since these activities result in repetitive patterns. The orange zone corresponds to the greatest power of the wavelet spectrum, that is, the moment of impact. The average variance in this zone is quite high, compared to the values of ADL (Fig. 5).

Figure 5 shows the results of the Morlet wavelet transform in ADL time series. The spectrum analyses show a clear difference among ADL and fall times series. The average

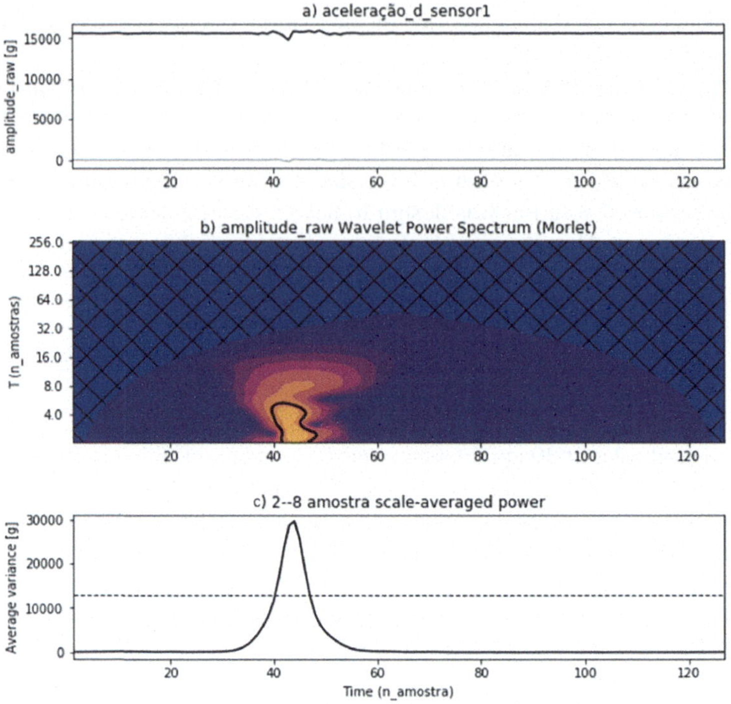

Fig. 4. Spectral analysis of fall time series. a) vectorial sum of acceleration, b) Morlet wavelet, c) average variance

variance of the fall time series is around 13,000 while ADL times series is around 260. Therefore, variance values equal to or above 13,000 can be set for fall detection, values from 260 to 13,000 can be set for medium activity detection, and values equal to or below 260 can be set for ADL detection.

Using a dataset with 80 samples (40 falls, and 40 ADL). The proposed model achieved an accuracy of 92.5%. However, the proposed classification model was able to correctly identify 100% of human falls and 85% of ADL events, meaning that there were false positives but there were not false negatives. This allowed the proposed solution to detect all human falls and warn users about 15% of the ADL events that raised above the Morlet wavelet transform triggers.

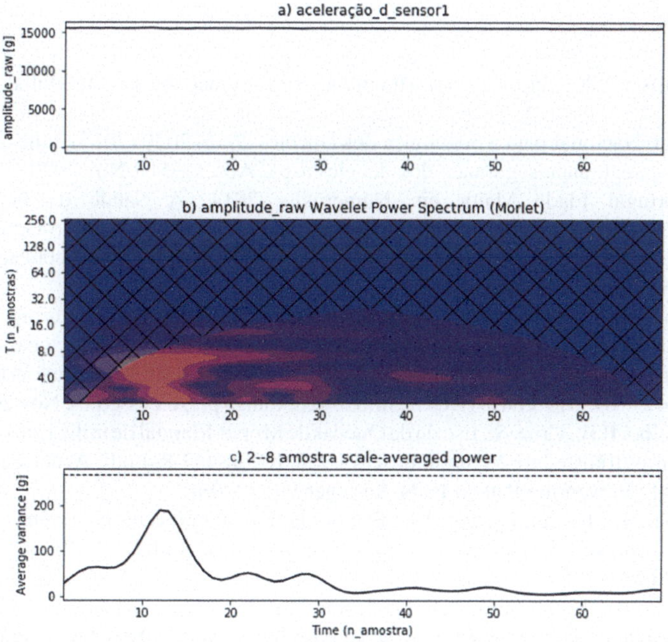

Fig. 5. Spectral analysis of ADL time series. a) vectorial sum of acceleration, b) Morlet wavelet, c) average variance

5 Conclusion

Human fall prevention is a utopia that can not be achieved with today's technology. However, the minimization of risks and aid time can be done to prevent long-term damage to people. The elderly population is the major group that is affected by this issue, where falls are recurrent and can become deadly. Therefore, society has the responsibility to address this issue.

This paper proposes an IoT-based solution that is deployed in the building, avoiding the needed maintenance of wearable devices. The paper proposed the IoT devices that need to be deployed in the room, the main architecture of the system, and a classification method using the Morlet wavelet.

The results were the result of the proposed solution using a human-body model to classify human falls in the fog-computing layer. The results are promising but not final. The proposed solution needs to be tested in a real environment and more developments should be made, such as the integration of new sensors in the spectral analysis, and a redundant classification method (e.g., using deep learning).

Acknowledgments. This work has received funding from FEDER Funds through COMPETE program and from National Funds through (FCT) under the project UIDB/00760/2020.

References

1. WHO: Falls (2020). https://www.who.int/news-room/fact-sheets/detail/falls. Accessed 02 Nov 2020
2. DGS: Plano Nacional para a Segurança dos Doentes 2015–2020 (2015). https://www.dgs.pt/qualidade-e-seguranca/seguranca-dos-doentes.aspx. Accessed 02 Nov 2020
3. DGS: Portugal Idade Maior em Números - 2014: A Saúde da População Portuguesa com 65 ou mais anos de idade (Tech. Rep.). Direção-Geral da Saúde (2014). https://www.dgs.pt/estatisticas-de-saude/estatisticas-de-saude/publicacoes/portugal-idade-maior-em-numeros-2014-pdf.aspx. Accessed 02 Nov 2020
4. INE: Mantém-se o agravamento do envelhecimento demográfico, em Portugal, que só tenderá a estabilizar daqui a cerca de 40 anos (Tech. Rep.). Instituto Nacional de Estatística (2017). https://www.ine.pt/xportal/xmain?xpid=INE&xpgid=ine_destaques&DESTAQUESdest_boui=277695619&DESTAQUESmodo=2&xlang=pt. Accessed 02 Nov 2020
5. Luís Carvalho, P. B., Cruz, S.: Escala de Quedas de Morse-Manual de utilização (2015). https://www.esenf.pt/fotos/editor2/i_d/publicacoes/978-989-98443-8-4.pdf. Accessed 02 Nov 2020
6. Morse, J.M.: Preventing Patient Falls. Springer Pub (2008)
7. Rajagopalan, R., Litvan, I., Jung, T.P.: Fall prediction and prevention systems: recent trends, challenges, and future research directions. Sensors (Switzerland) (2017). https://doi.org/10.3390/s17112509
8. Pang, I., Okubo, Y., Sturnieks, D., Lord, S.R., Brodie, M.A.: Detection of near falls using wearable devices: a systematic review. J. Geriatr. Phys. Ther. (2019). https://doi.org/10.1519/JPT.0000000000000181
9. Mozaffari, N., Rezazadeh, J., Farahbakhsh, R., Yazdani, S., Sandrasegaran, K.: Practical fall detection based on IoT technologies: a survey. Internet of Things (2019). https://doi.org/10.1016/j.iot.2019.100124
10. Bet, P., Castro, P.C., Ponti, M.A.: Fall detection and fall risk assessment in older person using wearable sensors: a systematic review. Int. J. Med. Inform. (2019). https://doi.org/10.1016/j.ijmedinf.2019.08.006
11. Ramachandran, A., Karuppiah, A.: A survey on recent advances in wearable fall detection systems. Biomed. Res. Int. (2020). https://doi.org/10.1155/2020/2167160
12. Vallabh, P., Malekian, R.: Fall detection monitoring systems: a comprehensive review. J. Ambient. Intell. Humaniz. Comput. **9**(6), 1809–1833 (2017). https://doi.org/10.1007/s12652-017-0592-3
13. Chaccour, K., Darazi, R., El Hassani, A.H., Andres, E.: From fall detection to fall prevention: a generic classification of fall-related systems. IEEE Sens. J. (2017). https://doi.org/10.1109/JSEN.2016.2628099
14. Nadeem, A., Mehmood, A., Rizwan, K.: A dataset build using wearable inertial measurement and ECG sensors for activity recognition, fall detection and basic heart anomaly detection system. Data Brief **27**, 104717 (2019). https://doi.org/10.1016/j.dib.2019.104717
15. Espinosa, R., Ponce, H., Gutiérrez, S., Martínez-Villaseñor, L., Brieva, J., Moya-Albor, E.: A vision-based approach for fall detection using multiple cameras and convolutional neural networks: a case study using the UP-Fall detection dataset. Comput. Biol. Med. (2019). https://doi.org/10.1016/j.compbiomed.2019.103520
16. Salman Khan, M., Yu, M., Feng, P., Wang, L., Chambers, J.: An unsupervised acoustic fall detection system using source separation for sound interference suppression. Signal Proc. 110, 199–210 (2015). https://doi.org/10.1016/j.sigpro.2014.08.021
17. Yacchirema, D., De Puga, J.S., Palau, C., Esteve, M.: Fall detection system for elderly people using IoT and Big Data. Procedia Comput. Sci. (2018). https://doi.org/10.1016/j.procs.2018.04.110

18. Aziz, O., et al.: Validation of accuracy of SVM-based fall detection system using real-world fall and non-fall datasets. PLoS ONE (2017). https://doi.org/10.1371/journal.pone.0180318
19. Ajerla, D., Mahfuz, S., Zulkernine, F.: A real-time patient monitoring framework for fall detection. Wirel. Commun. Mob. Comput. (2019). https://doi.org/10.1155/2019/9507938
20. De Miguel, K., Brunete, A., Hernando, M., Gambao, E.: Home camera-based fall detection system for the elderly. Sensors (Switzerland) (2017). https://doi.org/10.3390/s17122864
21. Lu, N., Wu, Y., Feng, L., Song, J.: Deep learning for fall detection: three-dimensional CNN combined with LSTM on video kinematic data. IEEE J. Biomed. Health Inform. (2019). https://doi.org/10.1109/JBHI.2018.2808281
22. Wang, Y., Wu, K., Ni, L.M.: WiFall: device-free fall detection by wireless networks. IEEE Trans. Mob. Comput. (2017). https://doi.org/10.1109/TMC.2016.2557792
23. Casilari, E., Santoyo-Ramón, J.A., Cano-García, J.M.: Analysis of public datasets for wearable fall detection systems. Sensors (Switzerland) (2017). https://doi.org/10.3390/s17071513
24. Casilari, E., Santoyo-Ramón, J.A., Cano-García, J.M.: UMAFall: a multisensor dataset for the research on automatic fall detection. Procedia Comput. Sci. **110**, 32–39 (2017). https://doi.org/10.1016/j.procs.2017.06.110
25. Sucerquia, A., López, J.D., Vargas-Bonilla, J.F.: SisFall: a fall and movement dataset. Sensors (Switzerland) (2017). https://doi.org/10.3390/s17010198
26. Torrence, C., Compo, G.P.: A practical guide to wavelet analysis. Bull. Am. Meteor. Soc. (1998). https://doi.org/10.1175/1520-0477(1998)079%3c0061:APGTWA%3e2.0.CO;2
27. Palmerini, L., Bagalà, F., Zanetti, A., Klenk, J., Becker, C., Cappello, A.: A wavelet-based approach to fall detection. Sensors (Switzerland) (2015). https://doi.org/10.3390/s150511575
28. Yazar, A., Keskin, F., Töreyin, B.U., Çetin, A.E.: Fall detection using single-tree complex wavelet transform. Pattern Recogn. Lett. (2013). https://doi.org/10.1016/j.patrec.2012.12.010
29. Hubbard, B.: The World According to Wavelets the Story of a Mathematical Technique in the Making. Universities Press (India) Pvt, Limited (1996)
30. Caramia, C., De Marchis, C., Schmid, M.: Optimizing the scale of a wavelet-based method for the detection of gait events from a waist-mounted accelerometer under different walking speeds. Sensors **19**(8), 1869(2019). https://doi.org/10.3390/s19081869
31. PAOS: Wave:Wavelets 2 (1998). https://paos.colorado.edu/research/wavelets/wavelet2.html. Accessed 02 Nov 2020
32. PyCWT: PyCWT: spectral analysis using wavelets in Python — PyCWT 0.3.0a22 documentation (2017). https://pycwt.readthedocs.io/en/latest/. Accessed 02 Nov 2020

Detection of Suboptimal Conditions in Photovoltaic Installations for Household-Prosumers

Dalberth Corrales[1], Leonardo Cardinale-Villalobos[1](✉), Carlos Meza[1], and Luis Diego Murillo-Soto[2]

[1] Electronics Engineering School, Instituto Tecnológico de Costa Rica, Cartago, Costa Rica
lcardinale@tec.ac.cr

[2] Electromechanical Engineering School, Instituto Tecnológico de Costa Rica, Cartago, Costa Rica

Abstract. Advancements in photovoltaics and information technology makes it possible for cities to produce a fraction, if not everything, of the electrical power they require. Power generation in cities can be mainly achieved by photovoltaic installations located in the roofs of houses and buildings unevenly distributed among large areas. Suboptimal conditions such as soling, partial shadowing and electrical faults, are common in PV installations, decreasing their efficiency. It is advisable to detect when these suboptimal conditions occur to correct them and improve the overall performance of the PV installation. Nevertheless, the majority of suboptimal-detection techniques reported in the literature only consider large PV systems, neglecting rooftop PV installations. In this paper a low-cost embedded system suitable for detecting suboptimal conditions in small PV installation is presented. This system was validated for the detection of partial shadow that may occur due to strange objects in a specific region of the PV array. The system obtained a sensitivity of 87%, specificity of 87.9% and accuracy of 87.6%.

Keywords: PV prosumers · Fault detection · Partial shading · Soiling · Embedded systems

1 Introduction

The typical central scheme in which large power stations provide electricity unidirectionally to residential, industrial, and commercial end-users is slowly but steadily changing into a decentralized structure [1]. Due to the cost reduction of photovoltaic (PV) modules and to the improvements in information and communication technologies (ICT), it is now possible that industrial, residential, and commercial end-users produce their own electricity. Indeed, the modern power system comprises prosumer agents, i.e., end-users that consume, produce, and store electricity [2]. The "new" power scheme not only has distributed power

J. M. Corchado and S. Trabelsi (Eds.): SSCTIC 2021, LNNS 253, pp. 26–36, 2022.
https://doi.org/10.1007/978-3-030-78901-5_3

generation but also distributed intelligence employing smart systems that decide when to take, deliver or store power from the power grid [3]. The prosumer generation scheme can potentially help to decarbonize the power grid while reducing the energy bill for the end-users.

The most widely used technology for prosumers in the residential sector is PV due to its modularity and steady cost reduction in the last decades. Several countries have already achieved grid-parity in self-consumption PV systems in comparison to energy purchased from the grid [4]. Also, many governments have put into place incentive policies for PV installations such as net metering, net billing, and feed-in tariff [3].

To achieve maximum utilization of PV installations, it is required to ensure that the system is free of suboptimal conditions (e.g. faults, soiling) for as long as possible. Several authors have developed fault detection methods, e.g., [5–12]. Fault and soiling detection techniques that use images taken by unmanned aerial systems (UAS) are useful for large ground-mounted and/or large PV installations given that visual data about the site can be taken in relatively short time and does not require additional circuitry or measurement devices [13]. Nevertheless, UAS PV fault-detection is not suitable for small scale PV installations used for self-consumption in households given that, normally, PV installations in houses' roof are widely spread around the city. In order to detect suboptimal conditions in small PV installations a low-cost and easy to install detection system is required. In this regards, the authors found that there is a lack of development of low-cost IoT-based systems [14] for PV fault detection.

The present work proposes a low-cost embedded system that is experimentally validate for partial covering due to soling or objects. The paper is structued as follows: Sect. 2 describes the type of suboptimal conditions evaluated, Sect. 3 presents the method used for detecting suboptimal conditions. The materials and methods used in this research are summarized in Sect. 4 and the main results and discussion are presented in Sect. 5. Finally, conclusions are presented in Sect. 6.

2 Related Work

A partial covering on a PV array surface creates a specific region in which the irradiance is lower, resulting in a decrease in the power of the entire array [14], which will depend on the size and degree of shading [15,16]. The presence of strange objects (such as dirt or leaves) in a specific region of the PV array yields in a suboptimal operation which results in a decrease in the energy generated by the system. The latter should not be confused with uniform dirt, which also causes a decrease in the incidence of irradiance to the PV array (reducing the power generated), however, given that it typically occurs in the form of dust over the entire array, it does not cause partial shading [17].

Different methods to detect partial shadowing have been proposed. For example, UAV detection has been proposed in [13] and [18]. Mekki et al. in [19] proposes a partial shadowing detection technique based on a neural network algortihm. Using the i-v curve of the PV array Ma et al. in [20] are able to detect partial shadowing. Even though the previous techniques are effective require special equipment (UAS), high computational power (neural networks) or to stop the PV installation production to get the i-v curves of the array. Those conditions are not possible nor desire in a roof-top PV installation. In this next section, a low-cost, easy to implement system is presented to detect suboptimal operation conditions such as partial shadowing.

3 Suboptimal Detection Through Efficiency Analysis

Some fault detection methods require the experimental calculation of internal parameters of the solar panels that are not provided by the manufacturer [21–23]. In [12] it is shown that using the expression (1) it is possible to obtain an estimate of the PV efficiency with an accuracy that makes it suitable to detect suboptimal conditions in the system, i.e., a decrease in the efficiency. For this, it is sufficient to estimate the constants K_1, K_2 and K_3 with the solar panel data sheet, measure the irradiance (G) and the cell temperature (T).

$$\eta_T(T, G) = K_1 \cdot T + K_2 \cdot G + K_3 \tag{1}$$

Given its simplicity and sufficient accuracy, expression (1) is used in the present paper to detect suboptimal conditions in the PV array.

4 Materials and Methods

4.1 The PV Installation Analyzed

The experimental setup used for validation comprises two PV arrays of 12 modules connected in series with a peak power of 3.36 kW. The PV modules are of the same technology (monocrystalline) and the same model. Each PV array has a dedicated inverter. The analyzed PV arrays are shown in Fig. 1.

Fig. 1. PV array 1 and 2 used for the experiment. Both arrays are identical, with the same configuration of 12 modules in series.

4.2 Electronic Suboptimal Condition Detection System

The experiment used a low-cost IoT solution as a suboptimal condition detection system. Using an embedded circuit in a Raspberry Pi 3 model B, all the information was collected and processed to perform the suboptimal condition detection. Analog sensors were used to measure cell temperature and estimate irradiance. Power was obtained from the power inverters through a wifi link. The results were displayed in real time remotely on the embedded system terminal and backed up in a text file. A block diagram of the implemented system is shown in Fig. 2. In order to achieve the correct functioning of the system, it was necessary to ensure that all communication links were stable; otherwise, the system would produce erroneous information.

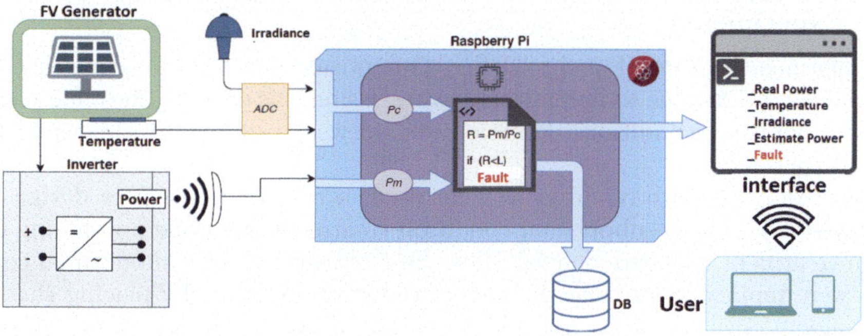

Fig. 2. Block diagram of the suboptimal condition detection system used in the investigation. It used a Raspberry Pi to collect, process and display the information.

The instrumentation used is described below:

- a low-cost SQ-110 photosynthetically active radiation (PAR) sensor with a correction factor applied to obtain equivalent insolation.
- PT-1000 thermocouple for measuring the temperature of the solar module.
- Sunny Boy 3.8-US inverter for the measurement of real power.

The temperature measurement was calibrated against a Southwire multimeter sensor making ambient temperature measurements and the irradiance was calibrated against a Vantage Pro2 weather station measurement.

4.3 Protocol for Suboptimal Detection

The method proposed by [12] was used to calculate the instant power. Considering the panel area of 1.64 m^2, 12 panels per array, STC efficiency of 17.11%, 15.43% at NOCT and 16.51% under low irradiance conditions (LIC). In addition, aging losses of 1.26% and 10% for the wiring were considered. This resulted in Eq. (2).

$$P_A(T,G) = -13,4 \times 10^{-3} \cdot T + 1.3 \times 10^{-4} \cdot G + 3.19 \quad (2)$$

The temperature was measured in a module of array 1. All modules of both arrays were considered to have the same temperature.

The electrical model generates an output power that was updated every 3 min, which was calculated from a rolling average that considers 3 measurements updated every minute. The system records the actual power produced by the PV array and compares it with the estimated theoretical power of the system under normal conditions. To identify a suboptimal condition, the system calculates the yield of the measured power versus calculated power ($Y = P_m/P_c$) and compares it to a yield limit that was determined empirically. If the yield is under this limit, the system is under-performing which may indicate possible suboptimal conditions, soiling or shadows.

4.4 Experiments

An experiment was developed to define a quantitative criteria for detecting suboptimal conditions due to partial shadowing. Using this criteria, the effectiveness of the suboptimal conditions detection system was evaluated by the replication of the experiment.

Since only two study subjects were used, a repeated-measures design was employed [24]. Three suboptimal condition treatments and one non-failing condition treatment (N) were defined. The three treatments with suboptimal conditions were replicated 12 times in each of the study subjects by placing shadows at a random position of the array. The non-failing condition treatment was replicated 12 times in each study subject. Thus, a total of 96 experimental units were generated. The failure treatments evaluated were: 1) FP: a fraction of a panel shaded (26.6% of its area), 2) 1P: a complete panel shaded and 3) 2P: two complete panels shaded. The treatments were applied without interaction between factors. Independence between treatments was assumed because there are no residual effects due to the treatments.

4.5 Description of Each Factor and the Levels of the Experiment

The factor evaluated in the experiment corresponded to partial shading, this was done by placing an opaque plastic screen over the PV array to achieve strong shading [16] (see Fig. 3.a) in a manner equivalent to how it has been done in previous research [19]. The FP, 1P, and 2P levels were defined by varying the area of partial shading, as has been done in other partial shading studies [14]. An example of the 2P treatment on one of the PV chains is shown in Fig. 3.b. Figure 3.c shows how the screen was place on a module for FP treatment.

4.6 Date of the Experiment

The experiment used to establish the yield limit criteria for suboptimal conditions detection system was carried out on October 24 and 25, 2019. Subsequently, on November 7 and 9 the experiment was replicated to evaluate the effectiveness of the system.

Fig. 3. a) Screens used for FP and 1P treatments, b) example of chain under 2P treatment, c) plastic screen for FP treatment on a module.

4.7 Data Processing

For the definition of the failure detection criterion, descriptive statistics with histograms [25] and inferential statistics with ROC curves [26] and confidence intervals [27] were used in order to define the yield point that could detect the greatest number of failures. A hypothesis test was performed to determine if there were significant differences between the failure and non-failure treatments using the ancova test [28]. Finally, to determine the effectiveness of the suboptimal detection method, ROC curves were used to calculate the sensitivity, specificity and accuracy [26]. The software used were Jamovi and easyROC.

5 Results and Discussion

5.1 Determination of the Power Yield Limit

The results of the experiment to determine the yield limit at which a suboptimal condition exists are shown in the Fig. 4. Three outliers were eliminated during data processing: two measurements from the SP treatment and one measurement from the 1P treatment (these can be identified in Fig. 4).

Using binary classifier analysis, the ROC curve shown in Fig. 5 was plotted. The area under the ROC curve (AUC) obtained was 0.944, which indicates that the system has a high accuracy [26] in identifying the state of the PV array. The ROC test indicates that 92.2% is the yield limit that best balances between sensitivity and specificity for discriminating between suboptimal conditions and normal conditions.

An analysis of covariance (ANCOVA) was conducted using the 93 trustworthy measurements of power yield. In this case the measured power yield is the continuous dependent variable which is related to the shading treatments (factor or categorical independent variable) and at the same time influenced by PV module temperature and irradiance (continuous independent variables). By applying a post-hoc test to this ANCOVA (see Fig. 6) it is determined that

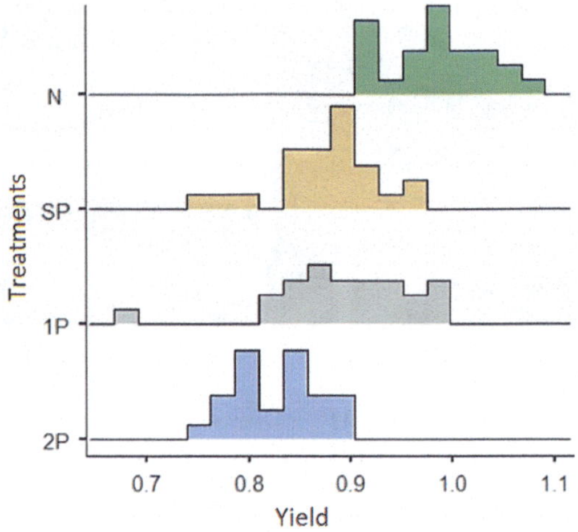

Fig. 4. Histogram of the yield measurements for the different treatments.

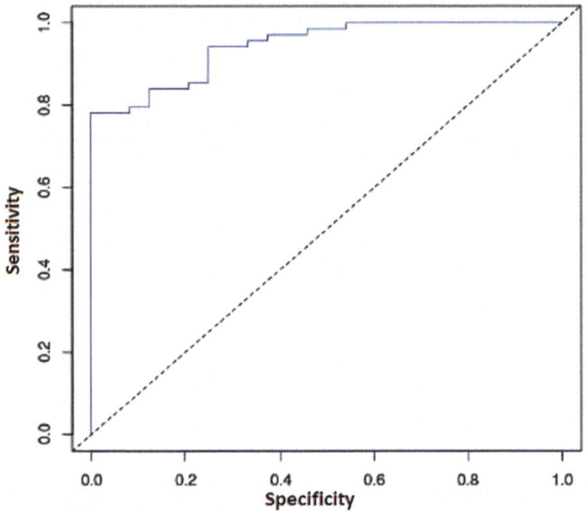

Fig. 5. ROC curve of the yield as an indicator used to detect suboptimal conditions.

between pairs of shading treatments there are significant differences in almost all cases ($p_{tukey} < 0.05$) [29]. The most important result is the significant differences between the absence of suboptimal conditions and each of the shadowing treatments. Therefore, this analysis indicates that performance is an appropriate variable to be used to detect this type of suboptimal condition.

Comparison							
Treatments			Mean Difference	SE	df	t	Ptukey
N	-	SP	0.1087	0.0121	83.0	9.02	< .001
	-	1P	0.0878	0.0117	83.0	7.53	< .001
	-	2P	0.1728	0.0120	83.0	14.40	< .001
SP	-	1P	-0.0209	0.0121	83.0	-1.73	0.316
	-	2P	0.0641	0.0119	83.0	5.40	< .001
1P	-	2P	0.0850	0.0118	83.0	7.19	< .001

Fig. 6. Post-hoc comparisons between the different levels of shading.

The results of the confidence intervals related to the marginal means obtained with the ANCOVA are shown in Fig. 7. In these results 0.944 is the highest confidence limit of all the failure treatments (specifically it's the upper limit for 1P) and it's under the lower limit for the no-failure treatment (N). So this means that 0.944 can be a yield limit that is significantly different between the absence of suboptimal condition and the shadowing; for this set of data. Comparing this power performance limit of 94.4% with the value of 92.6% found with the ROC test and knowing that the value generated by the ROC test can be varied as required [26]. It was decided to establish 94.4% as a criterion for the detection method, because it is of higher priority to detect the highest number of failures over the generation of false failure warnings.

			95% Confidence Interval	
T	Mean	SE	Lower	Upper
N	1.014	0.0116	0.991	1.037
SP	0.900	0.0115	0.878	0.923
1P	0.920	0.0120	0.896	0.944
2P	0.827	0.0116	0.804	0.850

Fig. 7. Estimated marginal means in the analysis of covariance.

5.2 Evaluation of the Suboptimal Conditions Detection System

To evaluate the effectiveness of the suboptimal conditions detection system, the same experiment for determining the yield limit was replicated; using the previously identified limit to detect suboptimal conditions. The measurements obtained in this case are plotted in Fig. 8, where, as seen in the box plot, seven measurements are outliers and thus eliminated for further analysis.

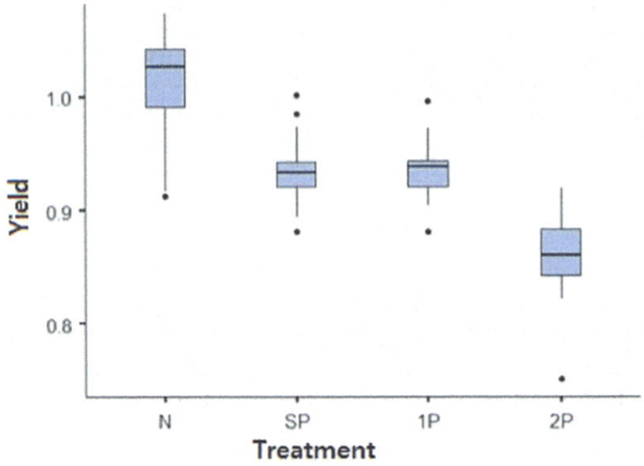

Fig. 8. Box plot of yield measurements for different treatments to evaluate the system.

Given that the proposed suboptimal conditions detection system is a binary classifier, it was evaluated using a confusion matrix as seen in Fig. 9, where N represents the absence of suboptimal conditions (positives) and F represents the occurrence of suboptimal conditions (negatives). There are 89 valid measurements of which 66 are actual suboptimal conditions (58 were correctly predicted) and the other 23 are actual absences of suboptimal conditions (20 were properly classified as such). Therefore, 20 measurements are true positives (TP), 58 are true negatives (TN), 8 are false positives (FP) and 3 are false negatives (FN). This results in a sensitivity of 87%, a specificity of 87.9% and an accuracy of 87.6%. Thus, of the 89 experimental units that included suboptimal conditions and normal condition treatments, 87% of the normal conditions were correctly classified as such, 87.9% of the suboptimal conditions were correctly detected, and in general, 87.6% of the cases were correctly identified.

		Actual		
		N	F	Σ
Prediction	N	20	8	28
	F	3	58	61
	Σ	23	66	89

Fig. 9. Confusion matrix obtained from the experiment for the evaluation of the suboptimal condition detection system.

6 Conclusions

Results show that it is possible to detect suboptimal conditions in photovoltaic installations with a low-cost IoT system, allowing to solve a need that is increasing with the growing development of distributed generation in many countries of the world. In this type of systems, the need for IoT solutions with stable communication systems were identified.

The methodology used allowed generating quantitative indicators related to the performance of suboptimal detection system, in this case in the presence of partial shadows, but it could be used for other subotpimal conditions.

The indicators obtained for sensitivity, specificity and accuracy were 87%, 87.9% and 87.6%, respectively. These results are associated with the experimentally estimated criterion that a yield less than or equal to 94.4% is associated with the presence of a partial shadow. Further experimental research remains to be done to increase the performance of the system

Acknowledgement. This paper is part of a project entitled: Fault Identification in Photovoltaic Systems, ID 5402-1360-4201, financed by the Costa Rica Institute of Technology.

References

1. Fina, B., Auer, H., Friedl, W.: Profitability of pv sharing in energy communities: Use cases for different settlement patterns. Energy **189**(116), 148 (2019)
2. Parag, Y., Sovacool, B.K.: Electricity market design for the prosumer era. Nat. Energy **1**(4), 1–6 (2016)
3. Li, Z., Ma, T.: Peer-to-peer electricity trading in grid-connected residential communities with household distributed photovoltaic. Appl. Energy **278**(115), 670 (2020)
4. González-Romera, E., et al.: Advantages of minimizing energy exchange instead of energy cost in prosumer microgrids. Energies **12**(4), 719 (2019)
5. Appiah, A.Y., Zhang, X., Ayawli, B.B.K., Kyeremeh, F.: Review and performance evaluation of photovoltaic array fault detection and diagnosis techniques. Int. J. Photoenergy **2019**, 1–19 (2019)
6. Tyutyundzhiev, N., Lovchinov, K., Martínez-Moreno, F., Leloux, J., Narvarte, L.: Advanced PV modules inspection using multirotor UAV. In: 31st European Photovoltaic Solar Energy Conference and Exhibition, Hamburg (2015)
7. Chaudhary, A.S., Chaturvedi, D.: Thermal image analysis and segmentation to study temperature effects of cement and bird deposition on surface of solar panels. Int. J. Image Graph. Signal Process. **9**(12), 12–22 (2017)
8. Köntges, M.: Reviewing the practicality and utility of electroluminescence and thermography (2014)
9. Leva, S., Aghaei, M., Grimaccia, F.: PV power plant inspection by UAS: correlation between altitude and detection of defects on PV modules. In: 2015 IEEE 15th International Conference on Environment and Electrical Engineering (EEEIC). IEEE (2015)
10. Cardinale-Villalobos, L., Rimolo-Donadio, R., Meza, C.: Solar panel failure detection by infrared UAS digital photogrammetry: a case study. Int. J. Renew. Energy Res. (IJRER) **10**(3), 1154–1164 (2020)
11. Figueroa-García, J.C., Garay-Rairán, F.S., Hernández-Pérez, G.J., Díaz-Gutierrez, Y. (eds.): WEA 2020. CCIS, vol. 1274. Springer, Cham (2020). https://doi.org/10.1007/978-3-030-61834-6
12. Murillo-Soto, L., Meza, C.: Fault detection in solar arrays based on an efficiency threshold. In: 2020 IEEE 11th Latin American Symposium on Circuits Systems (LASCAS), pp. 1–4 (2020)

13. Cardinale-Villalobos, L., Meza, C., Murillo-Soto, L.D.: Experimental comparison of visual inspection and infrared thermography for the detection of soling and partial shading in photovoltaic arrays. In: Nesmachnow, S., Hernández Callejo, L. (eds.) ICSC-CITIES 2020. CCIS, vol. 1359, pp. 302–321. Springer, Cham (2021). https://doi.org/10.1007/978-3-030-69136-3_21
14. Mellit, A., Tina, G., Kalogirou, S.: Fault detection and diagnosis methods for photovoltaic systems: a review. Renew. Sustain. Energy Rev. **91**, 1–17 (2018)
15. Mäki, A., Valkealahti, S.: Power losses in long string and parallel-connected short strings of series-connected silicon-based photovoltaic modules due to partial shading conditions. IEEE Trans. Energy Conv. **27**(1), 173–183 (2012)
16. Maghami, M.R., Hizam, H., Gomes, C., Radzi, M.A., Rezadad, M.I., Hajighorbani, S.: Power loss due to soiling on solar panel: a review. Renew. Sustain. Energy Rev. **59**, 1307–1316 (2016)
17. Javed, W., Wubulikasimu, Y., Figgis, B., Guo, B.: Characterization of dust accumulated on photovoltaic panels in doha, qatar. Solar Energy **142**, 123–135 (2017)
18. Quater, P.B., Grimaccia, F., Leva, S., Mussetta, M., Aghaei, M.: Light unmanned aerial vehicles (uavs) for cooperative inspection of pv plants. IEEE J. Photovolt. **4**(4), 1107–1113 (2014)
19. Mekki, H., Mellit, A., Salhi, H.: Artificial neural network-based modelling and fault detection of partial shaded photovoltaic modules. Simul. Model. Pract. Theory **67**, 1–13 (2016)
20. Ma, J., Pan, X., Man, K.L., Li, X., Wen, H., Ting, T.O.: Detection and assessment of partial shading scenarios on photovoltaic strings. IEEE Trans. Ind. Appl. **54**(6), 6279–6289 (2018)
21. Bastidas-Rodriguez, J.D., Franco, E., Petrone, G., Ramos-Paja, C.A., Spagnuolo, G.: Model-based degradation analysis of photovoltaic modules through series resistance estimation. IEEE Trans. Ind. Electron. **62**(11), 7256–7265 (2015)
22. Mellit, A., Chine, W., Massi Pavan, A., Lughi, V.: Fault diagnosis in photovoltaic arrays. In: 2015 International Conference on Clean Power (ICCEP) (2015)
23. Murillo-Soto, L.D., Figueroa-Mata, G., Meza, C.: Identification of the internal resistance in solar modules under dark conditions using differential evolution algorithm. In: 2018 IEEE International Work Conference on Bioinspired Intelligence (IWOBI), IEEE, pp. 1–9 (2018)
24. Tango, T.: Repeated Measures Design with Generalized Linear Mixed Models for Randomized Controlled Trials. Taylor & Francis Group, Tokyo Japan (2017)
25. Pyzdek, T.: Descriptive statistics. The Lean Healthcare Handbook. MP, pp. 145–149. Springer, Cham (2021). https://doi.org/10.1007/978-3-030-69901-7_12
26. Pintea, S., Moldovan, R.: The Receiver-Operating Characteristic (ROC) analysis: fundamentals and applications in clinical psychology. J. Cogn. Behav. Psychother. **9**(1), 49–66 (2009)
27. Tu, W.: Basic principles of statistical inference. In: Ambrosius, W. (ed.) Topics in Biostatistics, vol. 404, pp. 53–72. Humana Press (2007)
28. Huitema, B.: The Analysis of Covariance and Alternatives: Statistical Methods for Experiments, Quasi-Experiments, and Single-Case Studies, 2nd edn. John Wiley & Sons, Hoboken (2011)
29. Ruxton, G.D., Beauchamp, G.: Time for some a priori thinking about post hoc testing. Behav. Ecol **19**(3), 690–693 (2008)

Blockchain

RETRACTED CHAPTER: Blockchain Module for Securing Data Traffic of Industrial Production Machinery on Industrial Platforms 4.0

Akash Aggarwal[1], Yeray Mezquita[2(✉)], Diego Valdeolmillos[2], A. J. Gupta[1], Alfonso González-Briones[2,3], Javier Prieto[2], and Emilio S. Corchado[2]

[1] Department of Mathematical Science, IIT(BHU) Varanasi, Varanasi 221005, India
[2] BISITE Research Group, University of Salamanca, Edificio Multiusos I+D+i, 37007 Salamanca, Spain
yeraymm@usal.es
[3] Research Group on Agent-Based, Social and Interdisciplinary Applications (GRASIA), Complutense University of Madrid, Madrid, Spain
alfonsogb@ucm.es

Abstract. Industry 4.0 (the Fourth Industrial Revolution) is a concept devised for improving the operation of modern factories through the use of the latest technologies, under paradigms such as the Industrial Internet of Things (IIoT) or Big Data. One of such technologies is Blockchain, which is able to provide industrial processes with security, trust, traceability, reliability and automation. This paper proposes a technological framework that combines an information sharing platform and a Blockchain platform. One of the main features of the this framework is the use of smart contracts for validating and auditing the content received throughout the production process to ensure the correct traceability of the data. The conclusion drawn from this study is that this technology is under-researched and has significant potential to support and enhance the industrial revolution. Moreover, this study identifies areas for future research.

Keywords: Industrial environments · Security · Blockchain · IoT

1 Introduction

Awareness has begun to emerge regarding the importance of security in the industrial environment, not only on critical and unique facilities, but any other industrial facility which can be used to threaten third parties, and whose infrastructure and equipment contain critical information [32]. An increasing number

The original version of this chapter was retracted: The retraction note to this chapter is available at https://doi.org/10.1007/978-3-030-78901-5_31

J. M. Corchado and S. Trabelsi (Eds.): SSCTIC 2021, LNNS 253, pp. 39–47, 2022.
https://doi.org/10.1007/978-3-030-78901-5_4

of cyber-attacks occur against industrial plants [4,6]. As a recent and representative example, Win32/Stuxnet [19], a malware for Siemens SCADA PCS 7, WinCC and STEP 7 applications (the industrial control systems that will operate the centrifuges used by the uranium enrichment plants). This program led to an attack on Iran's first nuclear power plant in 2010 [20]. This shows that it is necessary to consider how to protect data, systems, and networks as a whole, not only at the logical or software level but also at the physical level and how security functionalities are implemented in production processes, pointing to mechanisms that provide aggregated security services such as confidentiality, authentication, integrity, authorization 31, non-repudiation and availability in the context of IoT [12,23,24].

The protection of data, systems and networks is very much taken into account when developing today's generic Industry 4.0 platforms [10]. On these platforms, security is managed on the basis of the framework proposed in [16] (which offers protection to a wide range of vertical solutions), through the design of a double layer of security, where the constructed ecosystem is based on a distributed ledger, allowing for the traceability of data originating from different sources (integrity of information and the source).

This article proposes a security framework that provides physical and logical protection, characterizing the behavior of systems and communications, going beyond the mere detection of known threats and allowing for the early identification of new threats, through the detection of variations. To this end, we propose the development of a platform based on blockchain that allows to validate the data received from its origin in the production line until its final treatment in the platform. For this purpose, a platform has been developed that, through the use of Smart contracts, ensures the correct traceability of the data.

A review of holistic security in Industry 4.0 is presented in Sect. 2. Section 3 describes the proposal. Section 4 presents the obtaining results. Finally, Sect. 5 presents the conclusions and future work.

2 Holistic Security in Industrial Environments

Today's industry 4.0 allows people to monitor and control equipment from anywhere, thanks to the Internet connection. Moreover, all the industrial processes are highly optimized thanks to the collection of massive data; together with the unsupervised communication between the machines, the use of multiagent systems and the application of automatic learning models; [5,9,11].

The current industrial revolution comes with a high risk as the equipment and machines connected to the Internet are vulnerable to attacks. Numerous studies have been conducted in the literature on the methods of detecting intrusions and attacks to the machines connected to internet [2,7]. However, one of the trends that is currently growing in popularity is the implementation of blockchain technology to record the activity of the machines and ensure secure communications between them [15].

In the literature, some designs and implementations have been proposed that use blockchain technology to secure the connections between devices, allowing for the traceability of data in supply chains, creating automatic and unsupervised markets [14,16,17,26]

Blockchain is a technology that makes use of an encryption mechanism through key pairs with which point to point connections are secured [31]. In addition, the use of blockchain technology allows for the use of smart contracts to automate processes [18].

In this case study, as in those cited in this section, blockchain technology has been selected due to its many advantages. The next section addresses the implementation of and we will study how the problem of implementation and the possibility of use of public or private blockchains has been addressed in the design of the proposal.

3 Proposal

Traditional industries are evolving into smart factories by incorporating sensors for data collection and optimization; this allows to automate processes. A sensor is a device that detects and responds to some type of input from the physical environment. These include but are not limited to light, heat, motion, moisture.

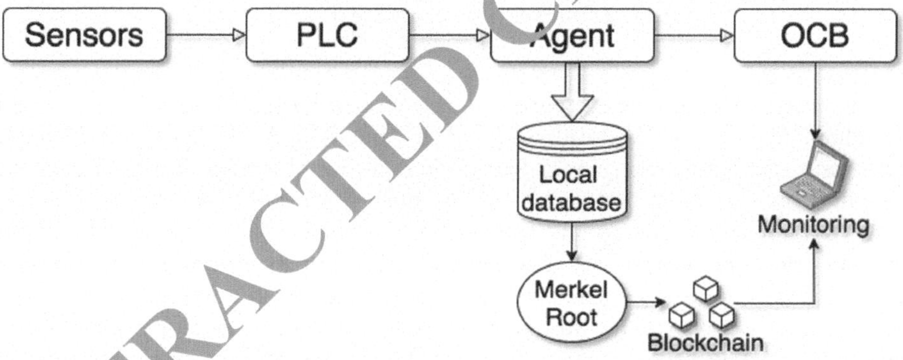

Fig. 1. A high level overview of the proposed model.

A signal comes out to be as output which is converted to human-readable display value [3,25]. In our model represented in Fig. 1, we consider a wide range of IoT data is gathered from the Sensors. The observed data goes through the Programmable Logic Controller (PLC) module, which is a ruggedized computer used for industrial automation.

The PLC receives data from connected sensors, processes the data, and triggers output based on pre-programmed parameters. PLC can monitor and record run-time data such as machine productivity or operating temperature, automatically start and stop processes, generate alarms if a machine malfunctions, and many more [8].

The processed output by the PLC go through the Agent module which helps in the concatenation to form a string (S) of the form ($S = Type\_of\_sensor + Value + PLC\_ID + Timestamp$) and to manage the local database. The formed string go through the Orion Context Broker (OCB) for real time monitoring which allows us to manage the entire process of the context information which includes but are not limited to updates, query, registration and subscription [30].

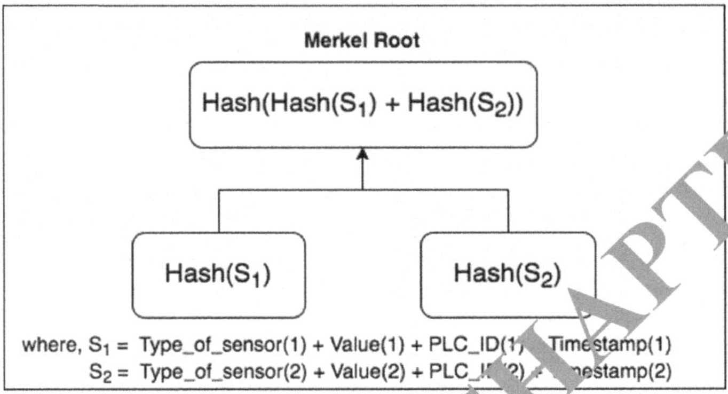

Fig. 2. Merkel tree for two leaf nodes.

The root of Merkel trees, created from each batch of data, is uploaded to the blockchain from the local database in each time slot. A Merkel tree is a hash-based data structure, where each leaf node contains the hash of a transactional block and each non-leaf node contains the hash of its children nodes.

A Merkle tree summarises all the verified transactions by repeatedly hashing the data and producing a new Merkle root [29]. For every fixed slot period, a hash value is generated for each string that is gathered in a local database, this hash value is added to the Merkle tree as a leaf node and the whole Merkle tree is updated leading to the new block in the blockchain as represented in Fig. 2. One of the main features of the proposed framework is the use of smart contracts which helps to validate and audit the content received throughout the production process. The validated data is monitored with the run-time stream of data which helps in the early identification of new threats, through the detection of variations.

Algorithm 1: PseudoCode: Storage in Smart Contract

```
public: setMeasurements (string _batchId, bytes32 _merkleRoot){
require(msg.sender == modules[_moduleId].plcAddress);
emit NewMeasurment(msg.sender, _batchId, _merkleRoot);
}
```

Every smart contract deployed in the blockchain e.g.- Ethereum, has its own personal database, known as smart contract storage. The public function described in the previous pseudocode algorithm helps maintain the smart contract storage which takes *_batchId*, the id of the batch of the stored data; and *_merkleRoot*, Merkle root observed for the given batch of data; as arguments.

Events and logs are important in a blockchain, where Ethereum is the most popular decentralized application, because it facilitates communication between smart contracts and their user interfaces.

When a transaction is verified, smart contracts can emit events and write logs to the blockchain that the frontend can then process. Whenever an event is emitted, the corresponding logs are written in the blockchain. Using an event to write logs in the blockchain is a cheaper form of storage [21, 22]. This is how a blockchain based framework is setup to ensure the correct traceability of the data and to provide physical and logical protection.

4 Results

In the proposed model, a wide range of IoT data is observed and manipulated according to the demand. One of the most significant concerns associated with such an interconnected heterogeneous network is the loss of personal information.

Public Blockchains are a form of peer-to-peer decentralized network that allows multiple nodes to participate in the network and perform operations without having to rely on a trusted third party. In permissioned Blockchains, written permissions are kept centralized at one entity or a group of entities whereas read permissions may be public or restricted to an arbitrary extent. Permissioned blockchains are faster, safer, and more efficient, however, they hold permissions centrally, which makes them lose the decentralization feature [27].

Advantages of Private Blockchain over Public Blockchain [1, 13]:

- It provides a greater level of privacy because of restricted read and write permissions.
- Can easily change the rules of blockchain, revert transactions, modify balances etc.
- Known validators restrict the addition of falsified blocks in the chain.
- Transactions are cheaper as verification is done by fewer nodes and less processing power is required.
- Well connected nodes help fix faults by means of a manual intervention and control the maximum block size.

In a public blockchain based on Ethereum, Gas is needed to maintain the network running. Gas is the Interface of the cyptocurrency native of Ethereum (Ether), which means that the value of Gas depends on Ether. This currency is paid by the user to the network when the user executes a function of a smart contract that changes its state [16].

The fixed amount of 21000 Gas is needed to execute a smart contract in the Ethereum network, in addition to the Gas needed to execute the operations of the smart contract [28].

In our aforementioned model, Algorithm 1 shows that there are mainly two operations involved in executing a smart contract.

- 1 transaction to call the function setMeasurements.
- 1 transaction to emit the measurements obtained in arguments.

So the total gas value is the addition of the gas value required to execute these two operations and the fixed amount required to execute a smart contract. So, the operation costs are negligible comparing with fixed amount of gas spent for executing the smart contract.

One of the most significant concerns right now regards the fact that the fees of public blockchains tend to exceed $0.01 per transaction. However, this may change in the future with scalable blockchain technology that has the potential to bring public-blockchain costs down to within one or two orders of magnitude of an optimally efficient private blockchain setup.

Ryan et al. in [28] calculated that, if the median gas price is 28 Gwei, then the total gas required for one transaction to be executed by one PLC on a public Ethereum Blockchain is: 21000 (Fixed amount) +1 (calling of function setMeasuremnts) +20000 (to emit 256-bit word to storage) $= 21001 Gas = 750$ Gwei. So the total Gas needed for 7 PLCs to execute during the entire day is $21000 + 7 * 1 * 24 + 7 * 20000 * 24 = 3380168 Gas =$ 120721[illegible] Gwei.

This section has considered the advantages and disadvantages of using a public or a permissioned blockchain network in this use case scenario. Thanks to the proposed framework, it is possible to run transactions on a public network, at an affordable cost.

5 Conclusions and Future Work

Industry 4.0 is not a far fetched concept, however more robust security solutions are required to deal with its complexity. To address security concerns regarding the transfer and logging of data transactions in an Industry 4.0 system, we have proposed a blockchain-based platform that allows for real-time monitoring and uses smart contracts to record sensor metadata in transaction logs.

Our model utilizes a blockchain based on Ethereum to test the smart contracts for the secure emission of the sensor data, overcoming issues such as: loss of personal information or manipulation of the data by third parties, which directly affects the performance of the organisation. Blockchain improves the security of communications, in addition to the fact that it gives us the possibility to trust the origin of the data. Hence, there are near-future prospects for integrating blockchain technology in Industry 4.0.

Future lines of research will focus on solving deployment problems for optimized performance, which includes, but is not limited to, the latency time of the sensor network and the traffic flow it supports. The authors will collaborate with a small factory operator to demonstrate the performance of the framework in a real-world environment.

Acknowledgements. Travel expenses were partially covered by the Travel Award sponsored by the open access journal Applied Sciences published by MDPI. On the other hand, the research of Yeray Mezquita is supported by a pre-doctoral fellowship granted by University of Salamanca and cofinanced by Banco Santander. Finally, this research has been supported too by the project "Intelligent and sustainable mobility supported by multi-agent systems and edge computing (InEDGEMobility): Towards Sustainable Intelligent Mobility: Blockchain-based framework for IoT Security", Reference: RTI2018-095390-B-C32, financed by the Spanish Ministry of Science, Innovation and Universities (MCIU), the State Research Agency (AEI) and the European Regional Development Fund (FEDR).

References

1. Alhadhrami, Z., Alghfeli, S., Alghfeli, M., Abedlla, J.A., Shuaib, K.: Introducing blockchains for healthcare. In: 2017 International Conference on Electrical and Computing Technologies and Applications (ICECTA), pp. 1–4. IEEE (2017)
2. Antón, S.D., Schotten, H.D.: Putting together the pieces: A concept for holistic industrial intrusion detection. In: ECCWS 2019 18th European Conference on Cyber Warfare and Security, p. 178. Academic Conferences and publishing limited (2019)
3. Bilger, B.: Occupancy sensor and method for home automation system. uS Patent 6,909,921 (2005)
4. Casado-Vara, R., Martin-del Rey, A., Affes, S., Prieto, J., Corchado, J.M.: Iot network slicing on virtual layers of homogeneous data for improved algorithm operation in smart buildings. Future Gener. Comput. Syst. **102**, 965–977 (2020)
5. Chamoso, P., González-Briones, A., De La Prieta, F., Venyagamoorthy, G.K., Corchado, J.M.: Smart city as a distributed platform: toward a system for citizen-oriented management. Comput. Commun. **152**, 323–332 (2020)
6. Consortium, I.I.: Industrial internet consortium, the industrial internet consortium's approach to securing industrial internet systems (2015). https://bit.ly/31Rxfep
7. Corchado, J.M., et al.: Deepint.net: a rapid deployment platform for smart territories. Sensors **21**(1), 236 (2021)
8. Erickson, K.T.: Programmable logic controllers. IEEE Potent **15**(1), 14–17 (1996)
9. Francisco, M., Mezquita, Y., Revollar, S., Vega, P., De Paz, J.F.: Multi-agent distributed model predictive control with fuzzy negotiation. Expert Syst. Appl. **129**, 68–83 (2019)
10. González Bedia, M., Corchado Rodríguez, J.M., et al.: A planning strategy based on variational calculus for deliberative agents (2002)
11. González-Briones, A., Castellanos-Garzón, J.A., Mezquita Martín, Y., Prieto, J., Corchado, J.M.: A framework for knowledge discovery from wireless sensor networks in rural environments: a crop irrigation systems case study. Wirel. Commun. Mob. Comput. **2018** (2018)
12. Group, B.R.: Doyfe project (2015). https://bisite.usal.es/es/apps/doyfe
13. Hjálmarsson, F.Þ., Hreiðarsson, G.K., Hamdaqa, M., Hjálmtýsson, G.: Blockchain-based e-voting system. In: 2018 IEEE 11th International Conference on Cloud Computing (CLOUD), pp. 983–986. IEEE (2018)
14. Javaid, A., Javaid, N., Imran, M.: Ensuring analyzing and monetization of data using data science and blockchain in loT devices. Ph.D. thesis, MS thesis, COMSATS University Islamabad (CUI), Islamabad 44000, Pakistan (2019)
15. Mezquita, Y., Casado, R., Gonzalez-Briones, A., Prieto, J., Corchado, J.M.: Blockchain technology in iot systems: review of the challenges. Ann. Emerg. Technol. Computi. (AETiC) **3**(5), 17–24 (2019)

16. Mezquita, Y., Gazafroudi, A.S., Corchado, J., Shafie-Khah, M., Laaksonen, H., Kamišalić, A.: Multi-agent architecture for peer-to-peer electricity trading based on blockchain technology. In: 2019 XXVII International Conference on Information, Communication and Automation Technologies (ICAT), pp. 1–6. IEEE (2019)
17. Mezquita, Y., González-Briones, A., Casado-Vara, R., Chamoso, P., Prieto, J., Corchado, J.M.: Blockchain-based architecture: a mas proposal for efficient agri-food supply chains. In: Novais, P., Lloret, J., Chamoso, P., Carneiro, D., Navarro, E., Omatu, S. (eds.) ISAmI 2019. AISC, vol. 1006, pp. 89–96. Springer, Cham (2020). https://doi.org/10.1007/978-3-030-24097-4_11
18. Mezquita, Y., Valdeolmillos, D., González-Briones, A., Prieto, J., Corchado, J.M.: Legal aspects and emerging risks in the use of smart contracts based on blockchain. In: Uden, L., Ting, I.-H., Corchado, J.M. (eds.) KMO 2019. CCIS, vol. 1027, pp. 525–535. Springer, Cham (2019). https://doi.org/10.1007/978-3-030-21451-7_45
19. Microsoft: Win32/stuxnet (2015). https://bit.ly/37JFQ40
20. News, B.: Stuxnet worm hits iran nuclear plant staff computers. nist, guide to industrial control systems (ics) security, special publication 800-82 (2010). https://goo.gl/A5e3G1
21. Ourad, A.Z., Belgacem, B., Salah, K.: Using blockchain for IOT access control and authentication management. In: Georgakopoulos, D., Zhang, L.-J. (eds.) ICIOT 2018. LNCS, vol. 10972, pp. 150–164. Springer, Cham (2018). https://doi.org/10.1007/978-3-319-94370-1_11
22. Pilkington, M.: Blockchain technology: principles and applications. In: Research Handbook on Digital Transformations. Edward Elgar Publishing (2016)
23. Pinzon, C.I., De Paz, J.F., Herrero, A., Corchado, E., Bajo, J., Corchado, J.M.: idmas-sql: intrusion detection based on mas to detect and block sql injection through data mining. Inf. Sci. **231**, 15–31 (2013)
24. Pinzón, C.I., De Paz, J.F., Tapia, D.I., Bajo, J., Corchado, J.M.: Improving the security level of the fusion@ multi-agent architecture. Expert Syst. Appl. **39**(8), 7536–7545 (2012)
25. Pottie, G.J., Kaiser, W.J.: Wireless integrated network sensors. Commun. ACM **43**(5), 51–58 (2000)
26. Qiu, H., Qiu, M., Memmi, G., Ming, Z., Liu, M.: A dynamic scalable blockchain based communication architecture for IoT. In: Qiu, M. (ed.) SmartBlock 2018. LNCS, vol. 11373, pp. 159–166. Springer, Cham (2018). https://doi.org/10.1007/978-3-030-05764-0_17
27. Rouhani, S., Deters, R.: Performance analysis of ethereum transactions in private blockchain. In: 2017 8th IEEE International Conference on Software Engineering and Service Science (ICSESS), pp. 70–74. IEEE (2017)
28. Ryan, D.: Calculating costs in ethereum contracts (2017). https://hackernoon.com/
29. Singh, M., Kim, S.: Blockchain based intelligent vehicle data sharing framework. arXiv preprint arXiv:1708.09721 (2017)
30. Sun, Y., Song, H., Jara, A.J., Bie, R.: Internet of things and big data analytics for smart and connected communities. IEEE Access **4**, 766–773 (2016)
31. Valdeolmillos, D., Mezquita, Y., González-Briones, A., Prieto, J., Corchado, J.M.: Blockchain technology: a review of the current challenges of cryptocurrency. In: Prieto, J., Das, A.K., Ferretti, S., Pinto, A., Corchado, J.M. (eds.) BLOCKCHAIN 2019. AISC, vol. 1010, pp. 153–160. Springer, Cham (2020). https://doi.org/10.1007/978-3-030-23813-1_19

32. Yigitcanlar, T., Butler, L., Windle, E., Desouza, K.C., Mehmood, R., Corchado, J.M.: Can building "artificially intelligent cities" safeguard humanity from natural disasters, pandemics, and other catastrophes? an urban scholar's perspective. Sensors **20**(10), 2988 (2020)

Distributed Decision Blockchain-Secured Support System to Enhance Stock Market Investment Process

Elena Hernández-Nieves(✉), José A. García-Coria, Sara Rodríguez-González, and Ana B. Gil-González

Bisite Research Group, University of Salamanca, Salamanca, SP, Spain
elenahn@usal.es

Abstract. One of the main purposes of financial technology solutions is to reduce the infrastructure costs of financial institutions. Adopting the use of FinTech technologies implies, in addition to a reduction in costs derived from physical entities, an improvement in user satisfaction, greater flexibility in terms of data access on different devices and more transparency in financial management. As a result, this paper proposes a blockchain-based scalable platform for investment recommendations. The platform proposed in this research serves as a decision support tool. A use case is included showing the process from the initial product search to the final investment recommendation that concludes in the optimal portfolio.

Keywords: Artificial intelligence · Big Data analytics · Blockchain · FinTech

1 Introduction

The worldwide financial sector has evolved thanks to the emergence of the Internet and the implementation of new technologies. This trend is reflected in the coining of the term FinTech (used to refer to software and other technologies that support or enable banking and financial services). The financial industry, aware of the need to apply technology to improve its activity, includes online credit, risk analysis and the processing of big data as FinTech developments [17]. The aim of this research is to improve customer experience by focusing and orienting itself entirely towards the customer. Currently, FinTech technologies are being developed in different startups that aim to sell a specific product (such as user experience in the case of Spring Studio or currency exchange in the case of Kantox). However, this research aims to create a blockchain-secured stock market investment process system, having open (or paid) data sources and allowing others to be aggregated (hence not explicitly defined). The key points of the contribution of this work can be summarized in:

J. M. Corchado and S. Trabelsi (Eds.): SSCTIC 2021, LNNS 253, pp. 48–60, 2022.
https://doi.org/10.1007/978-3-030-78901-5_5

1. It should serve as a decision support system.
2. It must incorporate a blockchain system for secure transactions.

The banking and insurance sector is undergoing a profound transformation process. An increasing number of online services are being offered in order to be more efficient and less costly for the company. This context is being exploited by many companies to create new value-added technological services that can be marketed to the end customer directly, or to the banking and insurance sector itself.

These tools currently require an analyst or expert to interpret the obtained data. It should be noted that platforms often incorporate a static reporting system without incorporating new visualisation mechanisms to facilitate data comprehension [4,5]. Therefore, it is up to an expert user to interpret the usefulness or not of the extracted data, and there is no intelligence that facilitates and specifies the information that the company needs.

The main motivation is to design a FinTech solution that allows to consult financial events in real time using big data and monitoring techniques. In addition, recommendation techniques according to the investor's profile and the algorithms that will help the user to learn how to invest depending on their profile are also investigated.

The paper is organized as follows: Sect. 2 overviews the current state of the art, Sect. 3 describes the FinTech platform architecture and the proposed model, including the modelling of the platform API module. Section 4 presents the obtained results. Finally, Sect. 4 covers the conclusions as well as future research.

2 State of the Art

2.1 Data Management Applied to Service Delivery

Approximately, 49% of companies are involved in Big Data projects or will be in the near future [26]. This is not surprising given its enormous advantages; business intelligence is fed with data in real time, while at the same time orienting the company towards the customer by providing information for predictive scenarios. The intensive use of Big Data is being made mostly in digital companies where analysis is critical to be able to reach the "virtual customer" [23]. These companies have achieved competitive advantages that have placed them in dominant positions in the market, the case of Amazon is an example [6].

Digital business is an increasingly time-consuming task. To anticipate the competitors, enterprises require knowledge of events, both inside and outside their organisation. Context is needed to support the demand of digital business. A highly dispersed and voluminous context, spread across hundreds of thousands of websites, devices and social media. A new category of cloud services needs to emerge to provide data to businesses, supporting their operations and decision-making processes [9,19]. The availability of consumer information (via mobile, social media, the cloud...) about its vendors will create a new style of valuation based on data from thousands of sensors embedded in products. This much more objective information that can be a differentiating factor for brands [3].

The research carried out aims to manage large amounts of data held by companies, providing a valuable tool to each user involved in the process. The technology derived from the use of Big Data makes it possible to observe and analyse the information, assisting the users of the service who may be interested in making decisions. In addition to Big Data, other technologies that have been considered are FinTech and Blockchain, which are explored further in the subsections that follow.

2.2 FinTech Technologies

FinTech is a new paradigm of recent implementation in which companies use ICT (information and communication technologies) to offer financial services efficiently [12]. The use of FinTech technologies often results in cost savings for the company that uses them. However, according to the FinTech Report published by the Spanish association of corporate financiers, only 46.2% of finance experts use some kind of FinTech solution in their work environment [1]. Even so, the financial sector is expected to see considerable growth in this type of tool in the coming years. Currently, different startups are being launched for different purposes at the national and international level (P2P payments, bitcoins, crowdlending, crowdfunding, investment tools, loans, currencies, etc.). Some outstanding startups are:

1. Adyen: Multi-channel payment service providing payment services to more than 3,500 companies, including Facebook, Uber and Airbnb.
2. eToro: A social investment network that offers access to global markets by allowing investors to follow other investors and invest by copying their trades [2].
3. Captio: Service that allows to report and manage the expenses and profits of a company [18].

One of the limitations of this technology is the relative lack of public awareness regarding its potential. This research aims to bring FinTech technologies closer to financial services companies, so they can become pioneers in the incorporation of technologies. Furthermore, it is intended that the FinTech technologies described in this research will be able to provide better efficiency and allow for the optimisation of resources.

2.3 Blockchain

Blockchain can be understood as a large set of records (called blocks) that are linked and encrypted to protect the security and privacy of transactions, not necessarily economic ones [24]. Each block contains a number of records that must be validated by sealing. This process is carried out by several users or nodes, and is called “block mining”, which involves a series of complex calculations that require time and (increasing amounts of) energy. When the process is finished, these blocks are permanently recorded and therefore one block cannot be modified without altering all the other blocks linked to it; an operation that would also require most nodes to validate it. There are several state-of-the-art technologies that support this procedure in a secure manner

1. Ethereum: One of the most widely used platforms due to its versatility and ease of use [28].
2. R3 Consortium: Financial institutions have created the R3 consortium in order to find out how to take advantage of blockchain in traditional financial systems [16].
3. TenX: It is a startup that has created a prepaid card that can be recharged with different cryptocurrencies. It can be used to pay in any establishment just like conventional cards [8].
4. Carsharing: Ernst & Young Global Ltd. subsidiary, EY, is developing a blockchain-based system that allows companies or groups of people to access a car-sharing service in a simple way [14].
5. Cloud storage: Storj wants to decentralise this service to improve security and reduce dependence on the storage provider [10].
6. Digital identity: recent massive security breaches and data theft have made the security of our identities a very real concern. Blockchain could provide a unique system for achieving irrefutable, secure and immutable identity validation. There are many companies developing services in this area, and all of them believe that applying blockchain technology for this purpose is an optimal solution [11].
7. Authorship management: There are platforms such as Ascribe, Bitproof, Blockai, Stampery that aim to attribute authorship to works through Blockchain [22].

Despite the enormous advantages of Blockchain technology in terms of security, confidentiality and data integrity, its penetration in the field of finance is still very limited. In addition, the removal of intermediaries provides very interesting cost savings for companies. This research aims to introduce Blockchain as an essential part of the services offered by the institutions.

3 FinTech Platform Architecture and Proposed Modelling

This research uses technology to make financial services available to customers that were previously only offered by banks. The ultimate goal of this research is a service development platform (PaaS) offering FinTech products. The proposed architecture involves the following key aspects (Fig. 1):

1. Access is facilitated through a web portal or a mobile application to enhance user-friendliness and socio-economic impact. A customisation layer of these services is integrated according to the target market niche.

2. The platform performs a Big Data analysis of the banking and insurance sector. Therefore, all the information required for stock market and financial analysis is considered. This analysis must be shown to end users in a graphical and intuitive way.
3. The platform is designed as a decision support tool. Therefore, it incorporates learning mechanisms based on the analysed data [15].
4. The platform is designed in a modular way. The decision support tool is the cornerstone on which these new services must be based.
5. On the basis of the created services, the platform should enable the establishment of smart contracts in blockchain.

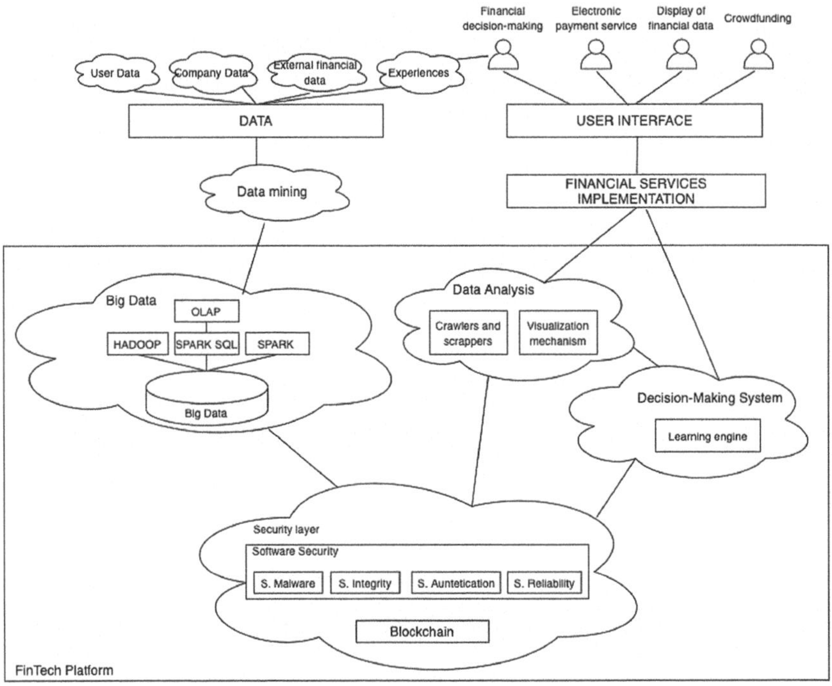

Fig. 1. Outline of the FinTech platform.

For the sectors covered here (banking and insurance), the immensity of the data to be processed makes it necessary to use Big Data-based technologies for information management and time constraint adjustment. In this regard, interoperability interfaces with external persistence systems have been designed to enable long-term storage and backup tasks. One of the cornerstones is the incorporation of the blockchain system for the security of the operations carried out within the platform. This system allows to validate transactions for the users connected to the platform, which decentralises the calculation process, saving computing time and improving efficiency, security and data integrity.

To be able to develop it, it is essential to design the information persistence in accordance with the storage system. On the basis of this data persistence format, decision making, data analysis, data mining, and Big Data processing modules have been defined, which can be synthesised and displayed according to the created service. These models are used for data visualisation by the companies involved. These companies will be able to transparently access the historical data and, in turn, incorporate new data that are relevant for the future use of the platform. The collected information will also be incorporated into a reasoning system (machine learning) that will be essential for the construction of a decision support tool.

3.1 Proposed Model

Due to the nature of the research, the optimal architecture is a Big Data model. In this section, we describe what the proposal consists, what problems it is targeted at and how they are solved. As a solution to the problem of processing large amounts of information due to the use of Big Data technology, it has been decided to approach the data computation methodology as stateless microservices. This allows to address the problem by completely eliminating the need for synchronisation between the different computation modules. At the same time, it allows for the scalability of the computation in a way that is linear to the number of nodes used for the computation. Thus, making use of this paradigm, an API is developed that allows for both, the collection of data and its processing.

The Stateless Microservices paradigm consists of dividing computational tasks into independent atomic units, not requiring communication or synchronisation with each other [25]. They must be timeless, i.e., their execution does not depend on the time of their execution. They are microservices in the sense that they do not consume computing resources beyond those used during the computation. This allows for a much more efficient use of processors and the main memory. They are stateless because operations are not required to maintain a constant state between steps. A complete procedure can execute each atomic operation in a different computation unit, thus allowing for a more efficient use of resources, and also, obtaining the maximum possible scalability. This is achieved, firstly, because all the information needed to perform an operation is passed as arguments to the entry point of that operation, without any other external state information being necessary. On the other hand, to perform the distribution of operations and to fully exploit this paradigm, some method of load balancing must be incorporated into the overall architecture of the platform.

Selection and Design of the Infrastructure. The Kubernetes has been chosen as the technology capable of meeting these requirements. Kubernetes, as mentioned above, is a deployment orchestration platform developed by Google, which makes use of Docker container technology. Its use makes it possible to encapsulate each component of the architecture in a container, as well as to specify in a file the architecture that the platform must have at the internal network level. It also specifies the individual deployment parameters of each module (the number of replicas, environment variables, exposed network ports, etc.) [20,21].

The ultimate goal is that the platform can be deployed on any server provider, whether it is a public cloud provider such as AWS (Amazon Web Service) or a private cloud, with the guarantee that the platform will work as expected without any type of configuration or adaptation to the underlying technology on which it is being deployed. Figure 2 illustrates the Kubernetes components used to deploy the platform in compliance with the established requirements, as well as the communication routes that the user requests would follow when making use of load balancing. In the Master node, the "Deployment" components are specified, is in charge of deploying the replicas of the components in the available nodes, trying to make the most of their computational capacity. Each node can host an indeterminate number of Front and API instances, depending on their computational characteristics. It is possible to deploy as many nodes as is required, and Kubernetes takes care of deploying the modules on them in an optimal way.

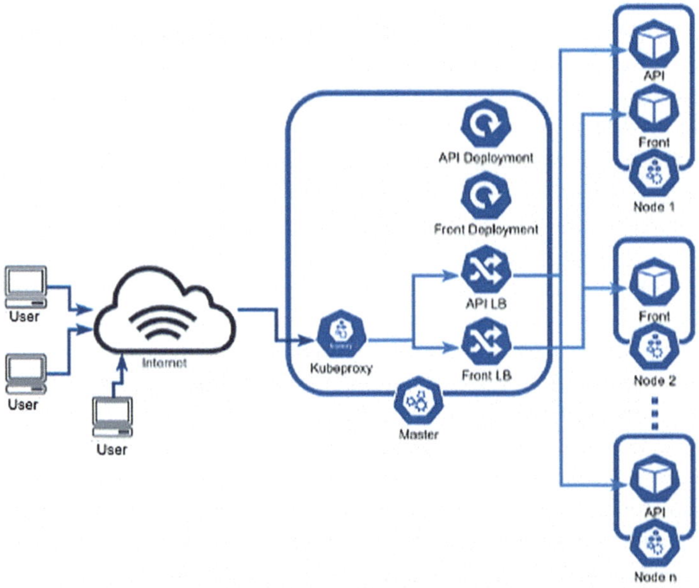

Fig. 2. Kubernetes architecture.

Regarding the deployment infrastructure in terms of hardware, a deployment model using the Amazon Web Services EC2 service has been proposed. In this model, the master node and the execution nodes must be in the same virtual private network. The master node is the only one that must have access to the internet, as the services hosted on the compute nodes are accessed through

Kubeproxy. The compute nodes are replicas of the same instance with Kubernetes installed and the Docker containers of the different services. The Master node then manages the deployment of these services on the nodes as appropriate.

3.2 Platform API Module

The API is designed to retrieve data from InvestPy [13] (a Python package to retrieve data from Investing.com, which provides data retrieval from up to: 39952 stocks, 82221 funds, 11403 etfs, 2029 currency crosses, 7797 indices, 688 bonds, 66 commodities, 250 certificates and 2812 cryptocurrencies). Also, it is designed to offer recommendation services on the platform. Python and Flask have been used to build it. Data extraction has been done through endpoints that obtain data for filtering, as well as endpoints that provide the data itself.

Regarding filtering we can find $get\_product\_types()$, which return the possible product types, which are stocks, certificates, etf, funds, index, bonds, cryptocurrencies, currency pairs and commodities. Of the aforementioned products, the first five use both the product name and the country to obtain the data, while in the remaining ones the product name is sufficient. Commodities stand out because they use groups instead of countries. To retrieve the filtering data available for each product, we have the method $get\_filtering\_data()$. This method obtains the type of product by argument and returns an object whose fields generally provide the available countries and names. If the historical record is chosen, a start and end date for the retrieval of the cases must be provided. This historical record is used later to train data for Machine Learning recommender models.

In addition, a number of indicators used to reinforce buying or selling decisions can also be obtained. To access these indicators, the same parameters as are used those used to identify the products above: name and, if necessary, country. In this way there are technical indicators, pivot points, and moving averages. The code of the functions does not differ much from the previous ones, as the bulk is still the control of arguments, the rest is reduced to the investpy call.

Black-Litterman Model. This model allows the investor's expectations to be included and, according to the confidence in these expectations, a greater or lesser weight is given to the asset within the portfolio [7]. The class computes the optimisation based on the Black-Litterman model, which assumes that the initial expected returns are those required for the equilibrium asset allocation to be equal to what we observe in the markets. The user only has to indicate how their assumptions about expected returns differ from the markets and state their degree of confidence in the alternative assumptions. Then, the Black-Litterman method calculates the desired asset allocation (mean-variance efficiency).

Markowitz Model. The Markowitz model is designed to find the optimal investment portfolio for each investor in terms of return and risk by making an appropriate choice of the assets that make up the portfolio [27]. The "Markowitz

Portfolio" class includes the procedure and methods necessary to find the optimisation for the portfolio entered by the user. The process of optimal portfolio selection can be summarised in three steps:

1. Fundamental analysis (macro and microeconomic aspects), as well as other aspects such as taxation, accounting regulations, entry and exit capacity, trading fees, price ranges.
2. Estimation of expected returns, variances and covariances, determination of the efficient frontier and the investor's indifference curves.
3. Choosing the optimal portfolio

Blockchain. This section contains the steps required to carry out a blockchain transaction and its subsequent storage in the database for its visualisation in a block explorer. To do this, the following steps are carried out:

1. The optimisation request is sent.
2. A tuple formed by the request and the response is generated.
3. Its hash is calculated.
4. The tuple and the hash are sent to the API endpoint for insertion in the blockchain.
5. The address returned by the API is stored together with the tuple in the database.

In addition, the following functions are available:

1. "*ADD_Recommendation*" function. In charge of generating a transaction, signing it and sending it to the blockchain.
2. "*GET_RECOMMENDATIONS*" function. In charge of retrieving all existing recommendations, or those corresponding to the index passed as an argument.
3. Block explorer function. It is a web application that provides access to the blocks. It works as a search engine connected to the blockchain. Its main function is to allow anyone with an Internet connection to follow in real time all the transactions carried out on the network.
4. "JS WEB" function. It is in charge of compiling, in a table, the data retrieved from the Back of the web. The table incorporates a link to the Ethrscan block explorer where all the information of each block can be visualised in detail.

Considering AI's current ability to ingest big data for exploratory studies and real-time decision-making [29], a distributed platform has been developed that provides a monitoring and decision support service for stock market investors, whether they are users or companies. To validate the system, a case study has been designed with several ETF products, making use of artificial intelligence techniques to improve the decision-making process of stock market investors. Figure 3 shows the results obtained by the platform for an ETF search. For each search, the country, the ETF name, its symbol, the values for the last day, the exchange rate and rotation, as well as the currency, are displayed. Figure 4

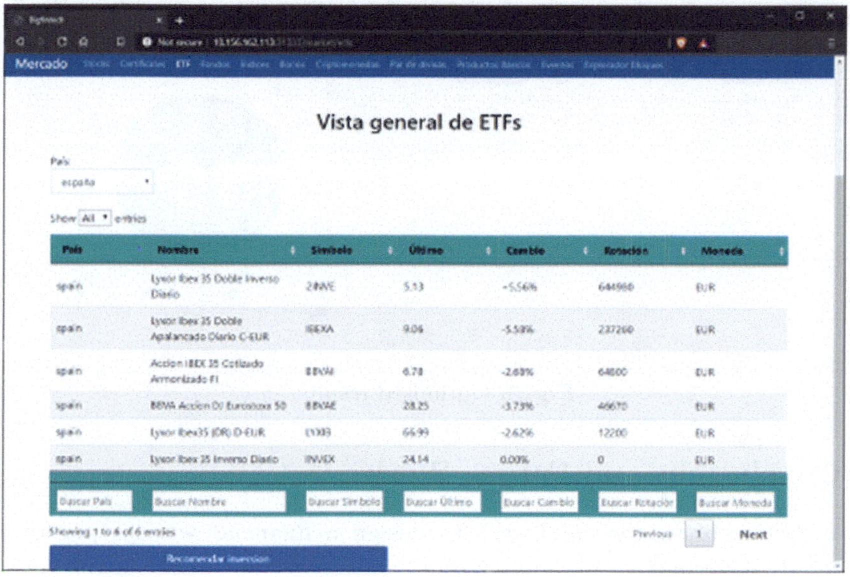

Fig. 3. ETFs general view.

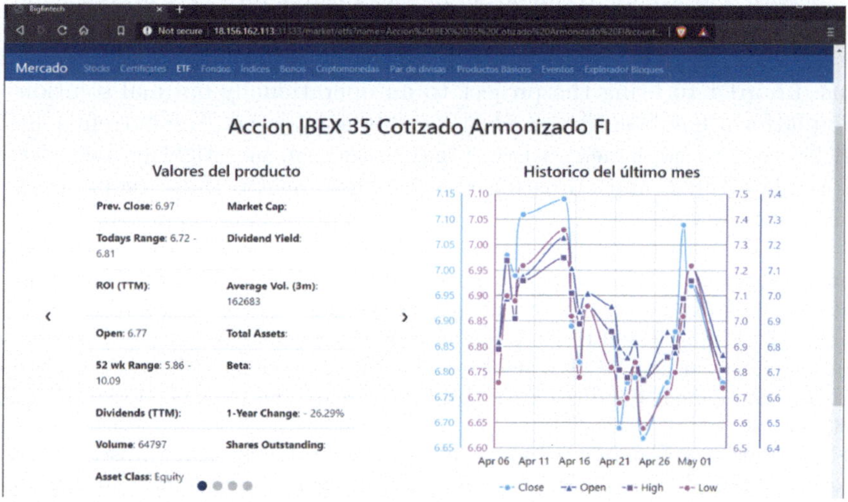

Fig. 4. Product values and last month's historical values for the IBEX 35.

shows the opening and closing values, the daily range and the last month's history for the harmonised IBEX 35. Figure 5 shows the optimised result. It shows the products, the country they come from and the recommended investment percentage. In addition, the privacy and security of all this data is ensured through blockchain technology.

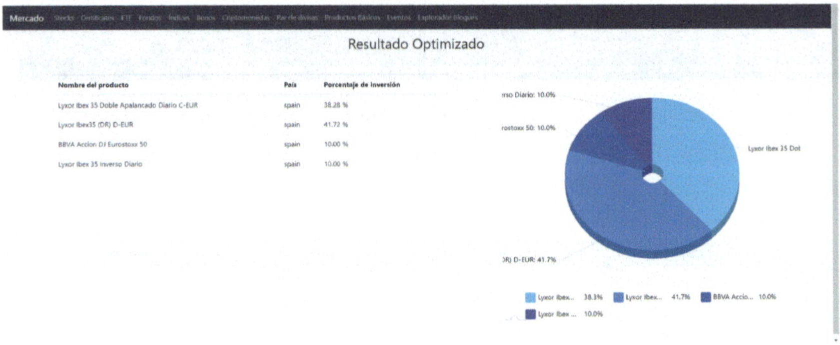

Fig. 5. Optimised results.

4 Conclusions and Future Work

The objective of this research was to design a financial services platform to support decision-making in banking and insurance markets. Being aware that a platform capable of implementing FinTech services would have to cover a much wider scope, the research focused on the design of an investment recommender secured with blockchain technology. In the future, further FinTech solutions could be designed and developed following the same procedure and making the necessary changes to the information search (sources) and decision support algorithms. In order to bring the project to an operationally optimal solution, the whole platform has been designed in a modular way, i.e., the current platform could be used to implement a new FinTech service, investigating new decision support algorithms, and adjusting the visualisations to those required by the new FinTech service.

Acknowledgments. This work has been supported by the project "Intelligent and sustainable mobility supported by multi-agent systems and edge computing (InEDGE-Mobility) (RTI2018-095390-B-C32), financed by the Spanish Ministry of Science, Innovation and Universities (MCIU), the State Research Agency (AEI) and the European Regional Development Fund (FEDER) and the research network Go2Edge (RED2018-102585-T).

The research of Elena Hernández-Nieves is supported by the Ministry of Education of the Junta de Castilla y León and the European Social Fund through a grant from predoctoral recruitment of research personnel associated with the University of Salamanca research project "ROBIN: Robo-advisor intelligent".

References

1. Arner, D.W., Barberis, J., Buckley, R.P.: The evolution of FinTech: a new post-crisis paradigm. Geo. J. Int. **47**, 1271 (2015)
2. Assia, Y.: eToro-building the world's largest social investment network. In: The FinTech Book: The Financial Technology Handbook for Investors, Entrepreneurs and Visionaries, pp. 196–197 (2016)

3. Carneiro, D., Ara'jo, D., Pimienta, A., Novais, P.: Real time analytics for characterizing the computer user's state. Adv. Distrib. Comput. Artif. Intell. J. (ADCAIJ) **5**(4), 1–18 (2016)
4. Casado-Vara, R., Martin-del Rey, A., Affes, S., Prieto, J., Corchado, J.M.: IoT network slicing on virtual layers of homogeneous data for improved algorithm operation in smart buildings. Futur. Gener. Comput. Syst. **102**, 965–977 (2020)
5. Chamoso, P., González-Briones, A., De La Prieta, F., Venyagamoorthy, G.K., Corchado, J.M.: Smart city as a distributed platform: Toward a system for citizen-oriented management. Comput. Commun. **152**, 323–332 (2020)
6. Chen, H., Chiang, R.H., Storey, V.C.: Business intelligence and analytics: from big data to big impact. MIS Q. **36**, 1165–1188 (2012)
7. Chen, S.D., Lim, A.E.: A Generalized Black-Litterman model. Oper. Res. **68**(2), 381–410 (2020)
8. Chen, Y.: Blockchain tokens and the potential democratization of entrepreneurship and innovation. Bus. Horiz. **61**, 567–575 (2017)
9. Corchado, J.M., et al.: Deepint.net: a rapid deployment platform for smart territories. Sensors **21**(1), 236 (2021)
10. Crosby, M., Pattanayak, P., Verma, S., Kalyanaraman, V.: Blockchain technology: beyond bitcoin. Appl. Innov. **2**, 6–10 (2016)
11. Cross, D.B., Hallin, P.J., Thomlinson, M.W., Jones, T.C.: U.S. Patent No. 7,703,128. U.S. Patent and Trademark Office, Washington, DC (2010)
12. Dapp, T., et al.: FinTech-the digital(r) evolution in the financial sector. Deutsche Bank Res. **11**, 1–39 (2014)
13. del Canto, A.B.: investpy-financial data extraction from Investing.com with Python (2020)
14. Dorri, A., Steger, M., Kanhere, S.S., Jurdak, R.: Blockchain: a distributed solution to automotive security and privacy. IEEE Commun. Mag. **55**(12), 119–125 (2017)
15. González Bedia, M., Corchado Rodríguez, J.M.: A planning strategy based on variational calculus for deliberative agents. Comput. Inf. Syst. **9**, 2–13 (2002)
16. Guo, Y., Liang, C.: Blockchain application and outlook in the banking industry. Finan. Innov. **2**(1), 1–12 (2016). https://doi.org/10.1186/s40854-016-0034-9
17. Hernández-Nieves, E., Hernández, G., Gil-González, A.B., Rodríguez-González, S., Corchado, J.M.: Fog computing architecture for personalized recommendation of banking products. Exp. Syst. Appl. **140**, 112900 (2020)
18. Mackenzie, A.: The FinTech revolution. Lond. Bus. Sch. Rev. **26**(3), 50–53 (2015)
19. Provost, F., Fawcett, T.: Data science and its relationship to big data and data-driven decision making. Big Data **1**(1), 51–59 (2013)
20. Riti, P.: Pro DevOps with Google Cloud Platform: With Docker, Jenkins, and Kubernetes. Apress (2018)
21. Shah, J., Dubaria, D.: Building modern clouds: using docker, kubernetes & Google cloud platform. In: 2019 IEEE 9th Annual Computing and Communication Workshop and Conference (CCWC), pp. 0184–0189. IEEE (January 2019)
22. Shrier, D., Wu, W., Pentland, A.: Blockchain & infrastructure (identity, data security). MIT Connection Science, pp. 1–18 (2016)
23. Snijders, C., Matzat, U., Reips, U.D.: Big Data: big gaps of knowledge in the field of internet science. Int. J. Internet Sci. **7**(1), 1–5 (2012)
24. Swan, M.: Blockchain: Blueprint for a New Economy. O'Reilly Media, Inc. (2015)
25. Tonelli, R., Lunesu, M.I., Pinna, A., Taibi, D., Marchesi, M.: Implementing a microservices system with blockchain smart contracts. In: 2019 IEEE International Workshop on Blockchain Oriented Software Engineering (IWBOSE), pp. 22–31. IEEE (February 2019)

26. Walker, S.: Big Data: A Revolution That Will Transform How We Live, Work, and Think (2014)
27. Wang, W., Li, W., Zhang, N., Liu, K.: Portfolio formation with preselection using deep learning from long-term financial data. Exp. Syst. Appl. **143**, 113042 (2020)
28. Wood, G.: Ethereum: a secure decentralised generalised transaction ledger. Ethereum Project Yellow Paper, vol. 151, pp. 1–32 (2014)
29. Yigitcanlar, T., Butler, L., Windle, E., Desouza, K.C., Mehmood, R., Corchado, J.M.: Can building "artificially intelligent cities" safeguard humanity from natural disasters, pandemics, and other catastrophes? An urban scholar's perspective. Sensors **20**(10), 2988 (2020)

Enhanced Cybersecurity in Smart Cities: Integration Methods of OPC UA and Suricata

David García-Retuerta[✉], Roberto Casado-Vara, and Javier Prieto

University of Salamanca, Patio de Escuelas Menores, 37008 Salamanca, Spain
{dvid,rober,javierp}@usal.es
https://bisite.usal.es/en/group/team/David

Abstract. The increasing urbanisation and digitalisation taking place all around the world has increased the interest in Smart Cities (SCs). The main goal of any SC is to provide a high quality of life to its citizens by enhancing the quality of the public and private services. However, mass-adoption of the IoT technology and IT technology creates a series of security challenges that must be addressed. One of the most popular machine-to-machine communication protocols is OPC UA, and a widely accepted intrusion detection and prevention system is Suricata. The goal of this research is to find theoretical solutions upon which a successful integration can be achieved. The advantages and disadvantages are examined, and alternative methods are proposed.

Keywords: Cybersecurity · Smart cities · OPC UA · Suricata

1 Introduction

We live in a world which is becoming increasingly urbanised. The United Nations (UN) has predicted that up to two thirds of the global population will live in cities by 2050 [1], subsequently consuming most of the resources generated in the world. Currently, cities are estimated to consume 75% of the world resources and energy, accounting for 80% of the released greenhouse gases [2]. Smart Cities (SCs) aim to increase the quality of life of its residents, generating wealth, optimising resource consumption and increasing public safety. Thus, efficient, cost-effective, innovative solutions are needed for a vast variety of aspects in every SC [3].

As SCs grow in size and importance, the risk of cyberattacks also increases [4]. The use of modern technologies to tackle urban challenges creates a dependency on several new network elements, which become another critical infrastructure to take care of [5]. Denial-of-service attacks targeted at industry or city management infrastructure can now greatly disturb the normal operation of a SC, affecting millions of people. Moreover, as technology continues to democratise, citizens become more exposed to insecure WiFi networks and fraudulent web services [6]. The Safe Cities Index has been proposed as a important measurement to quantify

J. M. Corchado and S. Trabelsi (Eds.): SSCTIC 2021, LNNS 253, pp. 61–67, 2022.
https://doi.org/10.1007/978-3-030-78901-5_6

the public safety of SCs. It focuses on four main aspects: digital security, health security, infrastructure security and personal security. Modern platforms such as UPC UA and Suricata have the ability to enhance infrastructure security easily, by improving the network security.

1.1 Network Security

Protecting the integrity, confidentiality and accessibility of computer networks and data is one of the main challenges of any Smart City around the world [7]. Its solutions can be hardware-based, software-based or a mixture of both. A cost-effective solution to this problem is to create a software application which monitors the whole network or various systems looking for malicious activity or policy violations—an intrusion detection system.

There are two types of intrusion detection systems:

- Host-based.
- Network-based.

The host-based intrusion detection system is a tool that resides on a network node. Its operation is similar to that of an antivirus. They send a periodic report to the central node from which the state of the network is diagnosed [8].

The network-based intrusion detection system is a device connected to the network that monitors all network traffic (sniffer) and when it detects an inclusion it sends an alarm signal to the central console to initiate the necessary measures [9]. This is the type of system found in Suricata.

1.2 Cryptographic Algorithms

Modern cryptography is based upon computational hardness assumptions. Algorithms are designed in such a way that any attacker would take an unpractical amount of time and resources to break into a secure transmission uninvited [10]. Algorithms are constantly being modernised due to the constant advancements in computer power and novel mathematical attack methods, with quantum computers being the most alarming concern at the moment [11].

In our case, OPC UA encrypts and signs the packets that are transferred between the various network modules using the Basic256SHA256 algorithm [12]. The signature is done using the famous SHA-2 algorithm, and the asymmetric encryption is done using 256-bit Basic which is based on the mathematical RSA encryption method.

Asymmetric encryption uses two keys, one for encryption and one for decryption [13]. Both keys are mathematically related and are generated at each of the nodes. The public key is transmitted to the central node and is used to encrypt messages, while the private key must never leave the node (terminal) and is used to decrypt messages.

To prevent a man-in-the-middle attack (Fig. 1), the following procedure is used: the terminal node (the one with the private key) encrypts some data with

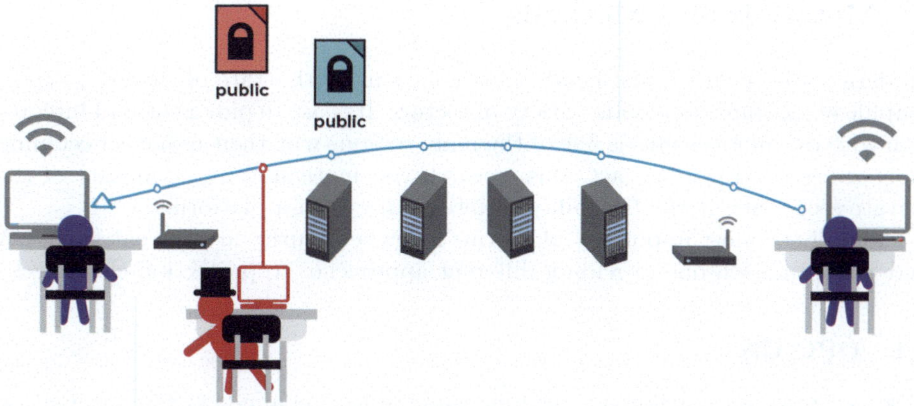

Fig. 1. Man-in-the-middle attack schema.

the private key and sends both the (blank) data and the encrypted data. Public key cryptography works in such a way that only the holder of the private key can efficiently perform this task, so if the central node has access to the public key, it can use it to decrypt the data. If it matches the data in the token, the recipient can be sure that it was generated by the private key holder (Fig. 2).

Fig. 2. Man-in-the-middle protection summary.

This paper is organised as follows: Sect. 1 presents the problem and makes a brief summary of the most important topics regarding cybersecurity in Smart Cities. Section 2 presents the programs which are the main focus of the current research: UPC UA and Suricata. Section 3 shows the possible integration methods of both technologies, as well as their advantages and disadvantages. Eventually, Sect. 4 shows the results of this work, their implication and proposes some alternatives.

2 Materials and Methods

Modern cybersecurity platforms offer solutions with state-of-the-art cryptographic algorithms for a wide variety of sectors. Its easy deployment and integration with existing systems is one of the main reasons why their usage is becoming more widespread [14]. In fact, this research was motivated by an inquiry of the private sector about the feasibility of integrating such a platforms.

Two of the most important platforms for cybersecurity are current UPC UA and Suricata, each one providing different approaches to protection challenges.

2.1 OPC UA

It is a protocol for machine-to-machine communication aimed at the automation of different industries. It allows the different machines in a factory to communicate with each other in a secure, scalable and standardised way. It uses the *Basic256SHA256* algorithm to guarantee the security of communications, signing the packets and encrypting them.

2.2 Suricata

It is an open source intrusion detection system capable of real-time detection, network security monitoring and inline intrusion prevention. It is widely popular and is appreciated by most security experts.

Part of its security analysis is based on applying a set of rules to ensure the trustworthiness of individual files. However, it is designed to ignore encrypted files in this analysis.

3 Integration

NOTE: The figures in this section show a simplified diagram of the network architecture. Actually, all the nodes communicate with each other, but for the sake of comprehension we represent all the packets as passing through a central node. How Suricata would fit into the connection is represented with the corresponding connection highlighted in red.

The possible integration methods of Suricata and UPC UA are as follows:

- **Sniffer.** Configure Suricata to intercept encrypted files, decrypt and analyse them (Fig. 3). Similar to an authorised sniffing attack.
  This configuration would require each node to transmit its own private key to Suricata, which is a serious threat to the security of the system. An authorised access to the keys' database would compromise the whole network. However, a fast performance is expected

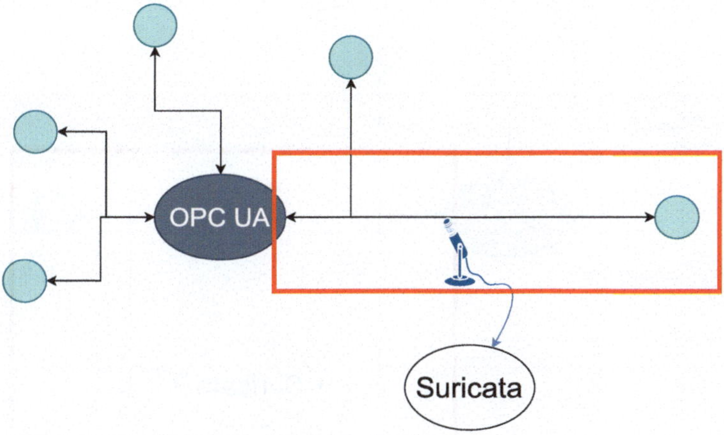

Fig. 3. Sniffer summary.

– **Man-in-the-middle.** Configure Suricata to intercept the encrypted files, decrypt, parse, and re-encrypt them and send them to the end node (Fig. 4). 256-bit Basic tool is unfortunately designed to prevent such attacks/configurations. Its implementation would require a lot of effort and modifications of the programs, with the basic security level of the network being lowered as a consequence. Moreover, the performance of the network is expected to be reduced as every single package must go through Suricata.

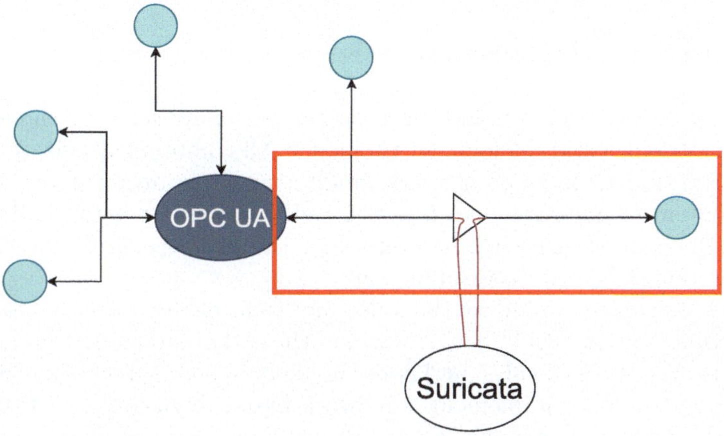

Fig. 4. Man-in-the-middle protection summary.

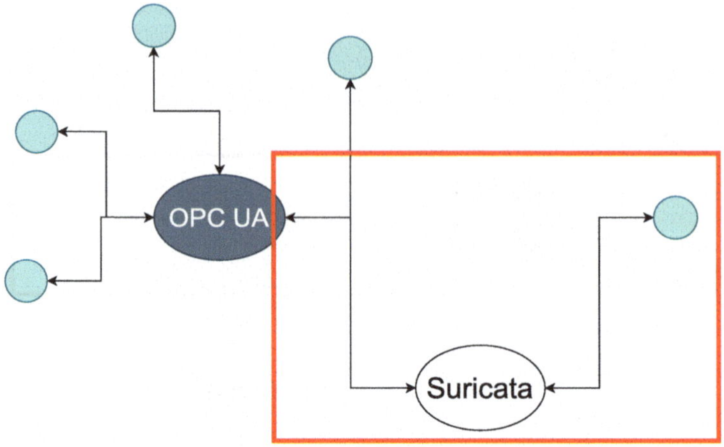

Fig. 5. Intermediary summary.

- **Intermediary.** Configure OPC UA so that the central node sends encrypted packets to Suricata, which decrypts, analyses and re-encrypts them to send them to the end node (Fig. 5).
  Its implementation would require a great effort, the lag in data transmission would be increased and a failure or an attack on Suricata would completely block the network.

None of these options would increase network security or reliability, which is precisely the goal of integrating Suricata with OPC UA.

4 Results and Discussion

The integration of UPC UA and Suricata creates many security challenges, as well as performance and deployment problems. All communications of UPC UA are safely encrypted, but any network monitoring program would need to read the raw content of packages, which would make the system highly vulnerable.

Thus, the new security threats outweigh the extra security layer, making integration pointless and disadvantageous.

A good alternative would be the following: to implement a host-based intrusion detection system with UPC UA. Each of the end nodes analyses its incoming packets based on a set of rules, and emits an alert in case it receives a suspicious packet. In addition, to implement a network-based intrusion detection system based on data flows. A statistical anomaly in the amount of data emitted by a node could raise the alarm to warn of a possible intruder. In case one of the end nodes is infected, in its attempt to replicate the virus, it will send packets to all available nodes. This unusual increase in the network packages can be detected by this method. Suricata can be used in this respect.

Acknowledgements. This research has been supported by the project "Intelligent and sustainable mobility supported by multi-agent systems and edge computing (InEDGE-Mobility): Towards Sustainable Intelligent Mobility: Blockchain-based framework for IoT Security", Reference: RTI2018-095390-B-C32, financed by the Spanish Ministry of Science, Innovation and Universities (MCIU), the State Research Agency (AEI) and the European Regional Development Fund (FEDER).

The research of Roberto Casado and Javier Prieto has been partially supported by the Salamanca Ciudad de Cultura y Saberes Foundation under the Talent Attraction Programme (CHROMOSOME project).

References

1. United Nations: World Urbanization Prospects: The 2014 Revision, Highlights. Department of Economic and Social Affairs. Population Division, United Nations 32 (2014)
2. Mohanty, S.P., Choppali, U., Kougianos, E.: Everything you wanted to know about smart cities: the Internet of Things is the backbone. IEEE Consum. Electron. Mag. **5**(3), 60–70 (2016)
3. Yigitcanlar, T., et al.: Artificial intelligence technologies and related urban planning and development concepts: how are they perceived and utilized in Australia? J. Open Innov. Technol. Mark. Complex. **6**(4), 187 (2020)
4. Lewis, J.A.: Assessing the Risks of Cyber Terrorism, Cyber War and Other Cyber Threats. Center for Strategic & International Studies, Washington, DC (2002)
5. Miller, B., Rowe, D.: A survey SCADA of and critical infrastructure incidents. In: Proceedings of the 1st Annual Conference on Research in Information Technology, pp. 51–56 (2012)
6. Khatoun, R., Zeadally, S.: Cybersecurity and privacy solutions in smart cities. IEEE Commun. Mag. **55**(3), 51–59 (2017)
7. Monzon, A.: Smart cities concept and challenges: bases for the assessment of smart city projects. In: 2015 International Conference on Smart Cities and Green ICT Systems (SMARTGREENS), pp. 1–11. IEEE (2015)
8. Ou, Y.j., Lin, Y., Zhang, Y.: The design and implementation of host-based intrusion detection system. In: 2010 3rd International Symposium on Intelligent Information Technology and Security Informatics, pp. 595–598. IEEE (2010)
9. Vigna, G., Kemmerer, R.A.: NetSTAT: a network-based intrusion detection system. J. Comput. Secur. **7**(1), 37–71 (1999)
10. Bellare, M., Rogaway, P.: Introduction to Modern Cryptography, vol. 207 (2005)
11. Mavroeidis, V., Vishi, K., Zych, M.D., Jøsang, A.: The impact of quantum computing on present cryptography. arXiv preprint arXiv:1804.00200 (2018)
12. Mühlbauer, N., Kirdan, E., Pahl, M.O., Waedt, K.: Feature-based comparison of open source OPC-UA implementations. In: INFORMATIK 2020 (2021)
13. Fujisaki, E., Okamoto, T.: Secure integration of asymmetric and symmetric encryption schemes. In: Wiener, M. (ed.) CRYPTO 1999. LNCS, vol. 1666, pp. 537–554. Springer, Heidelberg (1999). https://doi.org/10.1007/3-540-48405-1_34
14. Corchado, J.M., et al.: Deepint.dnet: a rapid deployment platform for smart territories. Sensors **21**(1), 236 (2021)

Review of Privacy Preservation with Blockchain Technology in the Context of Smart Cities

Yeray Mezquita[1(✉)], Alfonso González-Briones[1,2], Roberto Casado-Vara[1], Patricia Wolf[3], Fernando de la Prieta[1], and Ana-Belén Gil-González[1]

[1] BISITE Research Group, University of Salamanca. Edificio Multiusos I+D+i, 37007 Salamanca, Spain
yeraymm@usal.es
[2] Research Group on Agent-Based, Social and Interdisciplinary Applications (GRASIA), Complutense University of Madrid, Madrid, Spain
[3] Department of Marketing and Management, University of Southern Denmark, Campusvej 55, 5230 Odense M, Denmark

Abstract. Blockchain technology has credit for empowering smart cities. Due to the generation of huge amounts of data for direct transactions between devices, blockchain technology, by definition, must be used in this type of context. Although, due to the use of large amounts of multidimensional data, exists a risk that users' privacy may be violated. This privacy issue turns out to be of paramount importance, especially with new privacy regulations, like the General Data Protection Regulation (GDPR), that legalizes the requirement to protect the privacy of personal information. This work contributes with a review of the current state of the art in the context of blockchain-based smart cities and how it is managed the privacy-preserving issue.

Keywords: Blockchain · Smart city · Privacy · Review

1 Introduction

Blockchain technology has credit for empowering smart cities. Due to the generation of huge amounts of data for direct transactions between devices, blockchain technology, by definition, must be used in this type of context. Thanks to it, it is possible to obtain secure and automatic platforms, where devices can trust each other without the need for third parties.

Although, due to the use of large amounts of multidimensional data, users' privacy may be violated [4]. This privacy issue turns out to be of paramount importance, especially with new privacy regulations, like the General Data Protection Regulation (GDPR), that legalize the requirement to protect the privacy of personal information [29].

In the literature, this topic has been addressed numerous times, although all of them conclude that more research work is needed. This is because, in the

J. M. Corchado and S. Trabelsi (Eds.): SSCTIC 2021, LNNS 253, pp. 68–77, 2022.
https://doi.org/10.1007/978-3-030-78901-5_7

context of smart cities, data must be used and analyzed in near real-time, so due to the technical constraints that Internet of Things (IoT) devices have, it is not easy to implement complex cryptographic algorithms [6,12,13].

This work contributes with a review of the current state of the art in the context of blockchain-based smart cities and how it is managed the privacy-preserving issue. Urban researchers and practitioners alike will benefit from this work; along with a broad range of stakeholders who wish to understand the disruptive forces of Blockchain, as well as challenges in the context of Smart Cities.

The rest of the paper is structured in a theoretical background, Sect. 2 where some concepts of interest of smart cities and blockchain technology are described. Followed by the review of the literature in Sect. 3. Finally, in Sect. 4, it is discussed and concluded the results of the work.

2 Theoretical Background

Smart cities are those cities that, using technology and innovation, in addition to many other available resources, seek to increase the quality of life of the inhabitants of that city, promoting a more efficient and sustainable development, taking care of areas such as urban planning, accessibility, transportation, sustainability and energy saving, health and the environment. A smart city is a heterogeneous system that consists of a large IoT-based network of devices, offering various applications for citizens by collecting and analyzing real-time information [4].

Due to the high data exchange between smart city devices, blockchain technology is used to distribute the processes and secure the generated data [18]. The blockchain is an incorruptible digital distributed ledger of economic transactions that can be programmed to record not just financial transactions but virtually everything of value [28]. This ledger consists of a peer-to-peer (P2P) network of nodes that keeps the information stored redundantly. In an IoT system, BT can be used instead of traditional databases, helping on getting rid of centralized controllers such as banks, accountants, and governments [17].

Blockchain technology makes use of a public key signature mechanism; thanks to which it is possible to easily verify the source of the data generated, guaranteeing the integrity of the data generated. Due to the open, decentralized and cryptographic nature of a blockchain, we can list its main benefits [19]:

- Ensure data integrity, due to the use of consensus algorithms that keeps the data stored in the blockchain in a consistent state between the nodes of the network. This protocol provides the ledger with its immutable nature, making that everything written in there cannot be tampered or edited afterwards.
- Data signed by a public key mechanism. This feature allows the origin of any data stored in the blockchain to be easily verified.
- Dispose of some intermediaries that increase the price of the system's use while making it vulnerable to human errors in the operations of a system. This feature is due to the possibility of storing in the blockchain code that cannot be altered.

Smart contracts are a sequence of code stored in a blockchain that can be executed in a distributed way along with the network, reaching a consensus between the nodes on the result obtained from its execution. These programs facilitate, verify and enforce an agreement on a set of predefined conditions [7]. Smart contracts are self-execution and self-verification contractual agreements that automate the life cycle of a contract to improve compliance, mitigate risk and increase efficiency on any platform where entities with different interests have to interact with each other [20].

Every blockchain network is driven by a consensus algorithm. There is an increasing number of different types of consensus algorithms and their variations, although the most widespread algorithms are the ones that have proven to be effective in practice [32]:

- In the **Proof-of-Work** (PoW) algorithm, a node of the network must solve a cryptographic problem to be able to add a new block to the blockchain. The computational cost and the difficulty of solving the problem, the energy spent on searching for its solution (work), and the simplicity of verifying it, are sufficient reasons to discourage the nodes that add new blocks (miners) from performing illegal transactions.
- **Proof-of-Stake** (PoS) is a consensus algorithm, in which miners take turns adding new blocks to the blockchain. The probability that a miner gets its turn to add a block depends on the number of coins deposited for the miner as escrow (Stake). This algorithm assumes that a node is going to be honest to avoid losing the escrow. Some variations of this algorithm include the delegated Proof-of-Stake algorithm (dPoS), in which participants vote, in the base of the stake they have, the nodes that can add new data to the blockchain.
- In **Practical Byzantine Fault Tolerance** (PBFT), the process of adding a new block to the blockchain is called a round. In each round, a node is selected to propose a new block, then the block is broadcasted to the network, to let it be validated. The block is validated in each node of the network, obtaining a vote for each node that successfully validated it. When a block receives 2/3 of the votes of all the nodes in the network, is considered valid and added to the blockchain.

Currently, every consensus algorithm has its risks and vulnerabilities. E.g., PoW wastes a massive amount of energy to solve the cryptographic problems to produce new blocks. Besides, it is very limited in terms of scalability [2]. In the case of the PoS algorithm, its Nothing at a Stake theory causes forks to occur more frequently in the blockchain than with other consensus algorithms [15]. In the case of PBFT, the main risk is that it is a permissioned protocol and not a truly decentralized one [30].

3 State of the Art

Blockchain technology seems to greatly exploit possibilities within a smart city, allowing the development of urban applications in the field of logistics, supply chains, transport, and domains of governance [1]. Although numerous challenges have to been addressed to allow the widespread adoption of this technology within any smart city [1,3,13]:

- **Intelligent participatory sensing for smart cities**. The extended use of smartphones and cloud services allows the collection and analyze data that enables various smart city applications, such as energy controlling and healthcare. However, the actual infrastructure of any smart city is incapable of using these features. Design new frameworks that exploit participatory sensing to collect data from trusted authorities and perform the real-time analysis is mandatory.
- **Security and privacy**. Due to the interconnection of multiple devices within a smart city, it is needed a defense framework to ensure data is safe in the transmission. With the implementation of a blockchain framework, it is possible to ensure the provenance of the data, but it is a challenge to maintain privacy within it. Owing to the transparent nature of blockchain technology, the transactions are available and visible for all network participants [20].
- **Storage**. Although cloud storage seems a rather useful solution to the increasing amount of information generated within a smart city, it is slow and could compromise data integrity. Centralized data storage schemes have been proposed, but they are vulnerable to a single point of failure and DoS attacks. In this regard, there are blockchain-based frameworks, proposed as decentralized schemes [11]. Although this kind of system still has issues like lack of trust and lack of privacy and security.
- **Energy efficiency**. Energy efficiency needs to be considered seriously due to the rapidly rising energy costs in smart cities. Because PoW is a highly computational expensive consensus algorithm, it is not suitable for its use within a smart city context. Other consensus algorithms could be used within a smart city, like PoS or PBFT, and have been proposed new ones like Proof of Trust (PoT) [33]. But it is still research needed within this field to allow the use of a scalable and energy-efficient blockchain network.
- **Scalability and performance**. Blockchain based solutions in a smart city must meet demand of business and government-based sectors, especially regarding scalability and performance. Because of the need of consensus within the blockchain network, not all the consensus protocols could be used in this regard.
- **Incentive mechanism**. Nodes in smart cities are assumed to be self-interested and therefore incentive mechanisms are needed to motivate these nodes to contribute towards data verification. Besides, it also important to design a punishment mechanism to punish malicious nodes and prevent double-spending attacks. In this regard, none of the proposed solutions in the literature have been accepted, and it is needed more research on this topic [3].
- **Interoperability**. Implementing an interoperable system is a challenging task due to the wide range of data formats involved in various blockchain systems along with the lack of standards.

- **Regulation**. Governments are prudent towards the use of cryptocurrencies and the blockchain platforms face regularity issues. In [31] it is highlighted the major regulatory issues that have adverse impact on blockchains and innovative distributed technologies. Therefore, new industry and government regulations are needed in order to evade disputes among the transacting parties as there is no need for a trusted intermediary for a decentralized blockchain technology.

In this paper, we will study the state of the art on the use of blockchain technology in the field of smart cities, focusing on data privacy. This study aims to identify and compile the key aspects needed for the successful implementation of blockchain technology within a smart city. To this end, a search for papers was carried out in the ScienceDirect[1] database, obtaining 630 papers, of which 18 have been used in this article. Besides, from those 18 articles selected, it has been found within them more useful papers for this work. The following keywords and their regular expressions were identified for the search:

- **Blockchain**: ("distributed ledger technolog" OR blockchain*).
- **Smart city**: ("smart city").
- **Privacy**: (privacy).

Smart cities manage data by digitalizing all information, which leads to a high risk of security and privacy issues [4]. Personalized data acquisition plays a very important role in providing customized services in smart cities. Although, improving data availability combined with powerful analytic tools increases the risk of privacy violations [5]. Thus, control for sharing this sensitive data should be maintained by the data owner to ensure privacy, willingness, and level of access [13].

Authors from [5], state that there are two goals when employing privacy enhancing technologies in Smart Cities. The first goal is to protect the identity of each individual who is represented in the data so that no one may learn that they are part of the set. The second goal is to protect all sensitive attributes for each individual. Those attributes are defined in [16], where it is defined the concept of citizens' privacy as a model with five dimensions: identity privacy, query privacy, location privacy, footprint privacy and owner privacy.

It is studied in [4] a Secret Sharing algorithm to address personal information protection issues in external cloud services to improve data integrity and security in the design of blockchain-based distributed systems. Cloud Service Providers (CSP) are connected in a blockchain to validate user data integrity and provide easy data access. In the blockchain, it is stored users' information safely through the Secret Sharing algorithm as a distributed system with improved security of existing centralized systems.

In [8], it is built a general model for distributed transactions called PP-BCETS (Privacy-preserving Blockchain Energy Trading Scheme) to preserve

[1] https://www.sciencedirect.com/.

the privacy of users. Furthermore, it introduced Ciphertext-Policy Attribute-Based Encryption (CP-ABE) as the core algorithm to reconstruct the transaction model of the blockchain-based smart city.

Federated learning that shares model updates without exchanging raw data has been proposed in [24] as an efficient solution for achieving privacy protection. In that paper, the consortium blockchain-based federated learning framework is proposed to enable decentralized, reliable, and secure federated learning without a centralized model coordinator. Finally, a differential privacy method with a noise-adding mechanism is applied for the blockchain-based federated learning framework.

Authors in [6] propose a solution for distributed management of identity and authorization policies by leveraging the blockchain technology to hold a global view of the security policies within the system. While in [23], it is proposed a platform called Smart Agora using a consensus protocol based on the concept of proving witness presence. In which it can be done real-time collective measurements over citizens' choices in a fully decentralized and privacy-preserving way. In [21] it is proposed a tamper-proof, immutable, authentic, non-repudiable, privacy protected and easy to share blockchain-based architecture for secured sharing of students' credentials.

In [26] it is stated that Artificial Intelligence (AI) and blockchain technology is revolutionizing the smart city network architecture to build sustainable ecosystems. Following that lead, [12] presents a Trustworthy Privacy-Preserving Secure Framework (TP2SF) for smart cities. The framework consists of a three-module architecture:

1. The first module is called the trustworthiness module, where an address-based blockchain reputation system is designed.
2. In the two-level privacy module, a blockchain-based enhanced Proof of Work (ePoW) technique is simultaneously applied with Principal Component Analysis (PCA) to transform data into a new reduced shape for preventing inference and poisoning attacks.
3. In the intrusion detection module, an optimized gradient tree boosting system (XGBoost) is deployed.

Besides, it is designed a blockchain-IPFS integrated Fog–Cloud infrastructure called CloudBlock and FogBlock, which is in charge of deploy the proposed TP2SF framework in a smart city.

A privacy-preserving framework, based on blockchain technology and AI in a smart city is proposed in [25]. There, is used a neural network to keep the identity of the users secret. The neural weights codify the user information, although in case of a cybersecurity breach, the confidential identity can be mined and the attacker identified. Thus, it is enabled a decentralized authentication method.

[14] proposes a blockchain-based innovative framework for privacy-preserving and secure IoT data sharing in a smart city environment. The data privacy is preserved by dividing the blockchain network into various channels, where every channel comprises a finite number of authorized organizations and processes a specific type of data such as health, smart car, smart energy or financial details.

Access to users' data within a channel is controlled by embedding access control rules in the smart contracts. The experimental outcomes advocate that a multi-channel blockchain scales well as compared to a single-channel blockchain system.

Another example that makes use of artificial intelligence along blockchain technology is the proposal of [27]. There, it is proposed a Deep Learning-based IoT-oriented infrastructure for a secure smart city where Blockchain provides a distributed environment at the communication phase of a Cyber-Physical System. Besides, Software-Defined Networking (SDN) establishes the protocols for data forwarding in the network. To that end, it is designed three layers based IoT-oriented infrastructure for a smart city: i) connection, ii) conversion, and iii) application. Blockchain provides a peer-to-peer distributed environment, while SDN establishes the rules and regulations for data management in sustainable smart city networks.

As a summary, we could say that the decentralized nature of blockchain has resolved many security, maintenance, and authentication issues of IoT systems like smart cities. But there are still issues in other areas, specifically the implementation of blockchain in smart cities require proper privacy measures [10]. Privacy preservation with high efficiency remains an unsolved issue, which significantly constrains the practical deployment of a blockchain-based platform within a smart city case scenario [22]. Besides, it is worth noting that, although the increasing reliance of companies on advanced cryptography also entangles digital forensics; it improves users' privacy but makes it difficult to resolve legal disputes [9].

4 Conclusion

In this article, we have presented a study on the problems of implementing blockchain technology in a smart city, more specifically those related to user privacy. The study is far from perfect since it is limited to the use of a database together with the snowballing technique. However, we have obtained relevant information with which it has become clear that a large number of works motivate research in this area, as they consider the advances made to be of little relevance for the implementation of a real prototype.

In the literature, the blockchain network of a smart city is responsible for providing a mechanism for the verification of the data generated, never as a database in which they are stored. When performing information exchanges, works in the literature propose the use of cryptographic algorithms even together with machine learning techniques for user identity masking. Although the results of the solutions proposed in the literature make it clear that an improvement in data privacy is obtained, most of them are far from being implementable prototypes in a real smart city. Moreover, it should be taken into account that the privacy gain by users may come at the expense of a reduced ability to resolve legal disputes.

Acknowledgements. The research of Yeray Mezquita is supported by a pre-doctoral fellowship granted by University of Salamanca and cofinanced by Banco Santander.

Besides, this work has been supported too by the project "XAI - XAI - Sistemas Inteligentes Auto Explicativos creados con Módulos de Mezcla de Expertos", ID SA082P20, financed by Junta Castilla y León, Consejería de Educación, and FEDER funds.

References

1. Bagloee, S.A., Heshmati, M., Dia, H., Ghaderi, H., Pettit, C., Asadi, M.: Blockchain: the operating system of smart cities. Cities **112**, 103104 (2021)
2. Beikverdi, A., Song, J.: Trend of centralization in bitcoin's distributed network. In: 2015 IEEE/ACIS 16th International Conference on Software Engineering, Artificial Intelligence, Networking and Parallel/Distributed Computing (SNPD), pp. 1–6. IEEE (2015)
3. Bhushan, B., Khamparia, A., Sagayam, K.M., Sharma, S.K., Ahad, M.A., Debnath, N.C.: Blockchain for smart cities: a review of architectures, integration trends and future research directions. Sustain. Urban Areas **61**, 102360 (2020)
4. Cha, J., Singh, S.K., Kim, T.W., Park, J.H.: Blockchain-empowered cloud architecture based on secret sharing for smart city. J. Inf. Secur. Appl. **57**, 102686 (2021)
5. Curzon, J., Almehmadi, A., El-Khatib, K.: A survey of privacy enhancing technologies for smart cities. Pervasive Mob. Comput. **55**, 76–95 (2019)
6. Esposito, C., Ficco, M., Gupta, B.B.: Blockchain-based authentication and authorization for smart city applications. Inf. Process. Manage. **58**(2), 102468 (2021)
7. Gazafroudi, A.S., Mezquita, Y., Shafie-khah, M., Prieto, J., Corchado, J.M.: Islanded microgrid management based on blockchain communication. In: Blockchain-Based Smart Grids, pp. 181–193. Elsevier (2020)
8. Guan, Z., Lu, X., Yang, W., Wu, L., Wang, N., Zhang, Z.: Achieving efficient and privacy-preserving energy trading based on blockchain and ABE in smart grid. J. Parallel Distrib. Comput. **147**, 34–45 (2021)
9. Habibzadeh, H., Nussbaum, B.H., Anjomshoa, F., Kantarci, B., Soyata, T.: A survey on cybersecurity, data privacy, and policy issues in cyber-physical system deployments in smart cities. Sustain. Urban Areas **50**, 101660 (2019)
10. Hassan, M.U., Rehmani, M.H., Chen, J.: Privacy preservation in blockchain based IoT systems: integration issues, prospects, challenges, and future research directions. Futur. Gener. Comput. Syst. **97**, 512–529 (2019)
11. Kopp, H., Bösch, C., Kargl, F.: KopperCoin – a distributed file storage with financial incentives. In: Bao, F., Chen, L., Deng, R.H., Wang, G. (eds.) ISPEC 2016. LNCS, vol. 10060, pp. 79–93. Springer, Cham (2016). https://doi.org/10.1007/978-3-319-49151-6_6
12. Kumar, P., Gupta, G.P., Tripathi, R.: Tp2sf: a trustworthy privacy-preserving secured framework for sustainable smart cities by leveraging blockchain and machine learning. J. Syst. Archit. **115**, 101954 (2020)
13. Majeed, U., Khan, L.U., Yaqoob, I., Kazmi, S.A., Salah, K., Hong, C.S.: Blockchain for IoT-based smart cities: recent advances, requirements, and future challenges. J. Netw. Comput. Appl. **181**, 103007 (2021)
14. Makhdoom, I., Zhou, I., Abolhasan, M., Lipman, J., Ni, W.: PrivySharing: a blockchain-based framework for privacy-preserving and secure data sharing in smart cities. Comput. Secur. **88**, 101653 (2020)

15. Martinez, J.: Understanding proof of stake: the nothing at stake theory (June 2018). https://medium.com/coinmonks/understanding-proof-of-stake-the-nothing-at-stake-theory-1f0d71bc027. Accessed 09 Oct 2019
16. Martínez-Ballesté, A., Pérez-Martínez, P.A., Solanas, A.: The pursuit of citizens' privacy: a privacy-aware smart city is possible. IEEE Commun. Mag. **51**(6), 136–141 (2013)
17. Mezquita, Y., Casado-Vara, R., GonzÁlez Briones, A., Prieto, J., Corchado, J.M.: Blockchain-based architecture for the control of logistics activities: pharmaceutical utilities case study. Logic J. IGPL (2020)
18. Mezquita, Y., Gazafroudi, A.S., Corchado, J., Shafie-Khah, M., Laaksonen, H., Kamišalić, A.: Multi-agent architecture for peer-to-peer electricity trading based on blockchain technology. In: 2019 XXVII International Conference on Information, Communication and Automation Technologies (ICAT), pp. 1–6. IEEE (2019)
19. Mezquita, Y., González-Briones, A., Casado-Vara, R., Chamoso, P., Prieto, J., Corchado, J.M.: Blockchain-based architecture: a MAS proposal for efficient agri-food supply chains. In: Novais, P., Lloret, J., Chamoso, P., Carneiro, D., Navarro, E., Omatu, S. (eds.) ISAmI 2019. AISC, vol. 1006, pp. 89–96. Springer, Cham (2020). https://doi.org/10.1007/978-3-030-24097-4_11
20. Mezquita, Y., Valdeolmillos, D., González-Briones, A., Prieto, J., Corchado, J.M.: Legal aspects and emerging risks in the use of smart contracts based on blockchain. In: Uden, L., Ting, I.-H., Corchado, J.M. (eds.) KMO 2019. CCIS, vol. 1027, pp. 525–535. Springer, Cham (2019). https://doi.org/10.1007/978-3-030-21451-7_45
21. Mishra, R.A., Kalla, A., Braeken, A., Liyanage, M.: Privacy protected blockchain based architecture and implementation for sharing of students' credentials. Inf. Process. Manage. **58**(3), 102512 (2021)
22. Peng, L., Feng, W., Yan, Z., Li, Y., Zhou, X., Shimizu, S.: Privacy preservation in permissionless blockchain: a survey. Digit. Commun. Netw. (2020)
23. Pournaras, E.: Proof of witness presence: blockchain consensus for augmented democracy in smart cities. J. Parallel Distrib. Comput. **145**, 160–175 (2020)
24. Qi, Y., Hossain, M.S., Nie, J., Li, X.: Privacy-preserving blockchain-based federated learning for traffic flow prediction. Futur. Gener. Comput. Syst. **117**, 328–337 (2021)
25. Serrano, W.: The blockchain random neural network for cybersecure IoT and 5g infrastructure in smart cities. J. Netw. Comput. Appl. **175**, 102909 (2021)
26. Singh, S., Sharma, P.K., Yoon, B., Shojafar, M., Cho, G.H., Ra, I.H.: Convergence of blockchain and artificial intelligence in IoT network for the sustainable smart city. Sustain. Urban Areas **63**, 102364 (2020)
27. Singh, S.K., Jeong, Y.S., Park, J.H.: A deep learning-based IoT-oriented infrastructure for secure smart city. Sustain. Urban Areas **60**, 102252 (2020)
28. Valdeolmillos, D., Mezquita, Y., González-Briones, A., Prieto, J., Corchado, J.M.: Blockchain technology: a review of the current challenges of cryptocurrency. In: Prieto, J., Das, A.K., Ferretti, S., Pinto, A., Corchado, J.M. (eds.) BLOCKCHAIN 2019. AISC, vol. 1010, pp. 153–160. Springer, Cham (2020). https://doi.org/10.1007/978-3-030-23813-1_19
29. Voigt, P., von dem Bussche, A.: The EU General Data Protection Regulation (GDPR). A Practical Guide. Springer, Cham (2017). https://doi.org/10.1007/978-3-319-57959-7
30. Witherspoon, Z.: A hitchhiker's guide to consensus algorithms (November 2017). https://hackernoon.com/a-hitchhikers-guide-to-consensus-algorithms-d81aae3eb0e3. Accessed 09 Oct 2019

31. Yeoh, P.: Regulatory issues in blockchain technology. J. Financ. Regul. Compliance **25**, 196–208 (2017)
32. Zheng, Z., Xie, S., Dai, H., Chen, X., Wang, H.: An overview of blockchain technology: Architecture, consensus, and future trends. In: 2017 IEEE International Congress on Big Data (BigData Congress), pp. 557–564. IEEE (2017)
33. Zou, J., Ye, B., Qu, L., Wang, Y., Orgun, M.A., Li, L.: A proof-of-trust consensus protocol for enhancing accountability in crowdsourcing services. IEEE Trans. Serv. Comput. **12**(3), 429–445 (2018)

Edge and Fog Computing

Managing Smart City Power Network by Shifting Electricity Consumers Demand

Cátia Silva[1], Pedro Faria[1(✉)], and Zita Vale[2]

[1] GECAD - Research Group on Intelligent Engineering and Computing for Advanced Innovation and Development, Porto, Portugal
{cvcds,pnf}@isep.ipp.pt
[2] Polytechnic of Porto, Porto, Portugal
zav@isep.ipp.pt

Abstract. Demand Response (DR) concept, introduced by the Smart Grid paradigm, is presented as one of the main solutions to mitigate the effects of the intermittency of Distributed Generation sources in the network. With this, the consumer's role in the energy market is empowered and their flexibility is crucial. The technology advances allow bidirectional communication providing signals to the active consumers, given by the entity manager, participating in DR events to alleviate problems in the grid. The authors proposed a methodology, resorting to load shifting, to lessen voltage limit violations in some points of the grid. After the detention, the DR event is triggered and the small resources are scheduled, requesting a reduction to the active consumers. The authors believe that uncertainty in the response must be considered so, the willingness and availability from the active community are contemplated on the problem. The results show the necessity of pondering consumer behavior.

Keywords: Demand Response · Load shifting · Uncertainty · Remuneration

1 Introduction

The world's population is increasing, and the concept of smart cities was born to provide the answers for the problems that consequently arise – for instance, unsustainable energy and water use, pollution, or even environmental degradation. It is expected that, by implementing information and communication technologies, some of the consequences will be mitigated and will help to properly manage the communities. In this way, the consumer's role in the energy market is changing. Until now, low or no information was given to the end-users regarding the transactions in the market. With the increasing penetration of the Distributed Generation (DG) in the grid, namely solar and wind, the prior paradigm is no longer valid – the generation follows and satisfies the consumption needs. Instead, the demand side must adapt their needs to the available production since the behavior from the renewable-based technologies is intermittent [1–3].

The Demand Side Management solutions are considered, by the literature as the solution to avoid costly investments in the system restructure as well as aid on alleviate

J. M. Corchado and S. Trabelsi (Eds.): SSCTIC 2021, LNNS 253, pp. 81–91, 2022.
https://doi.org/10.1007/978-3-030-78901-5_8

the problems that will arise from demand peaks. Thus, Demand Response (DR) programs should be implemented with the help of bidirectional communication and Advanced Metering Infrastructures, including smart meters and smart appliances [4–6]. Having information and understanding the consumer load pattern, aid the entity manager in giving the right signals to change the consumption to alleviate any difficulties found in the grid, ensuring the security and reliability of the system [7, 8]. So, the role of small consumers in the energy market is changing. Yet, according to Iliana Ilieva, et al. [9] it will take time, education, and resources until taking rational decisions in the market transactions. However, the actual business models do not include or can deal with the uncertainty associated with these new resources. The unpredictable behavior, from both consumers and DG units, increase the complexity of managing the network. An entity was created to be responsible for the transactions in local communities with active players – the Aggregator. It is believed that developing smart buildings and smart homes, giving the right tools for the active consumers is a great step forward to the implementation in the real world of DR programs. However, other issues resulting from the use of such technologies, for instance, the large amounts of data, concerns regarding data security, safety, and privacy. The designed method to successfully implement DR in the real market should be able to manage these matters – artificial intelligence and machine learning are widely used in the treatment of big data.

Many models in the literature were created finding solutions to optimally manage and to aid and decrease the difficulty when dealing with the new players. Shaghayegh Zalzar and Ettore Francesco Bompard [10] modeled the demand flexibility using load shifting focusing and proposing an incentive-based settlement mechanism. The results were withdrawn, with high penetration of wind energy, indicate that monetary incentives are crucial for active consumer participation – without them, it is not profitable to provide such flexibility. Özge Okur et al. [11] evaluated the impact of DR for internal balancing to reduce the individual imbalances of an Aggregator regarding uncertain DG resorting to Model Predictive Control, reducing up to 30% in one of the case studies. However, the remuneration of the active consumers is not considered in their work.

The authors proposed a method to properly manage an active community in a smart city by developing a tool to aid the Aggregator on the complex task of handling uncertain resources. Being a continuity from other works [12–14], this paper it is presented a resolution for a voltage limit violation mitigated by triggering a DR event. After the detection of the problem, a DR event is triggered and flexibility from the active consumers is requested after the scheduling of the resources. Load shifting is applied, and the consumption is moved to another period. The authors consider the consumer perspective a crucial matter in developing DR solutions. In this way, the availability, and the willingness to contribute to such programs must be deemed. In this way, the case study is based on a smart city and how the Aggregator should manage the small resources associated, which can be a complex task due to the uncertainty of the active consumer response and the intermittent behavior of the DG resources.

The present paper is structured into five main sections. Firstly, a brief introduction to the topic, the proposed methodology, and the problem that the authors want to solve. After, in the Methodology section, the authors offer a detailed explanation of the solution presented. Next, the Case Study, where the characteristics of the dataset used as input, focusing on the community and the active consumers. In the Results and Discussion section, an analysis of the findings regarding the global perspective as well as the point of view of five selected consumers.

2 Methodology

In this section, the proposed methodology framework and assumptions are introduced. Since the focus is to optimally manage the participants in the DR event, the mathematical formulation respects the optimization scheduling to minimize the Aggregator's costs due to flexibility provision. By managing the availability of the local community, the entity manager will be able to mitigate limit violations by triggering DR events to reduce the tension on the faulty node. Figure 1 presents the proposed methodology.

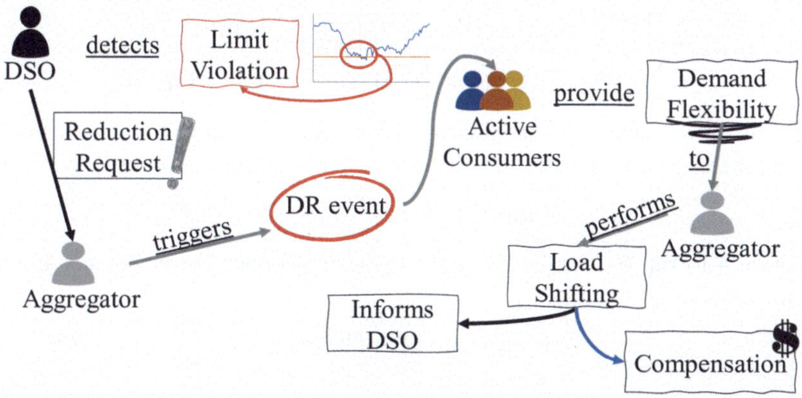

Fig. 1. Proposed methodology

Performed the power transit, as soon as the limit violation is detected, the DR event is triggered, and the participants are requested to reduce the contracted amount. It must be noticed that active consumers register to offer the flexibility voluntarily for the participation and receive compensation if there is a change in their baseline profile. The availability from the demand side will be crucial. The consumer has power over the appliances – even resorting to smart equipment, the user has always the freedom of choice. So, it is important to find ways to deal with the uncertainty. For instance, give the right compensation and reduce the discomfort that may incur from the event. In this way, the authors design the tool to aid the Aggregator in this complex task, being the focus of the proposed methodology. Triggered the DR event, the Scheduling phase starts. A linear approach is employed to optimally manage the smart city and the local community since can deal with several types of active small resources: consumers participating in DR programs (PDR), DG units (PDG), and even the joint of these two concepts. The DG is considered a priority in this approach but, when the amount is reduced, external suppliers are also considered. Two types of external suppliers were integrated: regular (PSUPR) and additional (PSUPA), considering the quantities contracted at distinct prices and where the last is only used in extreme cases (more expensive). Let p be the number of DG units, c the number of consumers, sr the number of regular suppliers, sa the number of additional suppliers, and t the actual period. The objective function is represented in Eq. 1 [14].

$$\begin{aligned} \text{Min. OF} = & \sum\left[P_{DG}(p, t)\, C_{DG}(p, t)\right] + \sum\left[P_{DR}(c, t)\, C_{DR}(c, t)\right] \\ & + \sum\left[P_{SUPA}(sa, t) C_{SUPA}(sa, t)\right] + \sum\left[P_{SUPR}(sr, t)\, C_{SUPR}(sr, t)\right] + P_{NSP}(t)\, C_{NSP}(t) \end{aligned} \tag{1}$$

The network balance constraint is presented in Eq. 2, achieving an equilibrium between consumption and generation where the sum of the requested reduction to the consumer initial load ($P_{initialLoad}$) should equal the total generation from DG units and external suppliers. In the extreme eventuality of demand not met by generation means, the variable from Non-Supplied Power (P_{NSP}) was added. The network is being well managed if the parameter PNSP is zero.

$$\begin{aligned} \sum\left[P^{initial}(c, t) - P_{DR}(c, t)\right] = & \sum\left[P_{DG}(p, t)\right] \\ & + \sum\left[P_{SUPA}(sa, t)\right] + \sum\left[P_{SUPR}(sr, t)\right] + P_{NSP}(t) \end{aligned} \tag{2}$$

The remaining constraints are presented between Eq. 3 and Eq. 10. In Eq. 3, the constraint for the maximum value of reduction for each consumer – contracted flexibility.

$$P_{DR}(c, t) \leq P_{DR}{}^{Max}(c, t) \tag{3}$$

Equation 4 to Eq. 6 describe DG units restricting upper, lower bounds and a total maximum value of generation.

$$P_{DG}(p, t) \leq P_{DG}{}^{Max}(p, t) \tag{4}$$

$$P_{DG}(p, t) \geq P_{DG}{}^{Min}(p, t) \tag{5}$$

$$\sum\left[P_{DG}(p, t)\right] \leq P_{DG}{}^{Total}(t) \tag{6}$$

Equation 7 to Eq. 10 refer to External Suppliers restricting them superiorly and total amount, for each type.

$$P_{SUPR}(sr, t) \leq P_{SUPR}{}^{Max}(sr, t) \tag{7}$$

$$\sum\left[P_{SUPR}(sr, t) \leq P_{SUPR}{}^{Total}(t)\right. \tag{8}$$

$$P_{SUPA}(sa, t) \leq P_{SUPA}{}^{Max}(sa, t) \tag{9}$$

$$\sum\left[P_{SUPA}(sa, t)\right] \leq P_{SUPA}{}^{Total}(t) \tag{10}$$

The following phase, Reduction Request, invite the active consumers to participate in the DR event and reduce the consumption to aid the Aggregator goal to mitigate the limit violation. As soon as the power transit is violation-free, the participants are remunerated with the proper compensation and the load is shifted to another period, convenient to both Aggregator and the active consumer reducing the discomfort caused.

3 Case Study

In this section, the information regarding the design of the case study is provided. The smart city has 96 consumers and, the grid is composed of 234 buses. The day is divided into periods of 15 min, so 96 periods were considered. On periods 38, 40, 42, 43, and 44 occurred voltage limit violation, as can be seen in Fig. 2. To solve the problem, DR programs are applied however only 64 consumers were willing to participate in DR events.

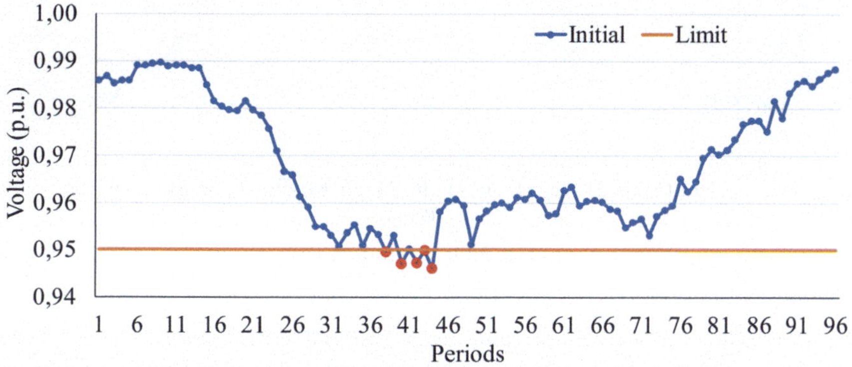

Fig. 2. Limit violation detection

The main goal of the present paper is assessment the limit from the proposed methodology when performing load shifting to mitigate threshold violation. Load shifting is applied to mitigate limit violation throughout the day – the programed consumption is moved forward in time to another period without causing more complications. It was assumed that shifted consumption should be scheduled between 5 h after the DR event but never 24 h after the same – a range of 19 h.

The DR event is trigged upon limit violation detection and participants should reduce the amount contracted at the time requested to be further scheduled according to the mentioned range. In this study, the availability of the consumer at the DR event time was also considered.

To avoid additional violations upon the event, the shifted load is assigned to different periods according to the consumer needs – preventing any discomfort. The remuneration is a monetary value of 0.22 m.u./kW. Understanding the impact on microgrids and five selected consumers, the next section will present and discuss the results. The chosen consumers are from different parts of the grid – 3 of them are in the same node as the limit violation and the remaining are from another distant part.

4 Results and Discussion

Throughout the present section, the results of the proposed methodology to properly manage the smart city by the Aggregator perspective – in this case, mitigate the limit violations. Firstly, from a global perspective, Fig. 3 shows the comparison between initial and final consumption after the application of the proposed method. The discrepancies between both curves are noticeable in the periods where the limits violations were detected – flexibility provided by active consumers.

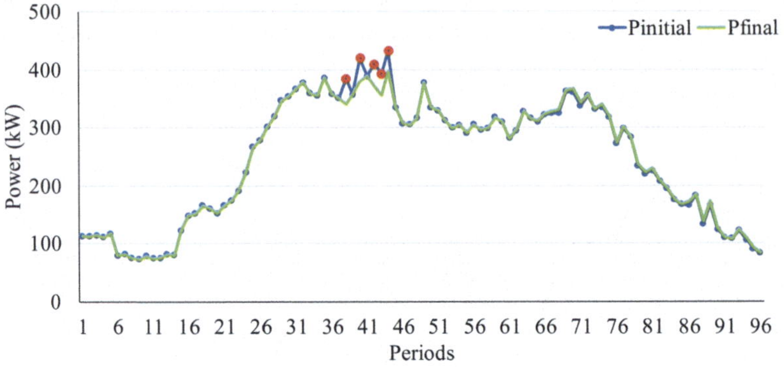

Fig. 3. Load shifting

With the reduction, the voltage values were restored to their proper place. As can be seen in Fig. 4, not only the values are above the limit, but further limit violations were created with this solution. Since the appliances from active consumers were shifted to different hours – according to their availability and to reduce the discomfort cause – the final curve follows the same line as the initial in the remaining periods.

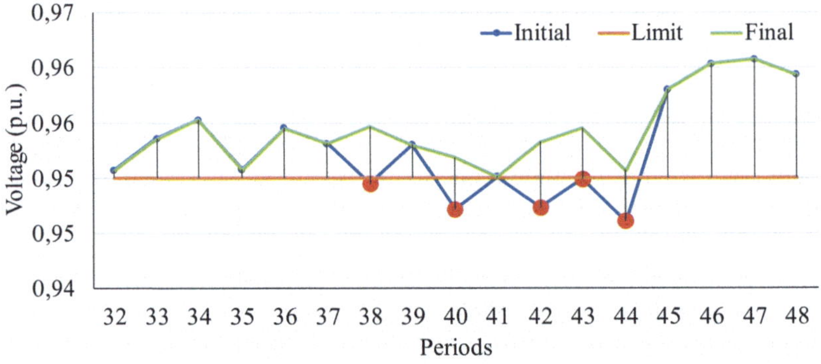

Fig. 4. Global perspective

From a more specific perspective, the five different consumers are confronted by the impact on the baseline profile as well as the remuneration granted from their participation. From the microgrid perspective, both Consumer 1 and 2 have their location far from the crisis point. Figure 5 and Fig. 8 show the comparison between initial and final consumption as well as the events in which the active consumer in question participate.

It is also highlighted where the consumptions were reallocated in the same days. May occur reallocations to the day after, according to the consumers' convenience. Also, the authors consider the willingness to participate in a DR event according to a certain context – period, since different consumers have different habits which should be considered when creating a solution for DR events.

First, Fig. 5 shows the results from Consumer 1, which participate in only one of the five events. The consumption was shifted from period 44 to period 83.

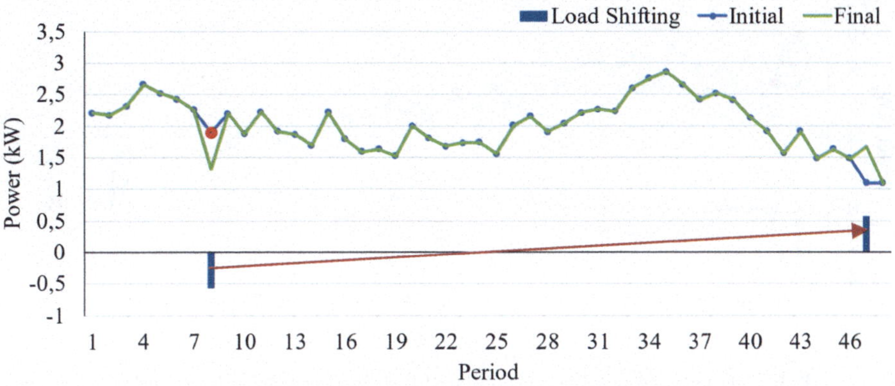

Fig. 5. Comparison between initial and final load profile – Consumer 1.

Regarding Consumer 2, Fig. 6 shows that participated in four of five events, although only three were scheduled for this day with high values of flexibility – between 2 and 2.5 kW.

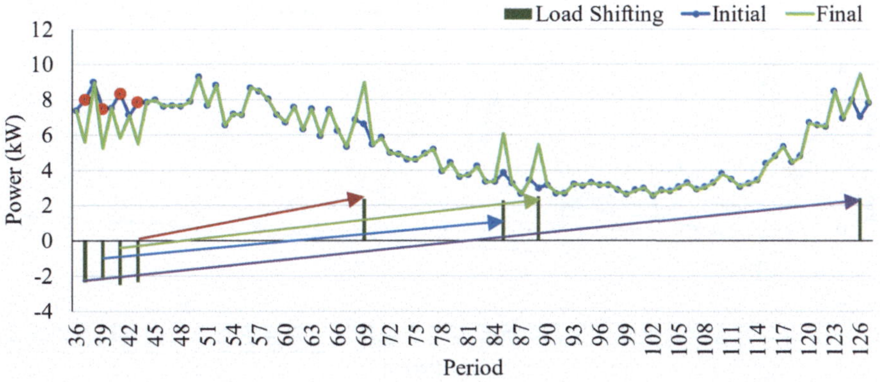

Fig. 6. Comparison between initial and final load profile – Consumer 2.

Consumer 3, Consumer 4, and Consumer 5 are from the same node where the limit violation was detected. Figure 7 presents the comparison between the initial and final load profiles from Consumer 3. This participant only takes part in two of five of the events which only one was scheduled to later that day. However, was the one, from the five selected that had a higher amount of reduction – almost 3 kW on period 40.

In respect to Consumer 4, although DR participant, the results show that this active consumer does not contribute to any event, as can be seen in Fig. 8. As referred earlier, participation is voluntary and depends on the willingness and availability of the user.

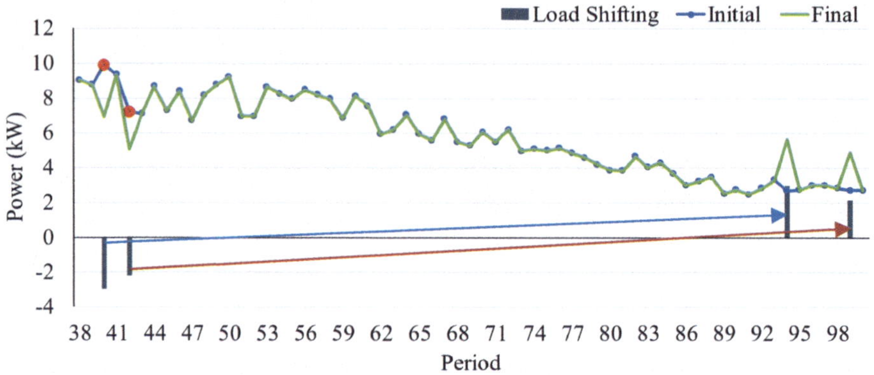

Fig. 7. Comparison between initial and final load profile – Consumer 3.

This is one of the examples where the uncertainty on the response must also contemplate the context of which the event is inserted. Finally, Consumer 5 participate in four out of five events, as can be seen in Fig. 8. Like Consumer 2, only three of the reductions were shifted to the same day.

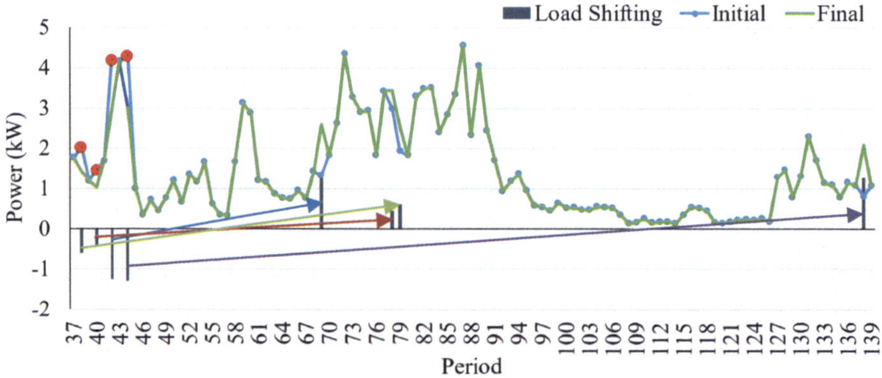

Fig. 8. Comparison between initial and final load profile – Consumer 5.

Table 1 presents the summary demonstrated earlier with the calculation of the remuneration received from each consumer in respect to their contribution to the management of the smart city. From this perspective, it is easier to see that none of the selected consumers participated in the mitigation of the limit violation of period 43. In period 38, only Consumer 2 and Consumer 5 provided flexibility with a total of 3 kW. In period 40, three consumers contributed with 5.65 kW but were on period 42 where these consumers reduce more – 5.93 kW. Finally, on period 44, Consumer 1, Consumer 2, and Consumer 5 join the event and reduce a total of 4.21 kW. Regarding the compensation, being the one with more contributions, Consumer 2 received 2.09 m.u./kW to shift the consumption to another period. After, Consumer 3 with 2.09 m.u./kW, Consumer 5 with 0.79 m.u./kW and finally, Consumer 1 with 0.13 m.u./kW.

Table 1. Reduction per period and remuneration values

Consumer ID	Reduction (kW)					Remuneration (m.u./kW)
	38	40	42	43	44	
1 – 41	0.00	0.00	0.00	0.00	0.57	0.13
2 – 48	2.39	2.24	2.50	0.00	2.35	2.09
3 – 59	0.00	2.97	2.17	0.00	0.00	1.13
4 – 60	0.00	0.00	0.00	0.00	0.00	0.00
5 – 51	0.61	0.44	1.26	0.00	1.29	0.79

From the study presented in this paper, the following conclusions can be withdrawn:

- The Aggregator must have powerful tools and make the right assumptions when dealing with the uncertainty on DR events to mitigate other problems on the management of the microgrid.
- When applying load shifting techniques, consider not only the grid issues but also the discomfort from the consumer perspective.
- The active consumers must be motivated to continuously contribute to the management of the grid.
- Uncertain responses must be dealt with the right motivations – remember that participation is voluntary, and the consumer has freedom of choice, even though penalties may be applied.

The authors believe that develop solutions where the consumer behaviors are considered, namely regarding the context of the event, should be a step forward to implement DR programs in the real market.

5 Conclusions

The growing concern regarding the climate change impacts on the so-called normal functioning of the world motivates the energy sector to exchange non-renewable solutions for technologies more environmentally friendly. Distributed Generation has more popularity but, to the system, the unstable behavior may cause problems with the security and the reliability. Now, instead of the generation following the consumption requests, the demand side must contribute with more flexibility – empowering the consumer role in the energy market. In this way, the consumer concept is changing – have more information and power to control their consumption with signals given by the energy market. Resorting to the technologies, smart appliances ease this task. For instance, giving the control to a managing entity to schedule the appliance according to the user preferences. When DR events are triggered, according to the contract with the entity manager, it is expected a response from the active consumer, rewarded with the right compensation.

In the present paper, the authors test the proposed methodology in a scenario where a voltage limit violation was detected, and a DR event must be triggered to deal with the problem. From the results, the solution was capable of mitigating the violation and not cause other problems after. However, when analyzing five different consumers, was more evident the importance of context and motivation. The consumer must be available and willing to participate, without that, the DR event becomes a more complex problem due to the uncertainty associated.

In future works, the authors intend to further study the consumer behavior, the factors that impact and motivate to take some decisions to create the right tool, resorting to context-aware methods, to aid the Aggregator on its difficulty of managing the active communities.

Acknowledgements. This work has received funding from the European Union's Horizon 2020 research and innovation programme under project DOMINOES (grant agreement No 771066), from FEDER Funds through COMPETE program and from National Funds through (FCT) under the projects UIDB/00760/2020, CEECIND/02887/2017, and SFRH/BD/144200/2019.

References

1. Hu, M., Xiao, F.: Quantifying uncertainty in the aggregate energy flexibility of high-rise residential building clusters considering stochastic occupancy and occupant behavior. Energy **194**, 116838 (2020). https://doi.org/10.1016/j.energy.2019.116838
2. Lokhande, S., Menon, V.P., Bichpuriya, Y.K.: Modelling of demand response for utility's LOAC forecasting. In: 2017 14th International Conference on the European Energy Market (EEM), June 2017, pp. 1–6 (2017). https://doi.org/10.1109/EEM.2017.7981985
3. Songyuan, Y., Fang, F., Liu, Y., Liu, J.: Uncertainties of virtual power plant: problems and countermeasures. Appl. Energy **239**, 454–470 (2019). https://doi.org/10.1016/j.apenergy.2019.01.224
4. Wang, B., Camacho, J.A., Pulliam, G.M., Etemadi, A.H., Dehghanian, P.: New reward and penalty scheme for electric distribution utilities employing load-based reliability indices. IET Gener. Transm. Distrib. **12**(15), 3647–3654 (2018). https://doi.org/10.1049/iet-gtd.2017.1809
5. Tiwari, S., Sabzehgar, R., Rasouli, M.: Load balancing in a microgrid with uncertain renewable resources and loads. In: 2017 IEEE 8th International Symposium on Power Electronics for Distributed Generation Systems (PEDG), April 2017, pp. 1–8 (2017). https://doi.org/10.1109/PEDG.2017.7972505.

6. Khan, Z.A., Jayaweera, D., Gunduz, H.: Smart meter data taxonomy for demand side management in smart grids. In: 2016 International Conference on Probabilistic Methods Applied to Power Systems (PMAPS), October 2016, pp. 1–8 (2016). https://doi.org/10.1109/PMAPS.2016.7764143
7. Zeng, B., Wu, G., Wang, J., Zhang, J., Zeng, M.: Impact of behavior-driven demand response on supply adequacy in smart distribution systems. Appl. Energy **202**, 125–137 (2017). https://doi.org/10.1016/j.apenergy.2017.05.098
8. Chen, S., Liu, C.-C.: From demand response to transactive energy: state of the art. J. Mod. Power Syst. Clean Energy **5**(1), 10–19 (2016). https://doi.org/10.1007/s40565-016-0256-x
9. Ilieva, I., Bremdal, B., Puranik, S.: Bringing business and societal impact together in an evolving energy sector. J. Clean Energy Technol. **7**(3), 42–48 (2019). https://doi.org/10.18178/JOCET.2019.7.3.508
10. Zalzar, S., Bompard, E.F.: An incentive-based settlement mechanism for participation of flexible demands in day-ahead markets. In: 2019 International Conference on Smart Energy Systems and Technologies (SEST), pp. 1–6 (2019). https://doi.org/10.1109/SEST.2019.8849038
11. Okur, Ö., Voulis, N., Heijnen, P., Lukszo, Z.: Aggregator-mediated demand response: minimizing imbalances caused by uncertainty of solar generation. Appl. Energy **247**, 426–437 (2019). https://doi.org/10.1016/j.apenergy.2019.04.035
12. Silva, C., Faria, P., Vale, Z.: Demand response and distributed generation remuneration approach considering planning and operation stages. Energies **12**(14), 2721 (2019). https://doi.org/10.3390/en12142722
13. Terras, J.M., Silva, C., Vale, Z., Faria, P.: Consumers modelling and clustering for the use of flexibility to mitigate network congestion. In: CIRED 2020 Berlin Workshop, no. September, pp. 22–23 (2020)
14. Silva, C., Faria, P., Vale, Z.: Rating the participation in demand response programs for a more accurate aggregated schedule of consumers after enrolment period. Electronics **9**(2), 349 (2020). https://doi.org/10.3390/electronics9020349

RETRACTED CHAPTER: An Edge-IoT Architecture and Regression Techniques Applied to an Agriculture Industry Scenario

María E. Pérez-Pons[1(✉)], Ricardo S. Alonso[1], Javier Parra-Domínguez[1], Marta Plaza-Hernández[1], and Saber Trabelsi[2]

[1] AIR Institute, IoT Digital Innovation Hub, Salamanca, Spain
eugenia.perez@usal.es

[2] Science Program, Texas A&M University at Qatar, P. O. Box 23874, Education City, Doha, Qatar
https://air-institute.org
https://air-institute.org

Abstract. The agricultural industry must adapt to todays market by using resources efficiently and respecting the environment. This paper presents the analysis of data and the application of the Internet of Things (IoT) and advanced computing technologies in a real-world scenario. The proposed model monitors environmental conditions on a farm through a series of deployed sensors and the most outstanding feature of this model is the robust data transmission it offers. The analysis of information collected by the sensors is measured using state-of-the-art computing technology that helps reduce data traffic between the IoT layers and the cloud. The designed methodology integrates sensors and a state-of-the-art computing platform for data mining. This small study forms the basis for a future test with several operations at the same time.

Keywords: Smart Farming · Data transfer · Eco-efficiency · Linear regression · Internet of Things · Edge computing

1 Introduction

When it comes to agricultural production, the disparity between developed and developing countries is becoming narrower, as the market is increasingly globalized and more competitive. The Common Agricultural Policy (CAP) [18] provides and manages resources of the EU budget, providing support to countries within the European Union (EU), in the form of income to farmers, market orientation and the environment. The CAP also regulated the evolution of quotas

The original version of this chapter was retracted: The retraction note to this chapter is available at https://doi.org/10.1007/978-3-030-78901-5_31

J. M. Corchado and S. Trabelsi (Eds.): SSCTIC 2021, LNNS 253, pp. 92–102, 2022.
https://doi.org/10.1007/978-3-030-78901-5_9

for agricultural industries. Although quotas have been eliminated, there are significant concerns among agricultural lobbies and the real time-changing effects of such quotas, as well as their economic impact within a market, over the years. However, the challenges facing the european dairy industry are also applicable to dairy producers around the world: the need to increase resource efficiency, to be more environmentally friendly and to apply the latest technological trends that allow to offer detailed information to the final consumer, while guaranteeing the safety and quality of the final product. The transformation of the agricultural sector is becoming a reality; the level of technology used in production is gradually increasing, as well as the challenges faced by producers [37]. The most recent scope of work for the CAP includes water resources and effective water management.

With technological advances, farms in developed and developing countries can benefit from the application of low-cost technologies [17]. In this regard, the *Internet of Things* (IoT) and, more specifically, the *Industrial Internet of Things* (IIoT), is presented as a key enabling technology for the implementation of resource monitoring and management solutions in various Industry 4.0 scenarios, including smart agriculture environments [41]. IIoT can be used in combination with other technologies, such as *cloud computing*, *big data*, *artificial intelligence*, or *distributed ledger technologies* (e.g., blockchain) to implement solutions that improve traceability and productivity of industrial processes [49]. However, when it comes to transmitting data to the cloud, several challenges arise in terms of data privacy, energy consumption or the costs associated with using cloud services [4]. In this regard, service providers charge fees depending on the amount of data that is transferred, stored and processed in the cloud [47]. By using *Edge Computing* technologies, it is possible to reduce the volume of traffic transferred between the IoT layer and the cloud [3]. In addition, this technology enables the execution of *machine learning* models at the edge of the network, reducing response time and providing a certain level of service even if communication with the cloud is interrupted; something common in scenarios where Internet connectivity is limited (for example, agricultural environments in rural areas) [5]. This research aims to develop a strategy on how farms can control their environmental efficiency and assess their chances of increasing profitability. Due to all the regulations and permitted production levels and their implications, this research presents an efficiency-oriented case study on a mixed farm in Spain.

The rest of this paper is organized as follows. Section 2 presents a review of the state-of-the-art literature regarding the economic and environmental effects of technology on the agricultural industry. Section 3 describes how IoT and EC help increase efficiency and profitability in Smart Farming scenarios. After that, Sect. 4 analyzes the profitability and environmental performance of an Edge-IoT platform in a Smart Farming scenario. Section 5 covers the experiment and initial results. Lastly, Sect. 6 draws conclusions and discusses future research.

2 Economic and Environmental Effects of Information Technology in the Agricultural Industry

There is a strong evidence that new technologies and incentives for sustainable production have had an economic impact on the agricultural industry, which affected production efficiency [7,34]. In recent years, companies have been focusing, not only on the efficiency of the product itself, but also on suppliers that meet sustainability requirements [20], and on the data they have [32]. Therefore, being a sustainable farm has become an added value, not only in terms of expenditure reduction, but also in terms of being suitable for new partners. In this sense, different methodologies are being developed to identify methods of assessing sustainable value chains [21].

The last European Horizon project launched by CAP focuses on setting *into legislation the political ambition of being the world's first climate neutral continent by 2050* [1]. To achieve this goal, there is a planned road-map, and this year the European Commission will launch the *European Climate Pact*, following its *Green Deal* strategy, which will be a lever for citizens to voice their opinions and play a role in designing new actions for Europe's environmental goals. Therefore, those incentives are at political and policy levels, even though companies are also moving towards environment management policies.

In [31], the different applications and benefits of the new *Precision Agriculture* (PA) concept have been reviewed. In this regard, the addition of *Decision Support Systems* (DSS) to PA represents the combination of data for optimal decision making [29]. While production levels and effectiveness at production frontiers incorporate variables more aligned with the environment and the macroeconomic situation, when it comes to measuring profitability, any economic instability is a determining factor in the equation, as studies such as [28]. Another approach that could be used to forecast optimal production levels is *short-term long memory* (LSTM), a type of artificial neural network (ANN). This solution has been proposed in studies such as [9], with good results.

In the mid-1990s authors such as [40] began to examine the terms and concepts of *eco-efficiency*. Eco-efficiency is the ability to produce more goods and services with less environmental impact and less consumption of natural resources [23]. The eco-efficiency ratio is normally measured as the ratio between the value added of what is produced (for example, GDP) and the added environmental impacts of the produced product or service (usually in terms of CO_2 emissions).

The next Section depicts some of the principal approaches where IoT and EC technologies have been applied in the field of Precision Agriculture and Smart Farming.

3 Industrial Internet of Things and Edge Computing Technologies in Smart Farming Scenarios

IoT offers many solutions to each of its application areas. Some of the most important functions are: multiple communication protocol management, data processing, real-time information and response, big data storage, data security and privacy [41], data visualization [13]. However, implementing these functions brings a number of challenges that need to be solved: Heterogeneity of data sources, security, privacy, latency, real-time response, use of shared computing resources, etc. [6]. Despite the fact that the use of IoT data ingestion layers enables the solution of the heterogeneity problem, other issues need to be addressed. One of these is the high volume of data that can be transmitted to the IoT platform by hundreds, thousands or even millions of devices. In this regard, solutions such as the *edge computing* paradigm have emerged in response to the need to reduce the amount of data traffic between the IoT layer and the cloud.

The Edge Computing paradigm reduces congestion due to the need for compute, network or cloud storage resources [49]. With this strategy, the computing and service infrastructure is closer to data sources and end users by migrating filtering, processing or storage from the cloud to the edge of the network [39]. In addition, there are a wide variety of scenarios in which IoT and Edge Computing based solutions are used. Among the most relevant applications are *Industry 4.0* [42], *smart energy* [43], *Smart cities* [11,48] or, in the case of this work, *smart agriculture industry* [5], as well as others, such as applications for tourism [19].

While there are scenarios where edge computing is being applied in a single environment, such as an *ad hoc* system, advances are also being made in the deployment of *edge as platform* functionalities. This increases the repeatability of the solution. On the other hand, there are reference architectures whose guidelines can be followed when designing and implementing systems and platforms based on edge computing [42]. *Precision agriculture* only considers the variability of the crop data. However, *Smart Farming* provides more comprehensive analytics, predictions and recommendations, as well as task automation, taking into account historical and real-time information about crops, machines, livestock or people [46].

There are specific use cases where IoT and Edge Computing are being applied to smart agriculture, such as the work of [2], which measures the quantity and quality of grain in silos; [33] which measured the efficiency on monthly basis, [8], which monitors bicarbonate irrigation for precision hydroponic farming; [12], which uses RFID sensors and egg detectors to process hen behavior and welfare locally and in the cloud; [15], which analyzes multispectral images for crop quality control; [25], which applies neural networks, SVM (Support Vector Machines) and electronic odor techniques to rapidly detect and identify moldy apples; as well as the work of [36], which proposes an insect monitoring system for open fields using vibroacoustic sensors (Fig. 1).

However, Edge-IoT based solutions are usually created by end-to-end platforms. In this sense, [27] proposed a multilayer architecture formed by a perception layer (i.e., data retrieval), a network layer, a middleware layer (i.e., service management), an application layer, and a business layer (i.e., management of the entire system). One of its major drawbacks is that it does not consider aspects such as security. Therefore, it may not be the best choice for managing data and applications for value chain traceability. [26] proposed a framework to

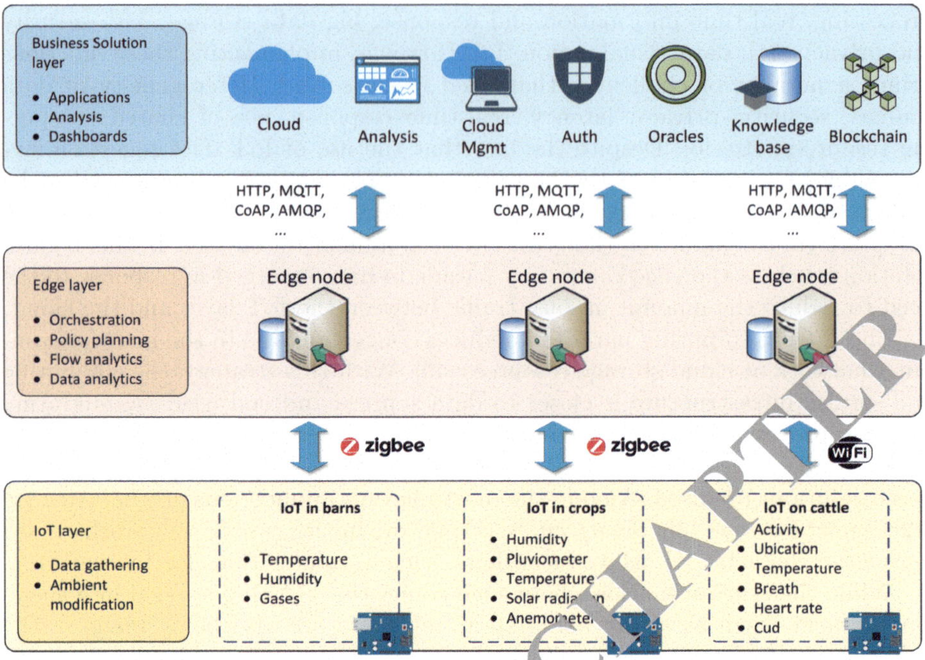

Fig. 1. SmartDairyTracer platform based on GECA that has been implemented in the analyzed scenario.

build intelligent farming systems. However, in their work, they do not present an implementation in a specific case study. On the other hand, [38] proposed a complete connected farm solution which used the IoT Service Server to create virtual IoT devices and a middleware management module installed on physical devices. Other authors such as [35] also proposed multi-tier platforms, while [10,44] introduced a basic IoT-Edge architecture to facilitate access to smart agriculture in developing countries. Finally, [30] proposed a scalable framework for data analytics, in which edge nodes preprocess and analyze the collected private data before sending the results to a remote server to estimate and predict the total crop yield.

Besides, there are various Edge Computing reference architectures used in industrial or Industry 4.0 environments, one of which is the Global Edge Computing Architecture (GECA) presented by [42], the Edge-IoT platform on which this work is based. GECA introduces Edge Computing functionalities that reduce the use of cloud computing, storage and network resources. Furthermore, it includes blockchain technologies that provide security and guarantee data integrity and traceability. This architecture, in turn, is the result of the analysis of four major reference architectures in the field of Edge-IoT in Industry 4.0: *FAR-Edge* [16], *INTEL-SAP* [24], the architecture of *Edge Computing Consortium* [14] and the *Industrial Internet Reference Architecture* (IIRA) [45]. GECA consists of three layers: the IoT layer, the edge layer, and the business solution layer. The above architecture was used to implement the *SmartDairyTracer* platform for process tracking in agribusiness.

The first stage of the SmartDairyTracer platform was implemented and tested in a real-world scenario by [5], which confirmed the benefits of Edge Computing, reducing data traffic by more than 46%, which translates into potential cost reductions. The experiments presented in [5] showed that it was possible to reduce the cost associated with data transfer between the IoT layer and the remote cloud by introducing reference architecture design principles, such as GECA, in an agro-industrial platform designed to monitor, track and optimize the resources and processes performed in the value chain in a mixed dairy production scenario. Additionally, the introduction of *Edge* nodes improved the reliability of communication with the cloud, reducing the number of missing values in the database.

In the next section, the cost and environmental efficiency of SmartDairyTracer is analyzed in the same mixed dairy production scenario.

4 Edge-IoT Platform in a Smart Farming Scenario

This research focuses on the use of IoT and advanced computing in a real-world farm scenario, with the purpose of increasing the profitability and sustainability of agricultural activity. The proposed system includes various sensors for monitoring multiple variables over a given period of time. These variables are rain, temperature and humidity, and are used to measure the *Environmental Performance Index*. Two main advantages are expected from the ability to track these variables, given that this is a mixed operation. On the one hand, traceability saves water costs, through more efficient use of water for irrigation and for livestock. On the other hand, temperature sensors allow the farmer to counteract inappropriate temperatures that can affect the health or welfare of cattle, which means lower milk production. GECA's Edge nodes filter and pre-process data from the IoT layer devices [42]. In addition, they are responsible for eliminating values that have been repeated as a result of frame retransmissions from the physical sublayers (i.e., ZigBee, Wi-Fi) to the *IoT* layer. They can also average and analyze regression data that occurs on the same layer *Edge*. In both cases, the amount of data and the cost of transmitting it to the cloud is reduced, reducing the cost of data traffic and the need for computing and storage in the cloud.

The experiments and results presented by [5] show that the application of the GECA reference architecture in the construction of the agri-food platform, and the addition of the Edge layer to it, reduces the total amount of data transferred to the cloud by 46.72% in a scenario with a mixed dairy farm, with the same usage conditions and sensors. This reduction may be even more significant in other scenarios where, given their characteristics, the filtering and/or pre-processing phases of the GECA architecture are used.

5 Experimentation and Initial Results

To conduct the experiment, three different variables (i.e., rain, temperature and air humidity) have been considered, among the ones gathered by a set of wireless agro-meteo stations installed on the farm [5]. With the average of each day, a linear regression model has been applied to forecast the evolution of the tracked values. For the regression model, the data has been split into test and train, where the values to test the model are the 25% of the entire set, as shown in Fig. 2.

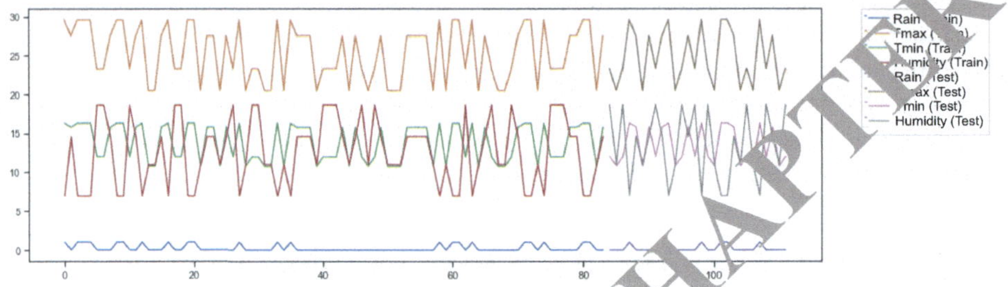

Fig. 2. The results correspond to the average values of each day in the period 01/09/2018 till 15/12/2018. Train and test data split for forecasting the evolution. The test has been split in the following way: 0.25 test and 0.75 train.

To make predictions and evaluate the model, the data has been decomposed into heteroscedasticity, tendency, and noise. To eliminate the heteroscedasticity, a logarithmic function has been applied. To understand the tendency of the current data, a linear regression model has been used. Then, the noise has been eliminated to resolve any data inconsistencies. The tendency of the train data when the linear regression model is applied is shown in Fig. 3.

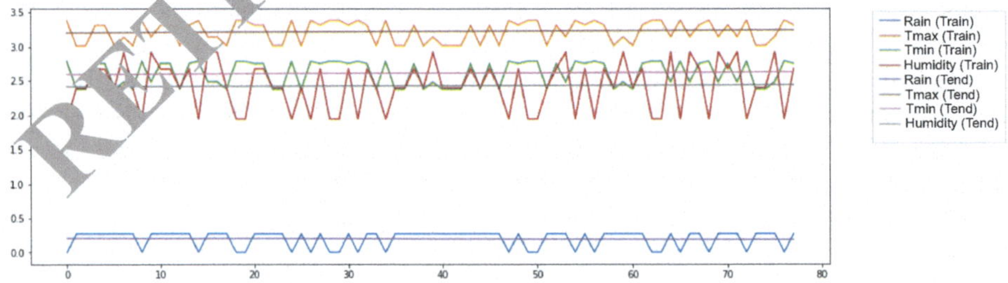

Fig. 3. Tendency of the different attributes and data collected from the sensors.

Figure 4 displays the results of the prediction tendency for each attribute. The graph presents the real data used as a test for the model and the forecast

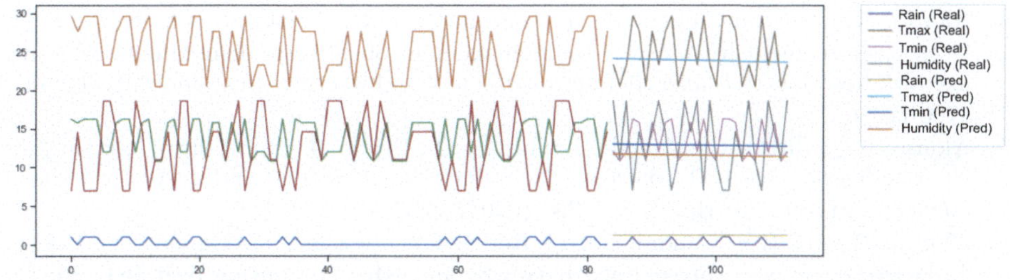

Fig. 4. Attribute tendency predictions made on the basis of the collected data.

on the tendency. The results are overlapped on the test data to assess whether the model could correctly generate the trend data.

6 Conclusions

This study provides the results of the monitoring of different variables in the Environmental Performance Index by means of real-time sensors. The application of an Edge-Computing platform reduces data traffic to the Cloud. This study shows that the application of cutting-edge techniques, such as IoT and Edge-Computing reduces the data traffic to the cloud and enables to monitor environmental variables and make decisions on the basis of projected trends. The agro-technological paradigm is leading to large-scale scenarios (that is, farms with millions of ha with a large number of sensors) which translates into an increase in data traffic to the Cloud. These innovative techniques are the result of adaptability to change and these techniques serve to reduce the volume of stored data and identify patterns and data tendencies.

Acknowledgments. This research was partially Supported by the project "Computación cuántica, virtualización de red, edge computing y registro distribuido para la inteligencia artificial del futuro", Reference: CCTT3/20/SA/0001, financed by Institute for Business Competitiveness of Castilla y León, and the European Regional Development Fund (FEDER). Authors would like to give a special thanks to Rancho Guareña Hermanos Olea Losa, S.L. (Castrillo de la Guareña, Zamora, Spain) for their collaboration during the implementation and testing of the platform.

References

1. European Commission Horizon 2050 (2019). https://ec.europa.eu/commission/presscorner/detail/en/IP_19_6691
2. Agrawal, H., Prieto, J., Ramos, C., Corchado, J.M.: Smart feeding in farming through IoT in silos. In: ISTA 2016. AISC, vol. 530, pp. 355–366. Springer, Cham (2016). https://doi.org/10.1007/978-3-319-47952-1_28
3. Ai, Y., Peng, M., Zhang, K.: Edge computing technologies for Internet of Things: a primer. Digital Commun. Netwo **4**(2), 77–86 (2018). https://www.sciencedirect.com/science/article/pii/S2352864817301335

4. Alonso, R.S., Sittón-Candanedo, I., Casado-Vara, R., Prieto, J., Corchado, J.M.: Deep reinforcement learning for the management of software-defined networks and network function virtualization in an edge-IoT architecture. Sustainability **12**(14), 5706 (2020)
5. Alonso, R.S., Sittón-Candanedo, I., García, Ó., Prieto, J., Rodríguez-González, S.: An intelligent edge-IOT platform for monitoring livestock and crops in a dairy farming scenario. Ad Hoc Netw. **98**, 102047 (2020)
6. Alonso, R.S., Sittón-Candanedo, I., Rodríguez-González, S., García, Ó., Prieto, J.: A survey on software-defined networks and edge computing over IoT. In: De la Prieta, F., et al. (eds.) PAAMS 2019. CCIS, vol. 1047, pp. 289–301. Springer, Cham (2019). https://doi.org/10.1007/978-3-030-24299-2_25
7. Balafoutis, A.T., et al.: Smart farming technologies – description, taxonomy and economic impact. In: Pedersen, S.M., Lind, K.M. (eds.) Precision Agriculture: Technology and Economic Perspectives. PPA, pp. 21–77. Springer, Cham (2017). https://doi.org/10.1007/978-3-319-68715-5_2
8. Cambra, C., Sendra, S., Lloret, J., Lacuesta, R.: Smart system for bicarbonate control in irrigation for hydroponic precision farming. Sensors **18**(5), 1333 (2018)
9. Cao, Q., Banerjee, R., Gupta, S., Li, J., Zhou, W., Jeyachandra, B., et al.: Data driven production forecasting using machine learning. In: SPE Argentina Exploration and Production of Unconventional Resources Symposium. Society of Petroleum Engineers (2016)
10. Casado-Vara, R., Martin-del Rey, A., Affes, S., Prieto, J., Corchado, J.M.: IoT network slicing on virtual layers of homogeneous data for improved algorithm operation in smart buildings. Future Gener. Comput. Syst. **102**, 965–977 (2020)
11. Chamoso, P., González-Briones, A., De La Prieta, F., Venyagamoorthy, G.K., Corchado, J.M.: Smart city as a distributed platform: toward a system for citizen-oriented management. Comput. Commun. **152**, 323–332 (2020)
12. Chien, Y.R., Chen, Y.X.: An RFID-based smart nest box: an experimental study of laying performance and behavior of individual hens. Sensors **18**(3), 859 (2018)
13. Corchado, J.M., et al.: Deepint.net: a rapid deployment platform for smart territories. Sensors **21**(1), 236 (2021)
14. Edge Computing Consortium, Alliance of industrial internet: edge computing reference architecture 2.0. Technical report, Edge Computing Consortium, November 2017. http://www.ecconsortium.net/Uploads/file/20180328/1522232376480704.pdf
15. ElMasry, G., Mandour, N., Al-Rejaie, S., Belin, E., Rousseau, D.: Recent applications of multispectral imaging in seed phenotyping and quality monitoring–an overview. Sensors **19**(5), 1090 (2019)
16. FAR-EDGE Project: FAR-EDGE Project H2020, November 2017. http://far-edge.eu/
17. Fleming, K., Waweru, P., Wambua, M., Ondula, E., Samuel, L.: Toward quantified small-scale farms in Africa. IEEE Internet Comput. **20**(3), 63–67 (2016)
18. Gardner, B.: European Agriculture: Policies, Production, and Trade. Psychology Press, Routledge (1996)
19. González Bedia, M., Corchado Rodríguez, J.M., et al.: A planning strategy based on variational calculus for deliberative agents. Comput. Inf. Syst. **9**, 2–13 (2002)
20. Gupta, M.C.: Environmental management and its impact on the operations function. Int. J. Oper. Prod. Manage. **15**(8) (1995)
21. Handfield, R.B., Walton, S.V., Seegers, L.K., Melnyk, S.A.: 'Green' value chain practices in the furniture industry. J. Oper. Manage. **15**(4), 293–315 (1997)

22. Humphreys, P., McIvor, R., Chan, F.: Using case-based reasoning to evaluate supplier environmental management performance. Expert Syst. Appl. **25**(2), 141–153 (2003)
23. Ichimura, M., et al.: Eco-efficiency indicators: measuring resource-use efficiency and the impact of economic activities on the environment. ESCAP, Bangkok (2009)
24. INTEL-SAP: IoT joint reference architecture from Intel and SAP. Technical report, INTEL-SAP, November 2018. https://www.intel.com/content/dam/www/public/us/en/documents/reference-architectures/sap-iot-reference-architecture.pdf
25. Jia, W., Liang, G., Tian, H., Sun, J., Wan, C.: Electronic nose-based technique for rapid detection and recognition of moldy apples. Sensors **19**(7), 1526 (2019)
26. Jones, J.W., et al.: Toward a new generation of agricultural system data, models, and knowledge products: state of agricultural systems science. Agric. Syst. **155**, 269–288 (2017)
27. Khan, R., Khan, S.U., Zaheer, R., Khan, S.: Future internet: the internet of things architecture, possible applications and key challenges. In: 2012 10th International Conference on Frontiers of Information Technology, pp. 257–260. IEEE (2012)
28. Machek, O., Špička, J.: Productivity and profitability of the Czech agricultural sector after the economic crisis. WSEAS Trans. Bus. Econ. **11**, 700–706 (2014)
29. McBratney, A., Whelan, B., Ancev, T., Bouma, J.: Future directions of precision agriculture. Precis. Agric. **6**(1), 7–23 (2005)
30. Park, J., Choi, J.H., Lee, Y.J., Min, O.: A layered features analysis in smart farm environments. In: Proceedings of the International Conference on Big Data and Internet of Thing, pp. 169–173, BDIOT 2017. ACM, New York (2017)
31. Pedersen, S.M., Lind, K.M. (eds.): Precision Agriculture: Technology and Economic Perspectives. PPA, Springer, Cham (2017). https://doi.org/10.1007/978-3-319-68715-5
32. Pérez-Pons, M.E., González-Briones, A., Corchado, J.M.: Towards financial valuation in data-driven companies. Orient. J. Comput. Sci. Technol. **12**(2), 28–33 (2019)
33. Pérez-Pons, M.E., Plaza-Hernández, M., Alonso, R.S., Parra-Domínguez, J., Prieto, J.: Increasing profitability and monitoring environmental performance: a case study in the agri-food industry through an edge-IoT platform. Sustainability **13**(1), 283 (2021)
34. Pérez-Pons, M.E., Parra-Domínguez, J., Chamoso, P., Plaza, M., Alonso, R.: Efficiency, profitability and productivity: technological applications in the agricultural sector. ADCAIJ: Adv. Distrib. Comput. Artif. Intell. J. **9**(4) (2020)
35. Popović, T., Latinović, N., Pešić, A., Zečević, Ž, Krstajić, B., Djukanović, S.: Architecting an IoT-enabled platform for precision agriculture and ecological monitoring: a case study. Comput. Electron. Agric. **140**, 255–265 (2017)
36. Potamitis, I., Rigakis, I., Tatlas, N.A., Potirakis, S.: In-vivo vibroacoustic surveillance of trees in the context of the IoT. Sensors **19**(6), 1366 (2019)
37. Reardon, T., Barrett, C.B., Berdegué, J.A., Swinnen, J.F.: Agrifood industry transformation and small farmers in developing countries. World Dev. **37**(11), 1717–1727 (2009)
38. Ryu, M., Yun, J., Miao, T., Ahn, I.Y., Choi, S.C., Kim, J.: Design and implementation of a connected farm for smart farming system. In: 2015 IEEE SENSORS, pp. 1–4. IEEE (2015)
39. Sánchez-Iborra, R., Sánchez-Gómez, J., Skarmeta, A.: Evolving IoT networks by the confluence of MEC and LP-WAN paradigms. Future Gener. Comput. Syst. **88**, 199–208 (2018)

40. Schmidheiny, S., Timberlake, L.: Changing Course: A Global Business Perspective on Development and the Environment, vol. 1. MIT Press, Cambridge (1992)
41. Sisinni, E., Saifullah, A., Han, S., Jennehag, U., Gidlund, M.: Industrial internet of things: challenges, opportunities, and directions. IEEE Trans. Ind. Inform. **14**(11), 4724–4734 (2018)
42. Sittón-Candanedo, I., Alonso, R.S., Corchado, J.M., Rodríguez-González, S., Casado-Vara, R.: A review of edge computing reference architectures and a new global edge proposal. Future Gener. Comput. Syst. **99**, 278–294 (2019)
43. Sittón-Candanedo, I., Alonso, R.S., García, Ó., Gil, A.B., Rodríguez-González, S.: A review on edge computing in smart energy by means of a systematic mapping study. Electronics **9**(1), 48 (2020)
44. Suma, N., Samson, S.R., Saranya, S., Shanmugapriya, G., Subhashri, R.: IoT based smart agriculture monitoring system. Int. J. Recent Innov. Trends Comput. Commun. **5**(2), 177–181 (2017)
45. Tseng, M., Canaran, T.E., Canaran, L.: Introduction to edge computing in IIoT. Technical report, Industrial Internet Consortium, November 2018. https://www.iiconsortium.org/IISF.htm
46. Wolfert, S., Ge, L., Verdouw, C., Bogaardt, M.J.: Big data in smart farming-a review. Agric. Syst. **153**, 69–80 (2017)
47. Wu, C., Toosi, A.N., Buyya, R., Ramamohanarao, K.: Hedonic pricing of cloud computing services. IEEE Trans. Cloud Comput. **6**(1), 182–196 (2018)
48. Yigitcanlar, T., Butler, L., Windle, E., Desouza, K.C., Mehmood, R., Corchado, J.M.: Can building "artificially intelligent cities" safeguard humanity from natural disasters, pandemics, and other catastrophes? An urban scholar's perspective. Sensors **20**(10), 2988 (2020)
49. Yu, W., et al.: A survey on the edge computing for the internet of things. IEEE Access **6**, 6900–6919 (2017)

Smart City Perspectives in the Context of Qatar

Reem Al Sharif[1] (✉) and Shaligram Pokharel[2]

[1] Engineering Management Department, Qatar University, Doha, Qatar
ra1900521@qu.edu.qa
[2] Industrial System Engineering, Qatar University, Doha, Qatar

Abstract. The smart city concept is being implemented in different countries. In order to make such a concept successful, countries need to invest heavily in information and communication technology and provide opportunity to utilize them for creating economic and social value. As the access to such technology, the funds required for investments, and the manpower needed to take such a concept forward, the right use of such technology becomes challenging. The challenge requires and understanding of dimensions and the risks associated with the investment and use of information and communication technology. The complexity of the technology and its integration also brings in to risks associate with it. When risks are not adequately understood and addressed, they may cause security and privacy issues, thereby affecting the very concept of smart city. This paper focuses on identifying smart city dimensions, and the technical and non-technical risks. Qatar's smart city project and the risks related to the information and communication technology applications used at the project are presented.

Keywords: Smart cities · Smart city dimensions · Technical risks · Non-technical risks · Qatar

1 Introduction

In 1990, the smart city concept was introduced to integrate advanced information and communication technology (ICT) based hardware and software in urban planning [1]. The smart city utilizes ICT to enhance people's quality of life, foster economy, simplify the process to resolve transport and traffic problems through proper management, encourage a clean and sustainable environment, and provide accessible interaction with the government's relevant authority [2]. The increased urban expansion and innovation in urban planning and ICT have encouraged planners to promote the smart city concept. Studies have suggested many dimensions in smart cities, such as smart mobility [3]; other dimensions are mentioned in [2].

Urban challenges such as congested transportation, high carbon energy network, infrastructure maintenance and repair, and urban security and policy should be considered the smart city concept [4]. Although cities like Dubai, Hong Kong, London, New York, Moscow, and Ottawa have adopted AI and robotics to develop smart applications to promote smartness in their cities, there are associated risks that can jeopardize the achievement of smartness [4].

J. M. Corchado and S. Trabelsi (Eds.): SSCTIC 2021, LNNS 253, pp. 103–113, 2022.
https://doi.org/10.1007/978-3-030-78901-5_10

Due to the complexity of systems needed for smart city development, smart city functionality can become vulnerable. Operational risks, strategy risks, and external risks may cause this vulnerability [5]. Techatassanasoontorn and Suo [6] divide risks into socio-political risks, approval risks, financial risks, technical risks, partnership risks, and resource management risks for smart cities. Risks are also associated with the security and privacy within the smart city systems [7].

Studies of smart city's risks may focus on an individual smart city system, such as energy systems risks highlighted by [8] and waste management system risks mentioned in [9]. However, studies should consider holistic risk assessment in smart city planning and operation. Assessment of risks can help mitigate the effects of technology, security, and privacy, political, environmental, managerial, and user trust and adoption. It is also mentioned that risk assessment can highlight the risk potential in different aspects of smart city design and operation [10].

A limited number of studies focus on governance policies, performance indicators, and standards in smart cities [11]; most studies highlight smart city themes, technologies, and innovations. Convenience research is desired to provide comprehensive information for government and policymakers about smart cities at a holistic level [11].

This paper provides a review of smart city dimensions and technical and non-technical risks. Qatar's smart city project (MDD) is investigated to identify developed smart applications about the dimensions and the corresponding risks considered during the implementation and highlight additional risks based on the reviewed literature.

Further discussion in this paper is organized as follows: literature review in Sect. 2. Qatar's smart city project details and project data are provided in Sect. 3. Discussions in Sect. 4 and conclusions are given in Sect. 5.

2 Literature Review

The content analysis method is used to review the literature related to the risks of smart cities. It is a quantitative method based on analyzing and categorizing the research topic's related text [12]. The method is used in studies by Al-Sobai, Pokharel, and Abdella, [13, 14] and Islam, Nepal, Skitmore, and Attarzadeh [15]. To collect literature databases available at Qatar University Library such as ScienceDirect, Taylor and Francis, and Wiley are used.

2.1 Smart Cities Dimensions

This section will present the smart city's dimensions regarding definitions, smart applications, and the dominant technology mentioned in the literature. This part of the literature review addresses the research question **RQ1**: What are Qatar's smart city dimensions and the associated applications?

The smart economy comprises guidelines and policies that inspire innovation and creativity in collaboration with scientific research, advanced technology, and the sustainability concept's attention to the environment [3]. Innovation, competitiveness, use of information and communication technologies in the overall aspect of the smart economy, and the socially responsible use of resources are considered as well ([3, 16, 17]).

Studies highlight forms of the smart economy, applications, and challenges that will lead to non-technical risks associated with each of the discussed applications. Studies mention platforms such as Amazon and Alibaba that work well for individual services but not complex products and suggest distributed marketplace models for smart cities aim, building a service ecosystem [18]. The usage of sharing economy concept is introduced in renewable energy to improve energy ingesting and support sustainable systems [19].

Smart cities' governance is associated with decision-making, public services, social services, transparent governance, and policies and strategies. Authors consider governance as the main building block in the collective efforts to develop effective interactions between all actors of smart cities ([20, 21]). Promoting open innovation requires interactive governance. The latter is facilitated through e-governance [2]. Cloud-based information services can help in decision-making by supporting participation, engagement, and information sharing [2]. Data privacy and security is the main challenge for governance applications, as mentioned in ([7, 10]).

Smart living is associated with the OECD Better-Life Initiative framework that considers the development and preservation of natural, economic, and human capitals as smart living elements [22]. Smart living can refer to smart buildings, education, and healthcare ([2],[49]), and social awareness [20]. Special care needs and emergency support enabled through the ICT are part of smart healthcare ([2, 23]). Smart living is an outcome of the smart economy [3]. ICT supports smart living through internet-enabled automated living space conditioning, lighting, and connected security systems [24]. Smart living applications are supported by technologies such as cloud storage and computing, AI, machine learning, data mining, and wireless sensor networks [25].

Smart mobility in terms of transportation systems and infrastructure is also studied by [26]. As per the study, common issues in smart cities are the traffic problems such as congestions, long queues, and delays. Smart systems should focus on vehicle usage and provide coordinated choices between the public and private vehicles for ease in their commuting. For example, in Singapore, measures have been established to make the real option choice between owning a vehicle (through a competitive bid for the certificate of entitlement) and utilizing the public system like taxi, bus, and rail for transportation. Real-time data on roads and routing and analysis to prospective travelers are collected by IoT technology [20]. A better-integrated transportation system for smart mobility is achieved by the widespread use of IoT in rural and urban areas [27].

Smart people form the social infrastructure of a smart city. It is related to human capital and social capital. The challenges of security and privacy of information and services provided to the people become more important in smart cities [28]. People's engagement with the government system through IoT is also important in smart cities [29], and functions like an e-government platform for information, access to and delivery of services through such a platform, and the service level agreements on such services promote smart city concept further.

Smart environment dimension includes enhancements in discarding waste; managing energy, smart grids, house and facilities, air and water quality; increasing green spaces, and; monitoring emissions [2]. Technology usage to maintain natural resources is also discussed in [30], which emphasizes that preserving natural resources requires sustainable methods to manage resources, protect the environment, and reduce pollution.

Internet of data and IoT are introduced as the used technologies to develop applications related to the smart environment [31]. These technologies use different sensors such as radio frequency identification (RFID), integrated circuits, optical sensors, and pressure sensors to manage a smart city environment. Such a provision helps the decision-makers to decide on the optimal use of resources on a real-time basis.

2.2 Risks Related to Smart Cities.

Smart city ecosystem includes all aspects of human life, such as transportation, logistics, education, and maintenance healthcare, computerized to be controlled and accessed through smart devices. Smart cities' development invites risks from multiple aspects [32]. The following paragraphs will demonstrate risks associated with smart cities from technical and non-technical perspectives. The research question being addressed for this part is **RQ2**: What are Qatar's smart city project's technical risks?

2.2.1 Technical Risks Associated with Smart Cities

Technical risks are related to technology and its implementation, such as AI, big data, and IoT risks. Technical risks are divided into three categories by [33]: network coverage in the city, choice of technology, and technology discontinuation. In addition, technical risks should consider security risks [32]. Risks include cybersecurity, interactions between devices, systems, the absence of supporting infrastructure, unorganized data management, and adaptation of different standards in technology and their integration. The following paragraphs will highlight the dominant technical risks found in the literature and, grouped based on the significant used technology.

Cybersecurity risk is a significant associated risk with IoT technology, blockchain, and AI. The increase of connected IoT devices is needed to support 'smart' ability in various sectors such as health, transportation, energy transmission, and others. Its vulnerability towards information hacking and misuse also increases. Therefore, the smart city concept should be supported with measures for cybersecurity risk management [34].

Authors ([35, 36]) show that cybersecurity is obtained by providing the requirements, privacy, and security. Breaches in cybersecurity can lead to fake alarms such as fires, earthquakes, or circuit breakdowns, which can endanger the public in the city [36].

The use of IoT should also be considered, but there are problems such as the interactions between devices and systems, absence of supporting infrastructure, unorganized data management, and absence of universal standards related to IoT ([35, 37]). Therefore, a holistic approach can be developed to include interdependence between all ICT-related actors: infrastructure, data space, and learning space to solve security risk [38].

Associated risks with AI applications mentioned in [39] are security and privacy risks. However, system complexity as an associated risk with artificial intelligence (AI) technologies is highlighted in [40]. Legal issues may be created because of these risks and require many verifications of compliance with existing laws related to fundamental rights protection.

AI's role in smart cities, businesses, and society are highlighted by [40] through applications such as data analytics in energy, education, health, security, transport, sustainable environment, and urban areas management. Also, the authors mention that a robust communication network monitoring system resulting from such an application can lead to early recognition of threats, frauds, crimes, fires, and accidents [40].

Risks in smart cities are not limited to technical risks [32]. The comprehensive analysis demonstrates that non-technical risks have a noticeable effect on the implementation and operation of smart cities. Therefore, the next section addresses the research question **RQ3:** What are the considered non-technical risks of Qatar's smart city project?

2.2.2 Non-technical Risks Associated with Smart Cities

Socio-economic risks include but not limited to the traditional mindset of stakeholders and decision-makers. The smart city concept implementation means handling multidisciplinary projects that require a considerable budget, trained personnel, and technology exposure of the citizens, decision-makers, and professionals. Social risks regarding the public proposition of specific technology are discussed [41]. Examples of this risk are Canada-Ontario's wind energy rollout and nuclear power in Germany. There are social struggles on these decisions as such systems involve different actors: regulators, customers, technology companies, and energy service providers for better efficiency, sustainability, and cost control [41], and the need for such alternate form of power in their locality.

Smart city projects face governance risks in terms of socio-political risks associated with policies, laws, rules, and political and social forces [42]. The risk of political leadership intrusion in solving technology involvement is highlighted in [43]. If the government wants to pursue a smart city, it should have a good governance system and minimum intrusions from the leadership.

When the strategic approach lacks the link between urban ICT development research agenda and sustainable development research plan is missing, the strategic risk emerges. City management needs to discuss strategic risks and challenges in strategy formulation and implementation [1].

Risk management in a smart city is not sufficiently addressed, probably due to time or financial restrictions [44]. Risk management includes risk identification, analysis, assessment, prioritization, and responses. The risk management process is important to enhance project performance. This enhancement is achieved by governing and monitoring the effects of uncertain and risky events on project objectives [15].

3 Research Design

3.1 Case Study: Qatar's Smart City - Mushaireb Downtown Development (MDD)

Qatar is evolving as one of the most dynamic and fastest-growing economies in the Middle East. The country is determined to innovate to improve its citizens' quality of life by preserving heritage and culture. Qatar's smart city project results from three years of research where experts of different disciplines are involved. Including architects, master planners, engineers, designers, and other experts (including specialists from Harvard, Princeton, Yale, and MIT to create a harmonized activity to combine the Qatari culture with the latest technologies. Construction began in 2010 and is scheduled in phases; the project includes more than 800 residential units, 10,000 parking spaces, and more than 100 buildings [45].

The MDD is studied in this paper as the project is in the final implementation stage. The MDD project aims to generate a sustainable downtown across 31-ha (76-acre) sites and bring people to discover a sense of community. The project has more than 100 buildings, including commercial and residential properties, retail, cultural, and entertainment areas. Sustainable designs are implemented to improve resource consumption, reduce waste generation, and decrease cost. These designs will achieve sustainability goals of reducing the carbon footprint. The project considers LEED (Leadership in Energy and Environmental Design) certification system developed by the US Green Building Council in building designs. The project uses the latest information and communication technologies to achieve smartness and sustainability [46].

3.2 Research Method

The research is completed through several stages, starting with the literature review, detailed in the previous section. Research questions are derived based on the reviewed literature and will be addressed by studying Qatar's smart city project (MDD). Readers may also want to review other methods such as PICO for this type of study. The PICO method focuses on the problem, investigation, developing viewpoints, and outcomes. However, the content analysis is used here as the focus here is more on the extraction of the information from the literature review. Content analysis also helps understand the viewpoints and methodologies, define the challenges, and develop the knowledge in the topic of discussion as mentioned in Rebeeh et al. [47]. The research method is given in Fig. 1.

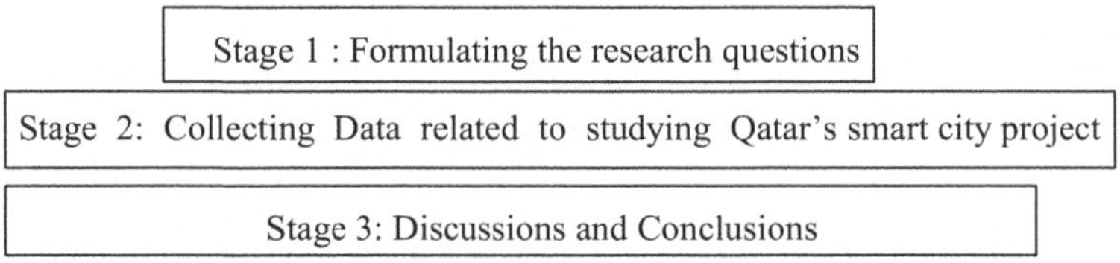

Fig. 1. Research method

3.2.1 Data Collection

Data from the MDD project are collected through a focus group with a representative from the project and the MDD team's information technology company. Focus group and document analysis are used to explore the research question by investigating different perspectives.

Focus group discussions, observational methods, and document analysis are qualitative analysis techniques. Combining two or more data collection methods, for instance, interviews and focus groups ('data triangulation'), enhances the study's credibility [46]. Focus group and document analysis are chosen to collect data from Qatar's different stakeholders' on the smart city project. The discussions are based on the three research questions mentioned in the previous section. Project documents and published brochures about (MDD) are studied to investigate risks, risk management strategies used within the project's life cycle and provided smart services at the smart city. The collected data concerning Qatar's smart city dimensions and applications are illustrated in Table 1.

Table 1. Qatar's smart city dimensions and applications data from the focus group and the project documents

No	Qatar smart city's services and applications (RQ1)	Dimension
1	Smart buildings and homes	Smart living
2	e-services portal	Smart governance
3	Smart energy management	Smart environment
4	Data analytics applications	Smart governance
5	People counting applications	Smart governance
6	Water conversations	Smart environment
7	Waste management applications	Smart environment
8	Central monitoring application	Smart governance
9	Smart transportation (internal tram network)	Smart mobility
10	Smart parking system	Smart mobility

The Collected data related to technical and non-technical risks associated with Qatar's smart city project addressing RQ2 and RQ3 are illustrated in Table 2 below.

Table 2. Technical and non-technical risks in Qatar's smart city project

No	Technical risks – from the literature review	Considered at Qatar's smart city project
1	Cybersecurity risk	Yes
2	Technical data and application risk	Yes
3	Network infrastructure risk	Yes
4	Data privacy and protection risk	No
5	Low productivity risk (related to blockchain technology)	N/A

(continued)

Table 2. (*continued*)

No	Non-technical risks – literature review	Considered at Qatar's smart city project
1	Policies, laws, and rules risks	Yes
2	Approvals and resource management risks	Yes
3	Legal issues related to data privacy	No
4	Strategic risks (integration between urban development ICT and sustainability)	Yes

4 Discussion

The planning and implementation of smart cities are measured as complex and multi-disciplinary projects where multiple dimensions must be considered. Not all smart city examples apply all the dimensions related to smart cities; they consider the significant impact of some of the dimensions to achieve their objectives. Qatar's MDD project emphasizes four dimensions: smart living, smart mobility, smart governance, and smart environment to provide quality of living through sustainability.

The paper also designates the difference in the level of technological applications. Due to ICT and related technologies, attention to the level of risks, especially related to security, privacy, and safety, becomes essential. In addition, a smart city faces both technical and non-technical risks. Therefore, Qatar's smart city project is in the final stage; it needs to review data privacy and protection risk, as they can bring a long-term impact and may jeopardize the value envisaged. The project should have continuous risk management and mitigation plans based on Qatar's policies, rules, and laws. Based on the application and the development elsewhere, the country may want to regulate some aspects of smart city projects and their operation in the country. As risk levels and complexity of their management changes over time because of technological development, MDD should also establish continuous risk assessment and monitoring.

The first research question to be addressed is related to the applicable smart city dimensions through different features. The examination of the MDD project shows that it uses smart living dimensions in terms of healthcare applications (e-health), education applications (e-learning), and smart building applications. For the smart mobility dimension, the project has considered transportation systems and smart vehicles. In terms of smart environment, it considers waste discarding, pollution control, energy management, smart grid quality of air and water, increases in green spaces, and monitoring emissions. For smart governance, e-government applications and services, and public participation platforms are considered. It is to note that the whole country is moving towards e-governance and e-health, for example, and therefore, eventually, the project may encompass other smart city dimensions in the future.

The second and third research questions are focused on the types of risks correlated with smart applications and the dominant types of risks. Qatar's smart city project considers most technical risks and some non-technical risks. The project considers cybersecurity risks, technical data and application risk, network infrastructure risk, data privacy and protection risk and energy consumption risk as technical risks and policies, rules and laws risks, approval and resource management, and strategic risks as non-technical risks. It has also established a risk management process to tackle the risks during and after the project.

5 Conclusions

The smart city concept is adopted to provide services through the efficient and effective use of technology. Different dimensions are considered for the development of smart cities. Some of these dimensions may require changes in the legal instruments in the country itself. For example, stricter rules are being developed on data privacy and cyber-security as these features may not be limited to one place and one country. The literature review shows that there are six dimensions established by the researchers based on the application of smart city concepts in different areas. However, it is also mentioned that a country may not adopt all the capabilities for one city or area due to overarching implications of some of the dimensions.

The study of the Msheirab smart city project shows that Qatar has used four dimensions, and there is an understanding of risks associated with these dimensions in the project. It also shows that within the dimensions, different factors can be considered to achieve smart living. While public transportation can be used as an option in some cases, in other cases, public transportation can be the option for smart mobility within the city. Therefore, the choice of factors and dimensions depends on a country's socio-cultural fabric, the level of the economic strength of the society, and the access to and usability of smart technologies.

References

1. Bibri, S.E., Krogstie, J.: Smart sustainable cities of the future: an extensive interdisciplinary literature review. Sustain. Cities Soc. **31**, 183–212 (2017)
2. Ismagilova, E., Hughes, L. Dwivedi, Y.K., Raman, K.R.: Smart cities: advances in research—an information systems perspective. Int. J. Inf. Manage. **47**(December 2018), 88–100 (2019)
3. Apostol, D., Bălăceanu, C., Constantinescu, E.M.: Smart – economy concept – facts and perspectives. HOLISTICA J. Bus. Public Adm. **6**(3), 67–77 (2015)
4. Golubchikov, O., Thornbush, M.: Artificial intelligence and robotics in smart city strategies and planned smart development. Smart Cities **3**(4), 1133–1144 (2020)
5. Mikes, A.: Managing risks: a new framework. Harvard Bus. Rev. Mag. (June) (2012)
6. Techatassanasoontorn, A.A., Suo, S.: Exploring risks in smart city infrastructure projects: municipal broadband initiatives. In: PACIS 2010 - 14th Pacific Asia Conference on Information Systems, pp. 13–24 (2010)
7. Čolić, N., Manić, B., Niković, A., Brankov, B.: Grasping the framework for the urban governance of smart cities in Serbia. The case of interreg SMF project clever. Spatium **4**(43), 26–34 (2020)
8. O'Dwyer, E., Pan, I., Acha, S., Shah, N.: Smart energy systems for sustainable smart cities: current developments, trends and future directions. Appl. Energy **237**(November 2018), 581–597 (2019)
9. Sharma, M., Joshi, S., Kannan, D., Govindan, K., Singh, R., Purohit, H.C.: Internet of Things (IoT) adoption barriers of smart cities' waste management: an Indian context. J. Clean. Prod. **270**, 122047 (2020)

10. Ismagilova, E., Hughes, L., Rana, N.P., Dwivedi, Y.K.: Security, privacy and risks within smart cities: literature review and development of a smart city interaction framework. Inf. Syst. Front. (2020)
11. Gupta, P., Chauhan, S., Jaiswal, M.P.: Classification of smart city research - a descriptive literature review and future research agenda. Inf. Syst. Front. **21**(3), 661–685 (2019). https://doi.org/10.1007/s10796-019-09911-3
12. Kohlbacher, F.: ePub WU institutional repository the use of qualitative content analysis in case study research. Forum Qual. Sozialforsch./Forum Qual. Soc. Res. **7**(1), 21 (2006)
13. Al-Sobai, K.M., Pokharel, S., Abdella, G.M.: Perspectives on the capabilities for the selection of strategic projects. Sustainability **12**(19) (2020)
14. Bouzguenda, I., Alalouch, C., Fava, N.: Towards smart sustainable cities: a review of the role digital citizen participation could play in advancing social sustainability. Sustain. Cities Soc. **50**(November 2018), 101627 (2019)
15. Islam, M.S., Nepal, M.P., Skitmore, M., Attarzadeh, M.: Current research trends and application areas of fuzzy and hybrid methods to the risk assessment of construction projects. Adv. Eng. Inform. **33**, 112–131 (2017)
16. Arroub, A., Zahi, B., Sabir, E., Sadik, M.: A literature review on smart cities: paradigms, opportunities and open problems. In: Proceedings of 2016 International Conference on Wireless Networks and Mobile Communication WINCOM 2016 Green Communication Networks, pp. 180–186 (2016)
17. Vinod Kumar, T.M., Dahiya, B.: Smart economy in smart cities. In: Advances in 21st Century Human Settlements. Springer, Singapore (2017). https://doi.org/10.1007/978-981-10-1610-3
18. Radonjic-Simic, M., Pfisterer, D.: Beyond platform economy: a comprehensive model for decentralized and self-organizing markets on internet-scale. Computers **8**(4), 1–44 (2019)
19. Akande, A., Cabral, P., Casteleyn, S.: Understanding the sharing economy and its implication on sustainability in smart cities. J. Clean. Prod. **277**, 124077 (2020)
20. Silva, B.N., Khan, M., Han, K.: Towards sustainable smart cities: a review of trends, architectures, components, and open challenges in smart cities. Sustain. Cities Soc. **38**(August 2017), 697–713 (2018)
21. Nilssen, M.: To the smart city and beyond? Developing a typology of smart urban innovation. Technol. Forecast. Soc. Change **142**(July 2018), 98–104 (2019)
22. Mizobuchi, H.: Measuring world better life frontier: a composite indicator for OECD better life index. Soc. Indic. Res. **118**(3), 987–1007 (2014)
23. Ben Atitallah, S., Driss, M., Boulila, W., Ben Ghézala, H.: Leveraging deep learning and IoT big data analytics to support the smart cities development: review and future directions. Comput. Sci. Rev. **38**, 100303 (2020)
24. Romero, M., Guédria, W., Panetto, H., Barafort, B.: Towards a characterisation of smart systems: a systematic literature review. Comput. Ind. **120** (2020)
25. Nitoslawski, S.A., Galle, N.J., van den Bosc, C.K., Steenberg, J.W.N.: Smarter ecosystems for smarter cities? A review of trends, technologies, and turning points for smart urban forestry. Sustain. Cities Soc. **51**(August), 101770 (2019)
26. Appio, F.P., Lima, M., Paroutis, S.: Understanding smart cities: innovation ecosystems, technological advancements, and societal challenges. Technol. Forecast. Soc. Change **142**(December 2018), 1–14 (2019)
27. Porru, S., Misso, F.E., Pani, F.E., Repetto, C.: Smart mobility and public transport: opportunities and challenges in rural and urban areas. J. Traffic Transp. Eng. (English edn.) **7**(1), 88–97 (2020)
28. Belanche-gracia, D., Casaló-ariño, L.V., Pérez-rueda, A.: Determinants of multi-service smartcard success for smart cities development: a study based on citizens' privacy and security perceptions. Gov. Inf. Q. **32**(2), 154–163 (2015)

29. El-haddadeh, R., Weerakkody, V., Osmani, M., Thakker, D.: Examining citizens' perceived value of internet of things technologies in facilitating public sector services engagement. Gov. Inf. Q. **36**(2), 310–320 (2019)
30. Staffans, A., Horelli, L.: Expanded urban planning as a vehicle for understanding and shaping smart, liveable cities, (November) (2014)
31. Perera, C., Zaslavsky, A., Christen, P., Georgakopoulos, D.: Sensing as a service model for smart cities supported. (September 2013), 81–93 (2014)
32. Ahad, M.A., Paiva, S., Tripathi, G., Feroz, N.: Enabling technologies and sustainable smart cities. Sustain. Cities Soc. **61**(March), 102301 (2020)
33. Singh, S., Sharma, P.K., Yoon, B., Shojafar, M., Cho, G.H., Ra, I.H.: Convergence of blockchain and artificial intelligence in IoT network for the sustainable smart city. Sustain. Cities Soc. **63**(April) (2020)
34. Lee, I.: Internet of Things (IoT) cybersecurity: literature review and IoT cyber risk management. Futur. Internet **12**(9), 157 (2020)
35. Mehmood, Y., Ahmad, F., Yaqoob, I., Adnane, A., Imran, M., Guizani, S.: Internet-of-things-based smart cities: recent advances and challenges. IEEE Commun. Mag. **55**(9), 16–24 (2017)
36. Sengan, S., Subramaniyaswamy, V., Nair, S.K., Indragandhi, V., Manikandan, J., Ravi, L.: Enhancing cyber–physical systems with hybrid smart city cyber security architecture for secure public data-smart network. Futur. Gener. Comput. Syst. **112**, 724–737 (2020)
37. Sovacool, B.K., Furszyfer Del Rio, D.D.: Smart home technologies in Europe: a critical review of concepts, benefits, risks and policies. Renew. Sustain. Energy Rev. **120**(May 2019), 109663 (2020)
38. Caviglione, L., Coccoli, M.: A holistic model for security of learning applications in smart cities. J. E-Learn. Knowl. Soc. **16**(1), 1–10 (2020)
39. Allam, Z., Dhunny, Z.A.: On big data, artificial intelligence and smart cities. Cities **89**(January), 80–91 (2019)
40. Yigitcanlar, T., Desouza, K.C., Butler, L., Roozkhosh, F.: Contributions and risks of artificial intelligence (AI) in building smarter cities: insights from a systematic review of the literature. Energies **13**(6) 2020
41. Meadowcroft, J., Stephens, J.C., Wilson, E.J., Rowlands, I.H.: Social dimensions of smart grid: regional analysis in Canada and the United States. Introduction to special issue of renewable and sustainable energy reviews. Renew. Sustain. Energy Rev. **82**(June 2017), 1909–1912 (2018)
42. Singh, P., Helfert, M.: Smart cities and associated risks: technical v/s non-technical perspective. In: CHIRA 2019 – Proceedings of 3rd International Conference on Computing Interaction Research and Applications, no. May, pp. 221–228 (2019)
43. Vidiasova, L., Cronemberger, F.: Discrepancies in perceptions of smart city initiatives in Saint Petersburg, Russia. Sustain. Cities Soc. **59**(May), 102158 (2020)
44. Pimchangthong, D., Boonjing, V.: Effects of risk management practice on the success of IT project. Proc. Eng. **182**, 579–586 (2017)
45. Msheireb. (2021)
46. Gill, P., Stewart, K., Treasure, E., Chadwick, B.: Methods of data collection in qualitative research: interviews and focus groups. Br. Dent. J. **204**(6), 291–295 (2008)
47. Rebeeh, Y.A., Pokharel, S., Abdella, G.M., Hammuda, A.S.: Disaster management in industrial areas: perspectives, challenges, and future research. J. Ind. Eng. Manag. **12**(1), 133–153 (2019)

Hybrid and Machine Learning Algorithms

Multi-subject Identification of Hand Movements Using Machine Learning

Alejandro Mora-Rubio[1(✉)], Jesus Alejandro Alzate-Grisales[1], Daniel Arias-Garzón[1], Jorge Iván Padilla Buriticá[1], Cristian Felipe Jiménez Varón[2], Mario Alejandro Bravo-Ortiz[1], Harold Brayan Arteaga-Arteaga[1], Mahmoud Hassaballah[3], Simon Orozco-Arias[4,5], Gustavo Isaza[5], and Reinel Tabares-Soto[1]

[1] Department of Electronics and Automation, Universidad Autónoma de Manizales, Manizales 170001, Colombia
alejandro.morar@autonoma.edu.co

[2] Department of Physics and Mathematics, Universidad Autónoma de Manizales, Manizales 170001, Colombia

[3] Department of Computer Science, Faculty of Computers and Information, South Valley University, Qena 83523, Egypt

[4] Department of Computer Science, Universidad Autónoma de Manizales, Manizales 170001, Colombia

[5] Department of Systems and informatics, Universidad de Caldas, Manizales 170002, Colombia

Abstract. Electromyographic (EMG) signals provide information about muscle activity. In hand movements, each gesture's execution involves the activation of different combinations of the forearm muscles, which generate distinct electrical patterns. Furthermore, the analysis of muscle activation patterns represented by EMG signals allows recognizing these gestures. We aimed to develop an automatic hand gesture recognition system based on supervised Machine Learning (ML) techniques. We trained eight computational models to recognize six hand gestures and generalize between different subjects using raw data recordings of EMG signals from 36 subjects. We found that the random forest model and fully connected artificial neural network showed the best performances, indicated by 96.25% and 96.09% accuracy, respectively. These results improve on computational time and resources by avoiding data preprocessing operations and model generalization capabilities by including data from a larger number of subjects. In addition to the application in the health sector, in the context of Smart Cities, digital inclusion should be aimed at individuals with physical disabilities, with which, this model could contribute to the development of identification and interaction devices that can emulate the movement of hands.

Keywords: Activity recognition · Computational modeling · Electromyography · Machine learning

J. M. Corchado and S. Trabelsi (Eds.): SSCTIC 2021, LNNS 253, pp. 117–128, 2022.
https://doi.org/10.1007/978-3-030-78901-5_11

1 Introduction

Electromyography (EMG) signals are generated by the electrical activity of skeletal muscle fibers and can be captured through two ways. First, invasive methods involve the insertion of a needle directly into the muscle; second, non-invasive methods rely on surface electrodes that are placed on the skin in areas close to the target muscle. In particular, surface EMG is the most suited for establishing human-machine interfaces of routine use [1]. These kind of signals are random, highly non-stationary [2], and display subtle variations associated with anatomical and motor coordination differences between subjects [3,4]. Therefore, different techniques of feature extraction and selection for time series [5] must be applied, which are also useful for developing optimal models of artificial intelligence (AI) [6].

These signals are used by health professionals for several purposes, including disease diagnosis and rehabilitation. Recently, with the increase number of automatic learning techniques, these signals are used in systems for movement classification [7,8], recognition of motor unit action potential (MUAP) [9], and diagnosis of neuromuscular diseases [10]. These tasks are achieved by measuring and studying features that can be extracted from EMG signals. Feature extraction can be done through different techniques, such as autoregressive models [11], signal entropy measurements [12], and statistical measurements of amplitude in the time and frequency domains [6]. This last technique includes the widely used Wavelet transform, which provides information in the frequency domain at high and low frequencies [13].

The feature extraction stage is followed by the classification stage, which relies on models that can learn from the extracted features [14]. Among computational models, AI is considered the main domain, which includes Machine Learning (ML) and Deep Learning (DL). AI refers to the process of displaying human intelligence features on a machine. ML is used to achieve this task, while DL is a set of models and algorithms used to implement ML [15]. In general, ML and DL technologies are powerful for extracting features and finding relationships between data; therefore, these approaches are suitable for tasks that rely on taking advantage of human experience [16,17]. Machine and Deep Learning algorithms applied to electrophysiological signals constitute a rapidly growing field of research and allow researchers to train computer systems as experts that can be used later on to support decision-making processes. From ML, the most commonly used models in this field are Support Vector Machine (SVM) [18], K-Nearest Neighbors (KNN) [9], and Ensemble classifiers [19,20]; in DL, different architectures are reported, including fully connected artificial neural networks (ANN) and convolutional neural networks (CNN) [21].

This paper proposes an approach to EMG signal classification, in which not all information of the time series is considered. Here, only raw data for hand gestures (i.e., the muscular activation values provided by each channel) are used to identify the patterns associated with six gestures, regardless of the subject performing the action. Therefore, we aimed to generalize among different subjects, which is a current issue in this classification task. The data used in this research consists of EMG recordings from 36 patients performing six different hand gestures. The classification accuracy obtained using this approach is over 90% for several models, demonstrating feasibility and outperforming state-of-the-art classification results involving several patients.

This paper is organized as follows: Sect. 2 presents the materials and methods, including the database and models used in the proposed framework, which are explained in detail in Sect. 3. Section 4 provides the experimental results and discussion. Finally, Sect. 5 concludes the paper and raises issues for further research.

2 Materials and Methods

2.1 Database

The information used to develop this research was retrieved from a free database called *EMG data for gestures Data Set.* available on the website of the UCI Machine Learning Repository. This database contains surface EMG signal recordings in a time interval where six different static hand gestures are executed, providing a total of 72 recordings from 36 patients. The static hand gestures included resting hand, grasped hand, wrist flexion, wrist extension, ulnar deviation, and radial deviation. These gestures were performed for three seconds, each with a three-second interval between gestures. The information was collected using the *MYO thalmic bracelet* device, which has eight channels equally spaced around the patient's forearm. Accordingly, each recording consists of an eight-channel time series, in which each segment is properly labeled with numbers zero to six, where zero corresponds to the intervals between gestures and numbers one to six refer to the gestures mentioned above, as reported in Table 1. This database was consolidated by [3] during their research on the factors limiting human-machine interfaces via sEMG.

Table 1. Labels of the gestures contained in the database.

Label	Gesture
0	Intervals between gestures
1	Resting hand
2	Grasped hand
3	Wrist flexion
4	Wrist extension
5	Ulnar deviation
6	Radial deviation

2.2 Machine Learning Models

The ML models implemented in the experiments included K-Nearest Neighbors (KNN), Logistic Regression (LR), Gaussian Naive Bayes (GNB), Multilayer Perceptron (MLP) using one hidden layer, Random Forest (RF), and Decision Tree (DT).

2.3 Deep Learning Models

Fully Connected Artificial Neural Network Architecture. The model described here corresponds to a network with three hidden layers and one output layer. The output layer has six units corresponding to the number of movements to identify. The activation function for the output layer was softmax and the loss function was categorical cross-entropy. The diagram of the ANN architecture is shown in Fig. 1.

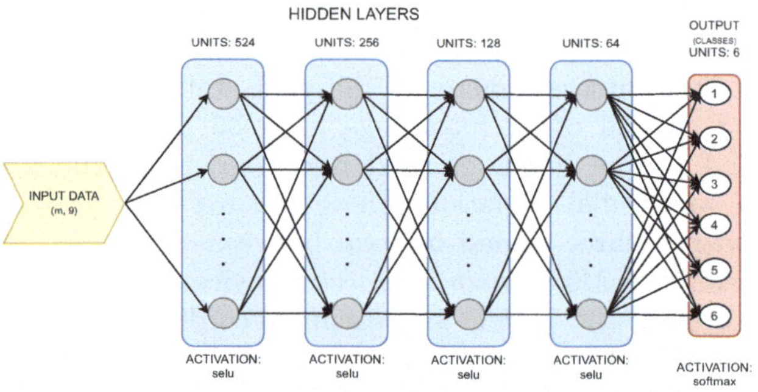

Fig. 1. Architecture of the fully connected artificial neural network used in this study.

Convolutional Neural Network Architecture. The structure of the model proposed here is a convolutional neural network with two 1D convolutional layers, one ReLU activation layer, three fully connected layers, and one output layer. The output layer has six units corresponding to the number of movements to identify. The activation function for the output layer was softmax and the loss function was categorical cross-entropy. The diagram for the CNN architecture is presented in Fig. 2.

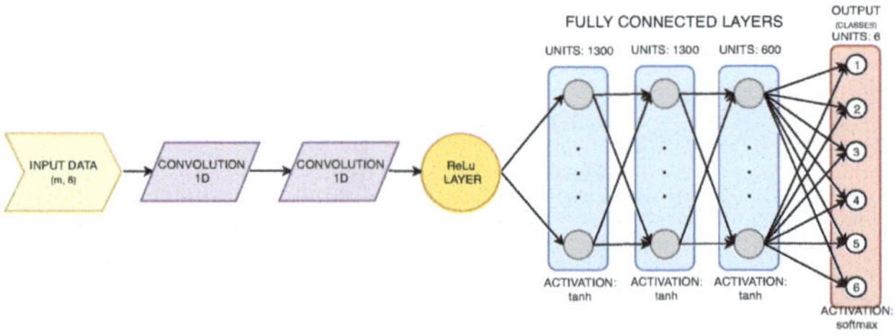

Fig. 2. Convolutional neural network architecture used in this study.

2.4 Tools

The complete experimental procedure is implemented using Google Colab platform and Python v. 3.7.3 programming language. Database import and preprocessing were done with Pandas library v. 0.24.2. The ML models mentioned in Sect. 2.2, in addition to GridSearchCV, principal component analysis (PCA), train-test split, and the scaling function were implemented using Scikit-learn library v. 0.21.2 [22]. The artificial neural network architectures and the corresponding training and validation processes were implemented using Keras library v. 2.2.4 [23]. Further, the resources used in this research are available in a public GitHub repository containing the EMG recordings files from the database, as well as a Jupyter notebook with the implemented source code. The repository is available here.

3 Proposed Framework

Figure 3 shows the steps of the proposed framework. Each step is explained in detail in the following subsections.

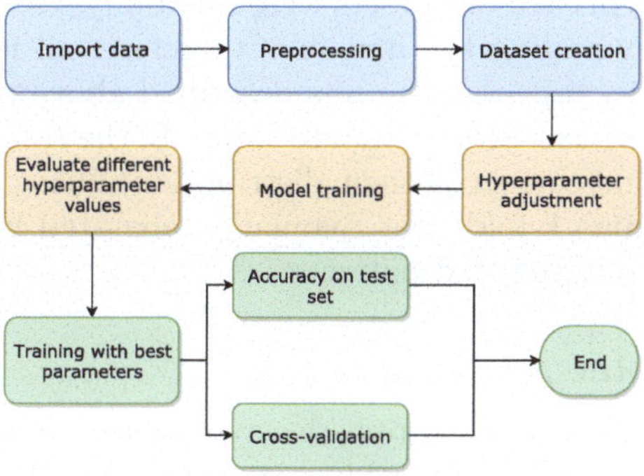

Fig. 3. Flowchart of the proposed framework.

3.1 Importing the Data

The data from the database is presented as a plain text file (.txt). The *read_csv* function from Pandas library was used to import these files.

3.2 Preprocessing

Data labeled as zero was removed since these corresponded to the resting intervals between gesture execution and do not contain relevant information to solve the problem. Therefore, only data captured during the execution of the six gestures of interest were kept. No feature extraction procedure was implemented in order to use the values of the eight channels provided by the device as discriminating information between gestures (i.e., raw data).

Considering the high susceptibility of EMG signals to noise and external artifacts, as well as the fact that relevant information of the signal is located in the frequency spectrum range from 0 450 Hz [24], we applied three different digital filters to eliminate or reduce noise as much as possible. The first filter consisted of a high pass filter with cut-off frequency 20 Hz, which eliminated most of the low-frequency noise while preserving the majority of the signal information [24]. The second and third filters were band pass with cut-off frequencies of 20–180 Hz and 10–480 Hz, respectively; this type of filter is commonly used to remove low-frequency noise and high-frequency information that may be irrelevant to the classification task [7,8].

3.3 Dataset Creation

For dataset creation, the patients were correctly distributed into training, validation, and test sets. We verified that each patient was present in only one set to avoid bias induced by including data from the same patient during training and testing. The training set comprised concatenated data from 30 patients, the validation set included three patients, and test set contained the remaining three patients. The patients in each of the sets were randomly selected. A second group of sets was created by scaling the first one; therefore, all features equally contributed to the training process. The sets described above were used during the hyperparameter adjustment process and to compute the *test set accuracy* metric. Finally, to conserve the correct patient distribution during the cross-validation process, each of the nine k-folds was manually generated by concatenating the data from four randomly selected patients.

3.4 Hyperparameter Adjustment

After dataset creation, we adjusted the parameters for each classification method. Given the number of variable parameters for ANN and CNN architectures, hyperparameter adjustment was done by tuning one or two parameters at a time using the GridSearchCV function implemented in Scikit-learn library. This method selects the best combination of parameters using the result of a cross-validation process. The parameters were adjusted in the following order: batch_size and epochs, optimization algorithm, network weight initialization, activation function, and neurons in hidden layers.

3.5 Model Training and Hyperparameter Optimization

For parameter optimization, we used an approach similar to the grid search. Given a set of parameters and value ranges, the model was trained and tested on the validation set using every possible combination. For traditional ML models, this process was implemented manually using *for* loops, whereas, for artificial neural networks, it was done using the GridSearchCV function from Scikit-learn library due to the high number of parameters.

3.6 Model Evaluation Using the Best Parameters

After performing hyperparameter optimization, we considered the best set of parameters for each model to generate the definitive evaluation metric. For this, cross-validation was used to measure model performance. This method uses the complete feature matrix and divides the dataset into k-fold parts. In this case, k-folds = 9 and each part contained data from four randomly selected patients. The data was iterated over these partitions, using eight for training and one for validation, until all of these passed through both states. This procedure was used to check the stability of the results. In addition to cross-validation, the test accuracy provided an estimate of generalization, given that the test set had not been used in any previous stage of the experiment. Finally, we computed precision, recall, and F1-score as per class performance measures.

To test for statistically significant differences in the cross-validation results applied to the raw and processed data among the models, we performed an analysis of one-way variance test (ANOVA) (i.e., under the assumption of normality) with the following structure:

$$H_0 : \mu_1 = \mu_2 = \mu_3 = \mu_4$$

$$H_A : \mu_i \neq \mu_j \; for \; i \neq j \wedge i = 1, ..., 4$$

Where μ_i is the mean of each dataset (raw data, 20 Hz, bandpass 20–180 Hz, bandpass 10–480 Hz).

4 Experiments and Discussions

4.1 Results

The best parameter values and datasets (normal/scaling) for each ML model were selected based on the results of the parameter optimization step (Table 2). Similarly, for ANN and CNN architectures, the optimized parameter values with GridSearchCV are shown in Table 3, and the neural networks were trained using these values. Performance accuracy and loss curves were generated in each training epoch to observe model behavior and determine the performance of the learning process. For CNN, the number of epochs was manually adjusted according to the model training curves; as a result, the most appropriate value found was Epochs = 100. Figure 4 shows the accuracy and loss curves during the training process for ANN.

The two methods mentioned in Sect. 3 for measuring model performance, namely cross-validation and test set accuracy, were implemented using the optimized parameters for each model. The cross-validation results for each model and filtered data are shown in Table 4, where STD stands for standard deviation.

Table 2. ML models

Models	Best parameter value
KNN	n_neighbors = 1 Dataset: Normal
LR	C = 0,3 Dataset: Scaling
GNB	N/A
MLP	hidden_layer_sizes = 800 Dataset: Scaling
RF	n_estimators = 71 Dataset: Scaling
DT	Default value: max_depth = None min_samples_split = 2 max_features = None Dataset: Scaling

Table 3. ANN and CNN

Parameters	Optimal values	
	ANN	CNN
Batch size	50,000	10,000
Epochs	120	100
Optimization algorithm	lr = 0.01	lr = 0.001
Weight initialization	Normal	Normal
Activation function	tanh	Linear tanh
Units in hidden layers	1,050	1,300, 1,300, 600
Filters	N/A	128
Kernel_size	N/A	3

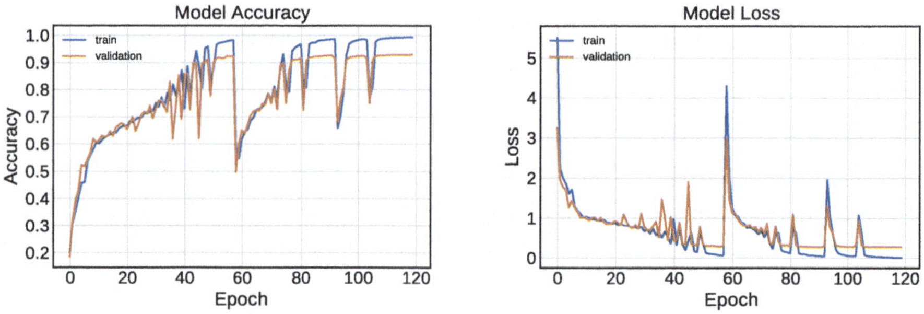

(a) (b)

Fig. 4. Training curves using raw data - ANN

In summary, the results from Table 4 indicate that LR, GNB, and MLP models display the lowest performance. On the other hand, the best results were achieved by KNN, RF, DT, ANN and CNN (93.22%, 95.39%, 93.29%, 93.73%, 94.77%, respectively). The ANOVA test for the cross-validation data did not show statistically significant differences between raw and processed data (P-value = 0.994 at a significance level of 5%). Test set accuracy was computed for the best-performing models using raw data and per class performance measures (e.g., precision, recall and F1-score). The resting hand gesture showed the best classification according to the established measures. The results per class for the top three models are reported in Table 5.

Table 4. Performance of the models based on cross-validation using different filters.

Models	Cross-validation (Accuracy [%] ± STD [%])			
	Raw data	Highpass 20 Hz	Bandpass 20–180 Hz	Bandpass 10-480 Hz
K-Nearest Neighbors	95.06 ± 1.98	94.94 ± 2.72	95.31 ± 2.69	95.03 ± 1.81
Logistic Regression	22.25 ± 3.34	16.81 ± 0.20	17.39 ± 0.20	17.62 ± 0.52
Gaussian Naive Bayes	63.27 ± 1.09	50.12 ± 1.31	58.46 ± 1.67	60.68 ± 1.22
Multilayer Perceptron	72.68 ± 1.11	62.95 ± 1.34	67.24 ± 1.32	68.40 ± 0.99
Random Forest	96.25 ± 1.63	95.78 ± 2.29	96.13 ± 2.32	96.02 ± 1.50
Decision Tree	94.59 ± 2.08	94.86 ± 2.86	95.13 ± 2.78	94.80 ± 1.89
Artificial Neural Network	96.09 ± 1.74	93.55 ± 3.85	95.59 ± 2.76	95.51 ± 1.64
Convolutional Neural Network	95.84 ± 1.81	94.64 ± 2.49	94.12 ± 1.06	93.82 ± 0.99

Table 5. Model performance based on class types.

Class	Model								
	RF			ANN			CNN		
	Precision	Recall	F1-score	Precision	Recall	F1-score	Precision	Recall	F1-score
Resting hand	0.93	1.00	0.96	0.93	0.99	0.96	0.94	1.00	0.97
Grasped hand	0.98	0.92	0.95	0.96	0.91	0.94	0.97	0.92	0.95
Wrist flexion	0.93	0.97	0.95	0.93	0.95	0.94	0.94	0.96	0.95
Wrist extension	0.97	0.96	0.96	0.94	0.94	0.94	0.95	0.95	0.95
Ulnar deviation	0.95	0.91	0.93	0.93	0.88	0.90	0.95	0.89	0.92
Radial deviation	0.96	0.96	0.96	0.94	0.94	0.94	0.94	0.95	0.95

4.2 Discussions

In this study, we explored the performance of different ML and DL techniques in the task of classifying six hand gestures from electromyographic signals. We found no statistically significant differences generated by using different preprocessing digital filters; thus, these filters did not benefit model performance compared to the raw data. We recommend working with raw data, without affecting performance, to save time and computational resources that would be invested in the preprocessing stage. Our reports demonstrate the feasibility of using the information provided by each of the channels as discriminant features between gestures. The best results described here are comparable to literature reports on similar problems and the same classification task. For example, [25] reported an accuracy of 99.2% in the diagnosis of neuromuscular disorders using data from 27 subjects, and a rotation forest classifier. Furthermore, [18] used EMG signals from lower limbs of two subjects to classify two different gait phases (e.g., swing and stance), reporting a classification accuracy of 96% using a SVM classifier. Both of these previous studies report a high classification accuracy; however, each considered the classification task of only two classes.

On the other hand, [21] reported an accuracy of 98.12% for the identification of seven hand gestures for a single subject using four time-domain features and stacked sparse autoencoders as a classifier. Similarly, [26] achieved a 96.38% accuracy using a RF model to identify three hand gestures in a subject-independent experiment with data from 10 subjects. Overall, these previous studies report a high classification accuracy; however, they also considered a small number of subjects. Furthermore, it is important to determine how the transition from single to multi-subject classification affects precision. Table 6 provides a summary of the works described.

Table 6. Performance comparison with other related works.

Methods	Task	Classes	Subjects	Accuracy (%)
Subasi et al. [25]	Neuromuscular disorders	2	27	99.2
Zieger et al. [18]	Gait phases	2	2	96
Zia ur Rehmann et al. [21]	Gestures	7	1	98.12
Wahid et al. [26]	Gestures	3	10	96.38
Rabin et al. [27]	Gestures	6	1–5	94.8–77.3
Our method	Gestures	6	36	96.25

In this context, we demonstrated the feasibility of traditional ML and DL models to identify different hand gestures from raw EMG signals. Furthermore, we achieved generalization among 36 subjects, while maintaining a high classification accuracy. Given our approach, each record can be used as an instance of the features matrix to obtain a more significant amount of labeled information for model training. Also, this approach avoids computational costs and reduces the time required to implement the different feature extraction methods. For instance, using Google Colab platform in a GPU environment, the KNN model takes less than a minute for training and testing; RF takes approximately seven minutes, and both DL models take nearly 20 min. These computational time metrics are not provided by other authors, which makes it difficult to determine if our approach is faster.

The accuracy obtained for Random Forest (96.25%), ANN (96.09%) and CNN (95.84%) (Table 4) is comparable to the literature reports previously discussed and outperforms the results presented as state-of-the-art for this classification task. Furthermore, our results correspond to a cross-validation procedure that provides robustness to the model and demonstrates the consistency of the results. The dataset used in this research (i.e., containing data from 36 patients) is bigger than those previously used in most multi-subject EMG signal classification tasks. Unlike other studies, we considered the correct distribution of the patients into the training, validation, and test sets to limit the information of a patient to only one dataset. This is an important point when designing and implementing ML models since these can memorize the information in the training set and, if the same information is in the validation or test set, biased performance measures may be generated.

5 Conclusions

The results confirm the feasibility of using machine learning and deep learning techniques to identify muscle activation patterns, specifically, hand movements.

Further, this paper presents a different approach to this classification problem using raw data provided by the *MYO thalmic bracelet* device to train models with high performance levels, regardless of the subject to which the signals belong. This allows overcoming different anatomical and biological factors between subjects, which translate into subtle differences in EMG signals that previously limited the performance of classification systems for different subjects. The models achieved high accuracy in multi-subject classification with 36 subjects. The best performances were obtained by implementing Random Forest with 96.25% accuracy, artificial neural network with 96.09%, convolutional neural network with 95.84%, K-nearest neighbor with 95.06%, and decision tree with 94.59%. Future work should be aimed to optimize and adapt the trained models for real-time classification tasks, such as motor control for prostheses.

References

1. Phinyomark, A., Khushaba, R.N., Scheme, E.: Feature extraction and selection for myoelectric control based on wearable EMG sensors. Sensors **18**(5), 1615 (2018)
2. Phinyomark, A., Campbell, E., Scheme, E.: Surface electromyography (EMG) signal processing, classification, and practical considerations. In: Naik, G. (eds.) Biomedical Signal Processing. SERBIOENG, pp. 3–29. Springer, Singapore (2020). https://doi.org/10.1007/978-981-13-9097-5_1
3. Lobov, S., Krilova, N., Kastalskiy, I., Kazantsev, V., Makarov, V.: Latent factors limiting the performance of sEMG-interfaces. Sensors **18**(4), 1122 (2018)
4. Molla, M.K.I., Shiam, A.A., Islam, M.R., Tanaka, T.: Discriminative feature selection-based motor imagery classification using EEG signal. IEEE Access **8**, 98255–98265 (2020)
5. Shakeel, A., Tanaka, T., Kitajo, K.: Time-series prediction of the oscillatory phase of EEG signals using the least mean square algorithm-based AR model. Appl. Sci. **10**(10), 3616 (2020)
6. Phinyomark, A., Phukpattaranont, P., Limsakul, C.: Feature reduction and selection for EMG signal classification. Expert Syst. Appl. **39**(8), 7420–7431 (2012)
7. Phukan, N., Kakoty, N.M., Shivam, P., Gan, J.Q.: Finger movements recognition using minimally redundant features of wavelet denoised EMG. Health Technol. **9**, 579–593 (2019). https://doi.org/10.1007/s12553-019-00338-z
8. Ji, Y., Sun, S., Xie, H.-B.: Stationary wavelet-based two-directional two-dimensional principal component analysis for EMG signal classification. Meas. Sci. Rev. **17**(3), 117–124 (2017)
9. Jahromi, M.G., Parsaei, H., Zamani, A., Stashuk, D.W.: Cross comparison of motor unit potential features used in EMG signal decomposition. IEEE Trans. Neural Syst. Rehabil. Eng. **26**(5), 1017–1025 (2018)
10. Diaz, M., Ferrer, M.A., Impedovo, D., Pirlo, G., Vessio, G.: Dynamically enhanced static handwriting representation for Parkinson's disease detection. Pattern Recogn. Lett. **128**, 204–210 (2019)
11. Koçer, S., Tümer, A.E.: Classifying neuromuscular diseases using artificial neural networks with applied autoregressive and cepstral analysis. Neural Comput. Appl. **28**(1), 945–952 (2016). https://doi.org/10.1007/s00521-016-2383-8
12. Chen, W., Wang, Z., Xie, H., Yu, W.: Characterization of surface EMG signal based on fuzzy entropy. IEEE Trans. Neural Syst. Rehabil. Eng. **15**(2), 266–272 (2007)

13. Gokgoz, E., Subasi, A.: Comparison of decision tree algorithms for EMG signal classification using DWT. Biomed. Sig. Process. Control **18**, 138–144 (2015)
14. Awad, A.I., Hassaballah, M.: Image Feature Detectors and Descriptors: Foundations and Applications. SCI, vol. 630. Springer, Cham (2016). https://doi.org/10.1007/978-3-319-28854-3
15. Bengio, Y., Goodfellow, I.J., Courville, A.: Deep learning. Technical report (2015)
16. Orozco-Arias, S., Isaza, G., Guyot, R., Tabares-Soto, R.: A systematic review of the application of machine learning in the detection and classification of transposable elements. PeerJ **7**, e8311 (2019)
17. Tabares-Soto, R., Raúl, R.P., Gustavo, I.: Deep learning applied to steganalysis of digital images: a systematic review. IEEE Access **7**, 68970–68990 (2019)
18. Ziegier, J., Gattringer, H., Mueller, A.: Classification of gait phases based on bilateral EMG data using support vector machines. In: Proceedings of the IEEE RAS and EMBS International Conference on Biomedical Robotics and Biomechatronics, Augus, vol. 2018, pp. 978–983. IEEE Computer Society, October 2018
19. Subasi, A., Yaman, E., Somaily, Y., Alynabawi, H.A., Alobaidi, F., Altheibani, S.: Automated EMG signal classification for diagnosis of neuromuscular disorders using DWT and bagging. Procedia Comput. Sci. **140**, 230–237 (2018)
20. Benazzouz, A., Guilal, R., Amirouche, F., Hadj Slimane, Z.E.: EMG feature selection for diagnosis of neuromuscular disorders. In: 2019 International Conference on Networking and Advanced Systems (ICNAS), pp. 1–5 (2019)
21. Rehman, M.Z.U., et al.: Multiday EMG-based classification of hand motions with deep learning techniques. Sensors **18**(8), 2497 (2018)
22. Pedregosa, F., et al.: Scikit-learn: machine learning in Python. J. Mach. Learn. Res. **12**, 2825–2830 (2011)
23. Chollet, F., et al.: Keras (2015). https://keras.io
24. De Luca, C.J., Gilmore, L.D., Kuznetsov, M., Roy, S.H.: Filtering the surface EMG signal: movement artifact and baseline noise contamination. J. Biomech. **43**(8), 1573–1579 (2010)
25. Subasi, A., Yaman, E.: EMG signal classification using discrete wavelet transform and rotation forest. In: Badnjevic, A., Škrbić, R., Gurbeta Pokvić, L. (eds.) CMBE-BIH 2019. IP, vol. 73, pp. 29–35. Springer, Cham (2020). https://doi.org/10.1007/978-3-030-17971-7_5
26. Wahid, M.F., Tafreshi, R., Al-Sowaidi, M., Langari, R.: Subject-independent hand gesture recognition using normalization and machine learning algorithms. J. Comput. Sci. **27**, 69–76 (2018)
27. Rabin, N., Kahlon, M., Malayev, S., Ratnovsky, A.: Classification of human hand movements based on EMG signals using nonlinear dimensionality reduction and data fusion techniques. Expert Syst. Appl. **149**, 113281 (2020)

An Open Domain Question Answering System Trained by Reinforcement Learning

Bghiel Afrae(✉), Ben Ahmed Mohamed, and Anouar Boudhir Abdelhakim

List Laboratory FSTT UAE Tangier, Tangier, Morocco
{mbenahmed,aboudhir}@uae.ac.ma

Abstract. Recently, Reinforcement learning has shown his ability as an effective approach for improving deep learning algorithms performance, moreover question answering problem are leveraged nowadays as competitive deep learning tasks. In this paper, we tried to realize a robust question answering model, by integrating a reinforcement learning such as combining a self-critical policy gradient for improving the seq2seq training, by the setup of a reward and baseline reward on the top of the model.

Keywords: RL · Chatbot · RNN · Policy gradient · Bleu score · Levenshtein distance · Self-critical

1 Introduction

The pervasive connectivity of the Internet provides the best mediation to recover information by hopping through documents.

We argue that, in order to further the ability of machine comprehension methods to recover answer about a question, we must move beyond a scenario where relevant information is coherently and explicitly stated within one click.

Traditionally, QA has attracted many AI researchers, and in recent years researchers have achieved considerable success applying neural network methods to question answering (QA).More recently, the question answering systems has become more and more popular given its superior performance without the demand of heavily hand-crafted engineering efforts.

Most methods for solving this task require training a ses2seq on a dataset (question, answer) pairs. The model is usually trained to maximize the log probability of a correct answer; [21] after training, those models are usually evaluated by computing of different metrics on test set, such as Bleu (papini et al.) [1] and rouge.

To address the inconsistency issue, reinforcement learning (RL) methods have been adopted to optimize sequence-level objectives [2]. The process of Reinforcement learning is inspired by the human learning process, by interacting with the surrounding environment where the learning process aims at maximizing a notion of cumulative reward. To train a dialogue system with reinforcement learning, the chatbot is put in use by the end users to become increasingly efficient throughout the conversations [3].

J. M. Corchado and S. Trabelsi (Eds.): SSCTIC 2021, LNNS 253, pp. 129–138, 2022.
https://doi.org/10.1007/978-3-030-78901-5_12

We try to fulfill these methods and study how to practically apply RL to evaluate the encoder-decoder model for obtaining a strong QA systems with quite competitive, so, in this paper, we propose an improvement to seq2seq model by integrating a self-critical policy gradient to train the model and optimize objectives.

In summary, we make the following contributions in this paper: in Sect. (2) we present a brief state of art, then in Sect. (3) we will discuss the related work to our model, and in Sect. (4) we present the dataset used, and in Sect. (5) we summarize the proposed model and technologies that we have used, finally, we present metrics and results to the conclusion and perspectives by the end.

2 Chatbot History

Throughout modern history, we've been obsessed with interacting with robots and artificial intelligence. But towards the latter half of the 20th Century, this dream started becoming a reality. Starting with ELIZA in the 1960s to Alexa and beyond, as we move further along the timeline, chatbot technology has exploded across social, administrable and business channel. chatbot origin belong to Alan Turing test by 1950, during World War II was one of the most prominent breakers of German code, issued an open challenge to the computer scientists of the world. It is due to this test that we can chronicle the history of chatbots, starting by Eliza in 1966, PARRING 6 years later, and the most advanced bot of its time ALICE.

The advent of the new millennium ushered in a lot of exciting, new chatbots. Evolving from pattern recognition technologies, these chatbots began to showcase machine learning and other advanced algorithms.

Due to this, they were able to learn from their interactions with humans. The history of chatbots took a whole new turn toward conversational agents beginning, to display a new kind of intelligence [4].

3 Related Works

The first chatbots were developed to perform improved notification processes. However, later, the new chatbots deployed respond to more complex requests such as financial advice, savings or meeting planning [5].

Actually, a lot of research has been done to make a developed chatbot since the very first one, and to explore new approaches; the researches has seen a highly improvement, where the most of them were based on encoder decoder architecture.

In [6], Sutskever and al, have used a multilayered Long Short-Term Memory (LSTM) to map the input sequence to a vector of a fixed dimensionality, and then another deep LSTM to decode the target sequence from the vector; where they found that the LSTM did not have difficulty on long sentences. they have discovered reversing the order of the words in all source sentences (but not target sentences) improved the LSTM's performance markedly, because doing so introduced many short term dependencies between the source and the target sentence which made the optimization problem easier.

Bghiel and al in their paper [7], have proposed another approach for the question answering problem with a grammatical correction first, where they have developed a set of deep learning models for natural language retrieval and generation—including recurrent neural networks, sequence-to-sequence models and attention mechanism—and evaluate them in the context of natural language processing.

They have used a BILSTM as a pre-attention RNN to encode inputs, then the result passed to attention mechanism for obtaining alphas results to generate a context vector, which they have passed to post attention LSTM and finally a softmax to obtain the answers distribution probabilities.

4 Dataset

WIKIPEDIA contains an abundance of human curated, multi-domain information and has several structured resources such as infoboxes and WIKIDATA (2012) associated with it.

WIKIPEDIA has thus been used for a wealth of research to build datasets posing queries about a single sentence (Morales et al., 2016; Levy et al., 2017) [8, 9] or article (Hewlett et al., 2016) [10]. However, no attempt has been made to construct a cross-document multi-step RC dataset based on WIKIPEDIA.

We have used only the pair questions and answers to train our model, where we have used the 43738 pairs questions and answer in open domain.

5 Background

In this section, we describe our proposed model and technologies used; Including data processing, seq2seq model and evaluator or improving model.

5.1 The Proposed Model

Reinforcement Learning refers to goal-oriented algorithms, which learn how to attain a complex objective (goal) or how to maximize along a particular dimension over many steps, so RL aims to enable to scientist to obtain the goal from deep learning algorithms training which is a the best results by tuning different hyper parameters.

The RNN can easily map sequences to sequences whenever the alignment between the inputs the outputs is known ahead of time, the simplest strategy for general sequence learning is to map the input sequence to a fixed-sized vector using one RNN, and then to map the vector to the target sequence with another RNN; this is why we have built our basic system based on this approach and improving it using RL.

Recently it has been shown that policy-gradient methods for reinforcement learning can be utilized to train deep end-to-end systems directly on non-differentiable metrics for the task at hand, here, we have employed an evaluator model based on reinforcement learning to improve our encoder decoder model performance by evaluating the hyper parameters to obtain the best model; we consider a RL model based on self-critical policy gradient that return each step a reward for optimizing our system by carefully optimizing our system using the test metrics (Fig. 1).

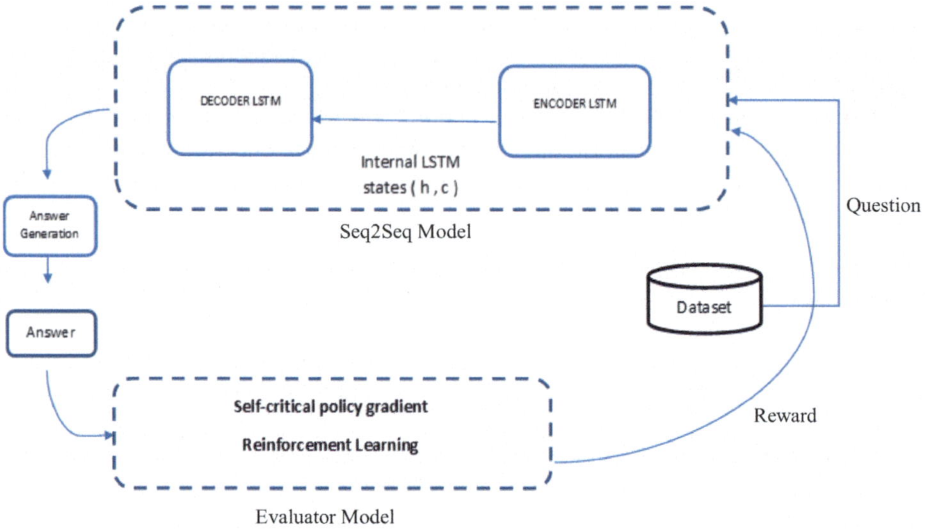

Fig. 1. The model architecture.

5.2 Basis Seq2Seq Model

In this section, we present our seq2seq based on the encoder-decoder network. In all our equations, x = {x1, x2,..., xn} represents the sequence of input (question) tokens, and y = {y1, y2,..., yn} the sequence of output (answer) tokens [11] (Fig. 2).

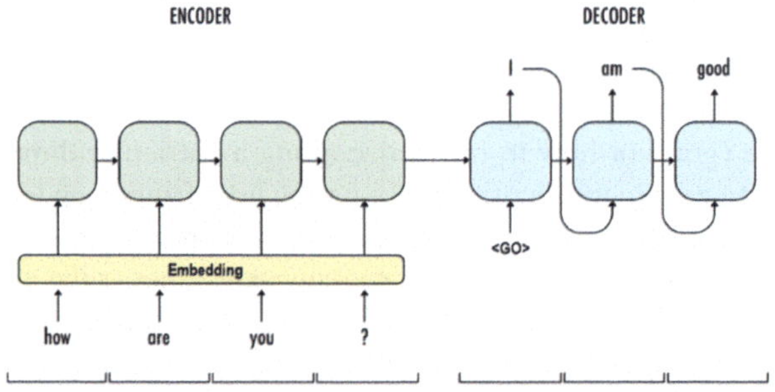

Fig. 2. Encoder decoder model.

Encoder Model

Encoder is simply an Embedding layer + LSTM, where the input is the padded sequence for source sentence, and the output is encoder hidden states. Here, Encoder is the a RNN with input as source sentences. The output can be the output array at the final time-step (t) or the hidden states (c, h) or both, depends on the encoder decoder framework setup.

The aim of encoder is to capture or understand the meaning of source sentences and pass the knowledge (output, states) to encoder for prediction.

Decoder Model

Decoder is another combining of Embedding layer and LSTM, where the input is encoder hidden states and input decoded sequence, and the output is the target decoded sequence. Here, the decoder is another RNN with input as target sentences. The output is the next token of target sentence. The aim of decoder is to predict the next word, with a word given in the target sentence.

Question Representation

we used integer to represent the each word in the sentence, so that we can use word embedding easily. Two separate corpus will be kept for source (questions corpus) and target sentences (answers corpus). To cater for sentence with different length, we capped the sentence at maxlen for long sentence and pad 0 for short sentence. Embedding layer is using low dimension dense array to represent discrete word token. In practice, a word embedding lookup table is formed with shape (vocab_size, latent_dim), then each individual word will look up the index and take the embedding array to represent itself (Fig. 3).

Given a vocabulary V, each individual word *wi* ∈ V is mapped into a real-valued vector (word embedding) w ∈ Rn where n is the dimension [11].

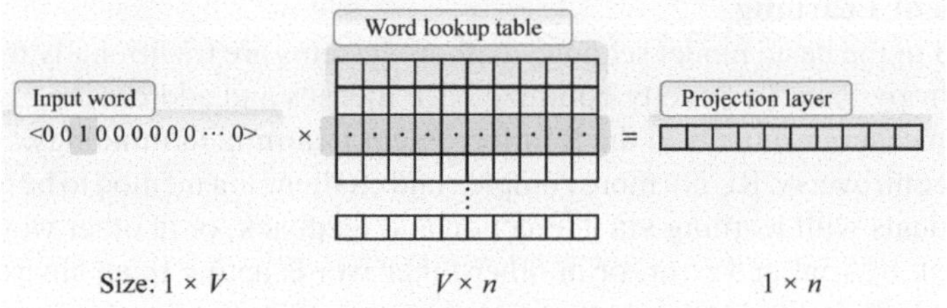

Fig. 3. Question representation.

LSTM

LSTMs are used for sequence modeling and long term dependency capture and designed to avoid the long term dependency problem. We have configured the model to produce only the last hidden state and the cell state. Long-term short-term memory networks are a special type of RNN, capable of learning long-term dependencies thanks to its ability to overcome the vanishing/exploding gradient problem. Remember information for long periods of time.

LSTM RNNs work by allowing the input xt at time t to influence the storing or overwriting of "memories" stored in something called the cell. This decision is determined by two different functions, called the input gate for storing new memories, and the forget gate for forgetting old memories. A final output gate determines when to output the value stored in the memory cell to the hidden layer. These gates are all controlled by the current values of the input xt and cell ct at time t, plus some gate-specific parameters. The image below illustrates the computation graph for the memory portion of an LSTM RNN [12] (Fig. 4).

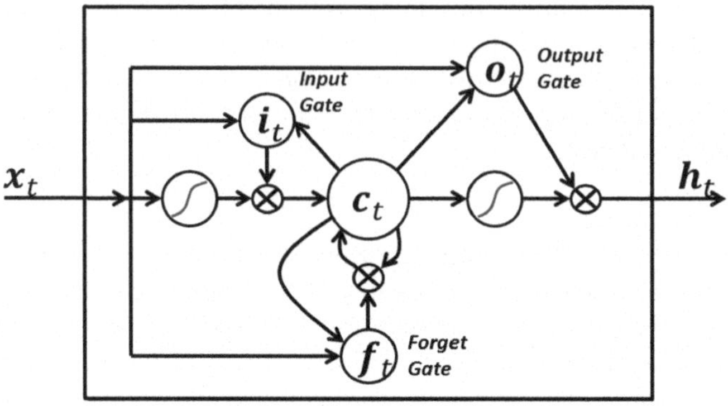

Fig. 4. LSTM unit.

5.3 Evaluator Model

Reinforcement Learning
As described in the basic model section, chatbots systems are traditionally trained using the cross entropy loss. To directly optimize NLP metrics and address the exposure bias issue, we can cast our models in the Reinforcement Learning terminology.

Beyond controversy, RL is a more complex and challenging method to be realized, but basically, it deals with learning via interaction and feedback, or in other words learning to solve a task by trial and error, or in other-other words acting in an environment and receiving rewards for it [13].

Our recurrent model basic model (LSTMs) introduced above can be viewed as an “agent” that interacts with an external “environment” (questions and answers). The parameters of the network, θ, define a policy pθ, that results in an “action” that is the prediction of the next word.

After each action, the agent (basic seq2seq model) updates its internal “state” (cells and hidden states of the LSTM, etc.). Upon generating the end-of-sequence (EOS) token, the agent observes a “reward” that is, for instance, the negative levenshtein distance or bleu score score of the generated sentence—we denote this reward by r. The reward is computed by an evaluation metric by comparing the generated sequence to corresponding real sequences. The goal of training is to minimize the negative expected reward [13] (Fig. 5).

Self-critical Policy Gradient
The goal of training is to minimize the negative expected reward:

$$\mathrm{L}(\theta) = -E_{ws\sim p\theta}[\mathrm{r}(w^s)] \tag{1}$$

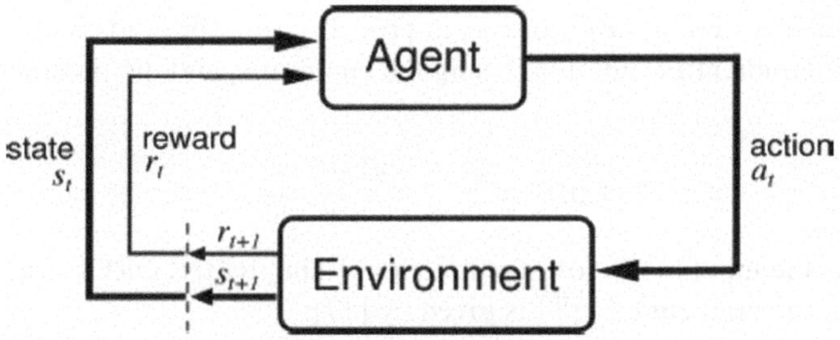

Fig. 5. RL basic model.

Where $w^s = (w_1^s, \ldots w_t^s)$ and w s t is the word sampled from the model at the time step t. In practice L(θ) is typically estimated with a single sample from pθ:

$$\mathrm{L}(\theta) \approx -\mathrm{r}(w^s),\, w^s \sim \mathrm{p}\theta \tag{2}$$

Policy Gradient with REINFORCE. In order to compute the gradient $\nabla_\theta \mathrm{L}(\theta)$, we use the REINFORCE algorithm [14] (See also Chapter 13 in [16]). REINFORCE is based on the observation that the expected gradient of a nondifferentiable reward function can be computed as follows:

$$\nabla_\theta \mathrm{L}(\theta)) = -E_{ws\sim p\theta}[\mathrm{r}(w^s)\ \nabla_\theta \log\ \mathrm{p}\theta(w^s)] \tag{3}$$

In practice the expected gradient can be approximated using a single Monte-Carlo sample $w^s = (w_1^s, \ldots w_t^s)$ from pθ, for each training example in the minibatch:

$$\nabla_\theta(\theta) \approx \mathrm{r}(w^s)\nabla_\theta \log p\theta(w^s) \tag{4}$$

REINFORCE with a Baseline

The policy gradient given by REINFORCE can be generalized to compute the reward associated with an action value relative to a reference reward or baseline b:

$$\nabla_\theta \mathrm{L}(\theta)) = -E_{ws\sim p\theta}[(\mathrm{r}(w^s) - \mathrm{b})\nabla_\theta \log\ \mathrm{p}\theta(w^s)] \tag{5}$$

The baseline can be an arbitrary function, as long as it does not depend on the "action" w s [15], since in this case:

$$\begin{aligned} E_{ws\sim p\theta}[\mathrm{b})\ \nabla_\theta \log\ \mathrm{p}\theta(w^s)] &= \mathrm{b} \sum \nabla_\theta \mathrm{p}\theta(w^s) \\ &= \mathrm{b}\nabla_\theta \sum{}_{ws} \mathrm{p}\theta(w^s) \\ &= \mathrm{b}\nabla_\theta 1 = 0 \end{aligned} \tag{6}$$

This shows that the baseline does not change the expected gradient, but importantly, it can reduce the variance of the gradient estimate. For each training case, we again approximate the expected gradient with a single sample w s ∼pθ:

$$\nabla_\theta \mathrm{L}(\theta) \approx -(\mathrm{r}(w^s) - \mathrm{b})\ \nabla_\theta \log\ \mathrm{p}\theta(w^s) \tag{7}$$

Note that if b is function of θ or t as in [16], Eq. (6) still holds and b(θ) is a valid baseline. Final Gradient Expression. Using the chain rule, and the parametric model of pθ we have:

$$L(\theta) = \sum_{t=1}^{T} \frac{\partial L(\theta)}{\partial st} \frac{\partial st}{\partial \theta},$$

Where st is the input to the softmax function. Using REINFORCE with a baseline b the estimate of the gradient of $\frac{\partial L(\theta)}{\partial st}$ is given by [17]:

$$\frac{\partial L(\theta)}{\partial st} \approx (r(w^s) - b)(p\theta(w_t|h_t) - 1w_s^t) \tag{8}$$

Self-Critical Sequence Training (SCST)

The central idea of the self-critical sequence training (SCST) approach is to baseline the REINFORCE algorithm with the reward obtained by the current model under the inference algorithm used at test time. The gradient of the negative reward of a sample w s from the model w.r.t. to the softmax activations at time-step t then becomes:

$$\frac{\partial L(\theta)}{\partial st} = (r(w^s) - r(\hat{w}))((p\theta(w_t|h_t) - 1w_s^t). \tag{9}$$

Where $r(\hat{w})$ again is the reward obtained by the current model under the inference algorithm used at test time. Accordingly, samples from the model that return higher reward than $\hat{w}$ will be "pushed up", or increased in probability, while samples which result in lower reward will be suppressed. Like BLEU [1], SCST has all the advantages of REINFORCE algorithms, as it directly optimizes the true, sequence-level, evaluation metric, but avoids the usual scenario of having to learn an estimate of expected future rewards as a baseline. In practice we have found that SCST has much lower variance, and can be more effectively trained on mini-batches of samples using ADAM. Since the SCST baseline is based on the test-time estimate under the current model, SCST is forced to improve the performance of the model under the inference algorithm used at test time [18].

Finally, SCST is self-critical, and so avoids all the inherent training difficulties associated with actor-critic methods, where a second "critic" network must be trained to estimate value functions, and the actor must be trained on estimated value functions rather than actual rewards. In this paper we focus on scenario of greedy decoding, where [18]:

$$\widehat{w_t} = \underset{w_t}{\arg\max}\, p(w_t|h_t) \tag{10}$$

6 Experiments

6.1 Metrics

The seq2seq model trained by evaluator model is evaluated by using different key performance measures such as levenshtein distance, and Blue Score where:

BLEU Score is a metric used to evaluate a sentence generated in a sentence of reference, proposed by Kishore Papineni, et al. in 2002. A perfect match gives a score of 1.0, whereas a perfect discordance gives a score of 0.0. The Blue Score allows us to automatically calculate a score that measures the resulting phrase quality [1].

And **Levenshtein distance (LD)** is a measure of the similarity between two strings, the source string (s) and the target string (t). The distance is the number of deletions, insertions, or substitutions required to transform s into t. The greater the Levenshtein distance, the more different the strings are [19].

6.2 Results

We have trained our models conducting several experiments for hyper-parameters automatically tuned where we have used a self-critic policy gradient to evaluate and train the basic model, where our results are summarized in Table1.

Table1. Table of results summarization

Metric	Mean Score	Graphics
LEVENSHTEIN DISTANCE (LD)	**0.302**	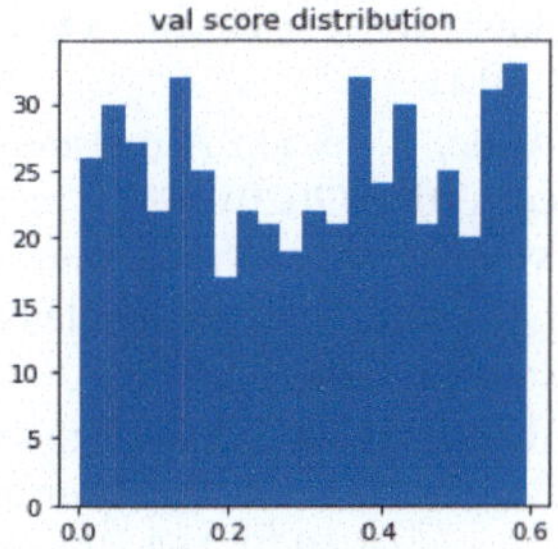
BLEU SCORE	**0.312**	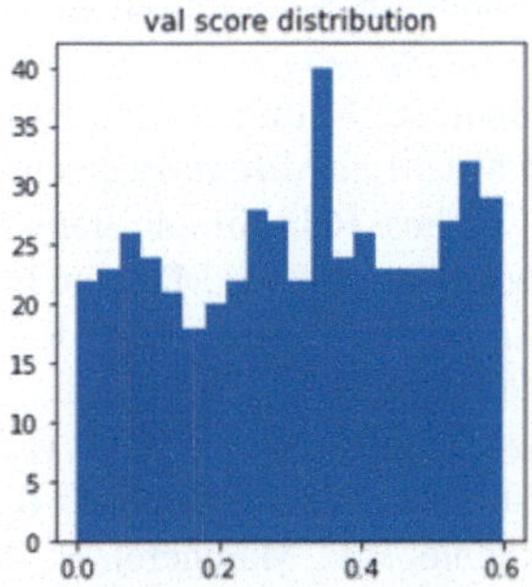

7 Conclusion

Trainable chatbots that exhibit fluent and human-like conversations remain a big challenge in artificial intelligence. Deep Reinforcement Learning (DRL) is promising for addressing this challenge, but its successful application remains an open question [20]. In this paper, we have tried to build a QA system trained by a self-critical policy gradient, which is likely to improve greatly the results of QA system.

References

1. Papineni, K., Roukos, S., Ward, T., Zhu, W.-J.: Bleu: a method for automatic evaluation of machine translation. (2001)
2. Wu, L., Tian, F., Qin, T., Lai, J, Liu, T.-Y.: A study of reinforcement learning for neural machine translation
3. Recent advances in conversational NLP: towards the standardization of Chatbot building: MaaliMnasri
4. Dsouza, J.: History of chatbots: what was, what is and what will. https://www.engati.com/blog/history-of-chatbots
5. https://fr.wikipedia.org/wiki/Chatbot
6. Sutskever, L., Vinyale, O., Le, Q.V.: Sequence to sequence learning with neural networks (2014)
7. Afrae, B., et al.: A question answering system with a sequence to sequence grammatical correction (2020)
8. Morales, A., et al.: Learning to answer questions from wikipedia infoboxes (2016)
9. Levy, et al.: Semantic relatedness of Wikipedia concepts – benchmark data and a working solution (2017)
10. Hewlet, D.: WikiReading: a novel large-scale language understanding task over wikipedia
11. Build a machine translator using Keras (part-1) seq2seq with LST. https://6chaoran.wordpress.com/2019/01/15/build-a-machine-translator-using-keras-part-1-seq2seq-with-lstm/
12. Hochreiter, S., Schmidhuber, J.: Long short_term_memory [LSTM]. Neural Comput. https://commons.wikimedia.org/wiki/File:Long_Short_Term_Memory.png. Accessed 4 Oct 2015
13. Moni, R.: Reinforcement learning algorithms— an intuitive overview. https://medium.com/@SmartLabAI/reinforcement-learning-algorithms-an-intuitive-overview-904e2dff5bbc
14. Williams, R.J.: Simple statistical gradient-following algorithms for connectionist reinforcement learning. In: Machine Learning, pp. 229–256 (1992)
15. Sutton, R.S., Barto, A.G.: Reinforcement Learning: An Introduction. MIT Press (1998)
16. Ranzato, M.A., Chopra, S., Auli, M., Zaremba, W.: Sequence level training with recurrent neural networks. In: ICLR (2015)
17. Zaremba, W., Sutskever, I.: Reinforcement learning neural turing machines. Arxiv (2015)
18. Rennie. S.J., Marcheret, E., Mroueh, Y., Ross, J., Goel, V.: Self-critical sequence training for image captioning. Watson Multimodal Algorithms and Engines Group IBM T.J. Watson Research Center, NY, USA (2017)
19. Haldar, R., Mukhopadhyay, D.: Levenshtein distance technique in dictionary lookup methods: an improved approach
20. Cuayáhuitl, H., et al.: Ensemble-based deep reinforcement learning for chatbots
21. Liu, S., Zhu, Z., Ye, N., Guadarrama, S., Murph, K.: Improved image captioning via policy gradient optimization of SPIDEr

Neural Network eXplainable AI Based on Paraconsistent Analysis - an Initial Approach

Francisco S. Marcondes, Dalila Durães(✉), Marco Gomes, Flávio Santos, José João Almeida, and Paulo Novais

Algorithm Center, University of Minho, Braga, Portugal
{francisco.marcondes,dalila.duraes,flavio.santos}@algoritmi.uminho.pt,
{marcogomes,jj,pjon}@di.uminho.pt

Abstract. Para-consistent logic is a non-classical logic whose foundations allow the treatment of contradictions without invalidating the conclusions. This paper presented an attempt of using Annotated Paraconsistent Analysis (APA) for supporting eXplained Artificial Intelligence (XAI) on Neural-Networks. For the study case, it was presented a situation where a binary classification model was able to correctly recognize one label but not the other with the form of the principle of explosion ($p \wedge \neg p$). By plotting the linear output max-min scaled data into the paraconsistent reticulate, it was possible to properly understand that the selected architectures were being able to predict. This result is early support for using APA for Neural-Network XAI, corroborating with this paper hypothesis.

Keywords: Paraconsistent logic · eXplainable AI · Neural network

1 Introduction

We live in a world full of uncertainties and contradictions. With the evolution of artificial intelligence (AI), we often expect machines to imitate the human brain and face the conflicts associated with this task [1]. Recently, Artificial Intelligence (AI) is democratized in our daily lives, with this proliferation having a significant impact on society. AI has already become ubiquitous, having become accustomed to making decisions for us in our daily lives. An example of this is the recommendation of products and films from Netflix and Amazon, suggestions for friends on Facebook, as well as personalized adverts on Google search results pages.

As Artificial Intelligence projects are becoming programming-in-the-large *cf.* [2], the importance of software engineering issues increases [3,4] yielding to the need of Software Engineering for Artificial Intelligence (SE4AI). Within the neural-network context, eXplainable AI (XAI) is cornerstone for SE4AI as it enables its issues to be addressed, providing model understanding by increased transparency [5].

J. M. Corchado and S. Trabelsi (Eds.): SSCTIC 2021, LNNS 253, pp. 139–149, 2022.
https://doi.org/10.1007/978-3-030-78901-5_13

Features engineering emerged as a science responsible for transforming raw data into relevant input information to overcome them [6]. In this way, it is possible to configure classifiers for the pattern recognition domain. This article addresses the problem of the quality of violence detection using a non-classical logical system capable of dealing with conflict situations. It is known as Paraconsistent Logic.

This paper hypothesis is that Annotated Paraconsistent Analysis (APA) *cf.* [7] can be applied for XAI. This short paper objective is to perform a first evaluation of the hypothesis based on a unique, yet meaningful, case study.

Accordingly, the remainder of the paper is organized as follows. Section 2 presents a preliminary background. Section 3 exhibits the methodology and methods, with the theoretical framework, the case study description, and the data format used in the analysis. Section 4 discusses research directions and open problems that we gathered and distilled from our analysis. Finally, Sect. 5 concludes this paper.

2 State of Art

2.1 Dataset

For this study it was used the RLVS dataset (Real Life Violence Situation) [8] composed with two thousand videos, a thousand classified as violence videos and a thousand classified as non-violence videos. Since it was not possible to find a an audio-based violence dataset, the audios from that data set was extract for building a derived violence audio dataset. However, some videos of RLVS have no sound and therefore the result was a smaller RLVS dataset: non-violence 193 audios and violence 745.

Based on this result the RLVS dataset was reconstituted aiming thousand videos for each category but all with audio. This derived data set is then named as RLVSwA (Real Life Violence Situations with Audio). Notice that the same criteria of good resolution clips (480p – 720p), variety in gender, race, age and several indoor and outdoor environments, as proposed in [8], was kept for RLVSwA in order to continue able of performing generalizations. Non-violence scenes in RLVS contain human actions as sports, eating, *etc.* The additional videos are all English speech real-life videos collected from the YouTube trimmed to three to seven seconds of duration.

2.2 Paraconsistent Logic

Roughly, paraconsistent logic handle inconsistency given by the principle of explosion by adding an additional dimension [7]. For a reference, Fig. 1 presents the paraconsistent reticulate. Therefore, situations that a value belongs to two sets is handled as $\top$ and that belongs to none set is handled as $\bot$. In addition it allows the computation of certainty ($\mu + \lambda - 1$) and uncertainty ($\mu - \lambda$) of belonging for a value [9].

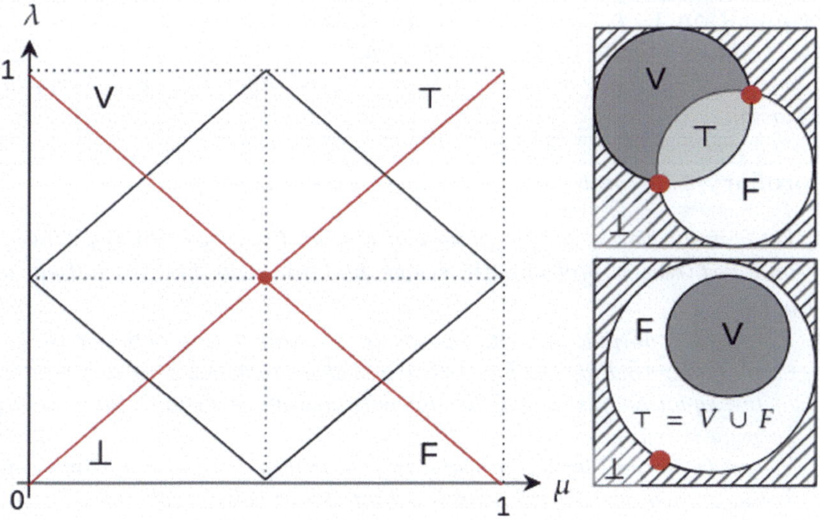

Fig. 1. Paraconsistent reticulate (with analogous set representations): V = True (0,1); F = False (1,0); $\top$ = Inconsistency (1,1); $\perp$ = Para-completeness (0,0); and the red dot = indefiniteness (0.5,0.5) [9]. The line [(1,0), (0,1)] is probability and the line [(0,0),(1,1)] is pertinence.

2.3 Data Privacy

Environmental surveillance can identify not only concerns but also trigger alerts. Surveillance and detection of violence can identify situations in the workplace and lead to regulation. Surveillance can help to create responsible institutions by providing information about scenes of violence and their causes. It can provide an evidence base for establishing and evaluating violence policy [10]. Surveillance, for example, will be central to the achievement of the United Nation's Sustainable Development Goals.

However, surveillance has been the subject of controversy at times bitter. Public surveillance can limit not only privacy but also other civil liberties. When surveillance involves name-based reporting, it can, to the extent that populations are informed, generate deep concern about invasions of privacy, discrimination and stigmatization [11].

3 Methodology and Methods

In this section, it will present our theoretical framework, the case study description, and the data format used in the analysis.

3.1 Theoretical Framework

Based on subsection II.B, the paraconsistent logic, we had applied some theoretical concepts, to create our framework.

Proposition 1. *Let p and $\neg p$ be two classes of a binary classification. The excluded middle $(p \vee \neg p)$ holds if p.*
However, if $\neg p$, $\neg p \vee \neg\neg p \Leftrightarrow \neg p \vee p \therefore \neg p$ is either $p \wedge \neg p$ yielding the explosion.

Corollary 1. *Probabilistic scaler loses Paraconsistent information*

1. *Let $\dot{a}$ and $\dot{b}$ be the output values of a neural network Binary Classifier and* `softmax` *a probabilistic scaler function that fits the output into $\dot{a} + \dot{b} = 1$.*
2. *The scaled output, if plot, results in a straight line between* $(0, 1)$ *to* $(1, 0)$ *(refer to Fig. 1); for each i added or subtracted in one left term of the equation results in the opposite operation in the other term for keeping it true.*

$\therefore$ *As can be noticed by the probability line in Fig. 1, this line complement is the paraconsistent information that is not being explored.*

Example 1. From Corollary 1, suppose a binary classifier that retrieved an output placed in full $\perp$ (0,0) or $\top$ (1,1) area. The `softmax` would reduce both to the same *indefinite* (0.5, 0.5) point for the output. However, neither $\perp$ or $\top$ are indefinite in the sense that $\perp$ implies that the result is in neither of these classes and $\top$ that the result is in both classes. For violence recognition $\perp$ may be a scene without humans and $\top$ with a scene with both people being violent and other not; therefore not *indefinite* results.

3.2 Case Study Description

The detection of violence through audio is a particular obstacle in the significant problem of surveillance. The purpose of audio violence detection is to automatically and effectively determine whether violence occurs or not in a short period by analyzing the audio signal. In recent years, this type of automatic recognition has become increasingly important for applications such as video surveillance and human-computer interaction [12–14].

Usually, building machine learning models to classify, describe, or generate audio concerns modelling tasks where the input data are audio samples. These audio samples are usually represented as time series. Dataset pre-processing is a step one must take to extract information from the underlying data. Data in a machine learning context should be useful for predicting a sample's class or the value of some target variable.

The audio samples were represented by images, spectrograms (by applying a non-linear transformation of the frequency scale called Mel Scale). A spectrogram is a visual representation of the spectrum of a signal's frequencies as it varies with time. Our main objective was to classify audio data in violence or non-violence. Therefore, using images as our audio data, we choose to apply Convolutional Neural Networks (CNNs). The reason to do it was because, in recent years, significant results have been achieved in generating and processing images with this kind of neural networks. This can partly be attributed to the great performance of deep CNNs to capture and transform high-level information in images.

The RLVSwA dataset was used for training the VGG19 [15] and the CNN6 [16] architectures for binary classification (Violence/Non-violence) whose evaluation resulted into the plots in Fig. 2.

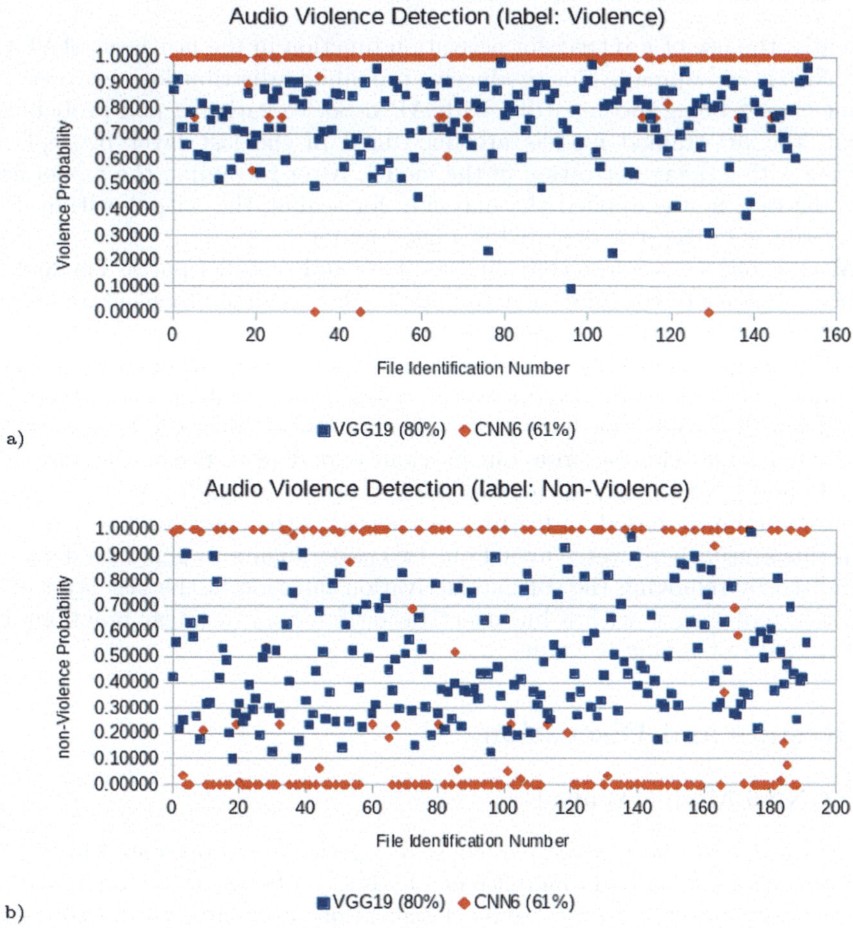

Fig. 2. Violence classification probabilities for the RLVSwA dataset

Notice that a pattern emerges, putting into words, means that *these architectures are able to classify violent audio as violence but they are not able to identify if a non-violent scene is or not violence.* In other words, by considering the negation of a class on its own, enables that each file to be classified as anything through the principle of explosion (see proposition 1). That aids on understanding the good performance on violence labeled movies (Fig. 2a) in opposition to a quasi-random performance for non-violence (Fig. 2b). Highlight these models have distinct accuracy yet an analogous issue. Nevertheless, for going further on explaining such behaviour a proper theoretical is required, APA (paraconsistent analysis) is a natural candidate that may provide additional information (Corollary 1).

3.3 Data Format Used in the Analysis

Currently, the use of `softmax` for activation function in the last layer of VGG19 and CNN6 is widespread. This produced a probability distribution over predicted output classes being incompatible with APA. So, to gather a non-probabilistic output, the pre-trained models are fine-tuned in the last layer by replacing `softmax` with `linear` activation in the model. After retraining the model using audio dataset, it was applied the max-min for scaling the output within [0..1]. The reesult is in the plot depicted in Figs. 3 and 4.

We can split a CNN into convolutional base and classifier parts. The first one comprises a stack of convolutional and pooling layers aimed to generate features from the image, referring to general features (problem independent). The second is usually composed of fully connected layers, and its primary goal is to classify the image based on the detected features, referring to specific features (problem-dependent). It is only the final layers of this network that the layers learn to identify particular classes. From our previous experiences, the two architectures (VGG19 and CNN6) used softmax as the last layer's activation function (which classified the image based on features extracted by the convolutional part).

In this sense, there were carried out two experiments to generate data non-normalized by removing the softmax activation function in the last layer of the networks, replacing it with a linear activation function (a linear function of its inputs), and re-training it again.

4 Results and Discussion

4.1 CNN6 Model Analysis

The plot of CNN6 data in the paraconsistent reticulate is presented in Fig. 3.

From this plot, no real difference can be noticed between the labels showing that model accuracy is not due to its classification capability but due to another factor. As a consequence that model's performance is untrustworthy since the classification is depending on an unknown factor. A possibility for this instance is to consider the dataset biased. Highlight that biased are not easy to identify since the feature that causes the bias is often concealed in the dataset and remains hidden after the training process due to the model's opacity. Therefore, a first finding is *(1) APA aids on identify potentially biased datasets.*

The concentration around-under the pertinence line shows that this model is not being capable on deciding to which class a particular feature set suits yet presents a slight tendency of classifying instances of both labels as non-violence.

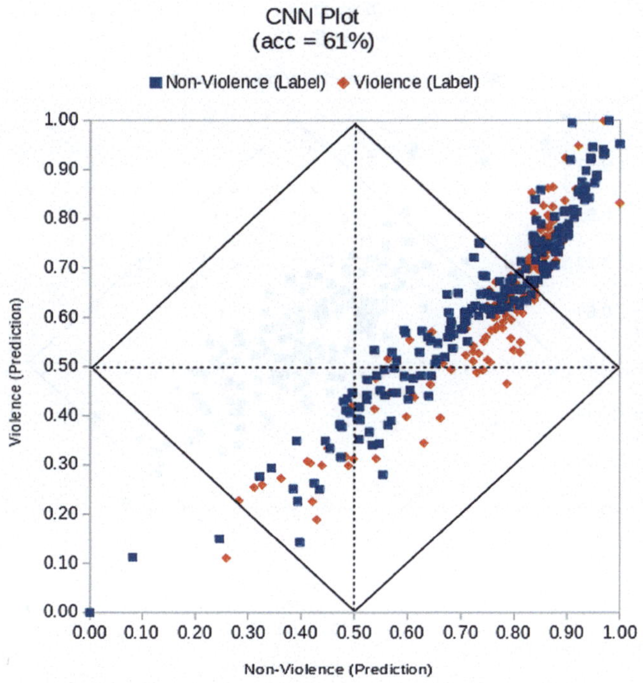

Fig. 3. CNN6 plot in the paraconsistent reticulate

Therefore, this model presents a high ambiguity level. Ambiguity means that a same set of features suits for both classes, therefore, that this feature set is not a convenient selection for discriminating this pair of classes. As a consequence, or another the class pair must be chosen or another feature set must be considered. In other words, as a second finding, is *(2) APA may help on identify an unsuitable feature set or class pair*, that may yield an enhanced re-framed model.

However, the concentration on the ambiguity quadrant suggests that feature set is succeeding for classification yet cannot defining the actual class. However, for some instances this feature set is not being able to classify on neither of these classes (the complement of these classes). Remark that those points may be due to an inaccuracy forced by the max-min approximation to be addressed in a future work. Nevertheless, supposing this is accurate, this would suggest the existence of third class suggesting that binary classification does suits for this dataset or feature set. For an instance, considering violence detection, an audio file with a TV hiss would fall in this quadrant. This reinforces finds (1) and (2).

4.2 VGG19 Model Analysis

The plot of VGG19 data in the paraconsistent reticulate is presented in Fig. 4.

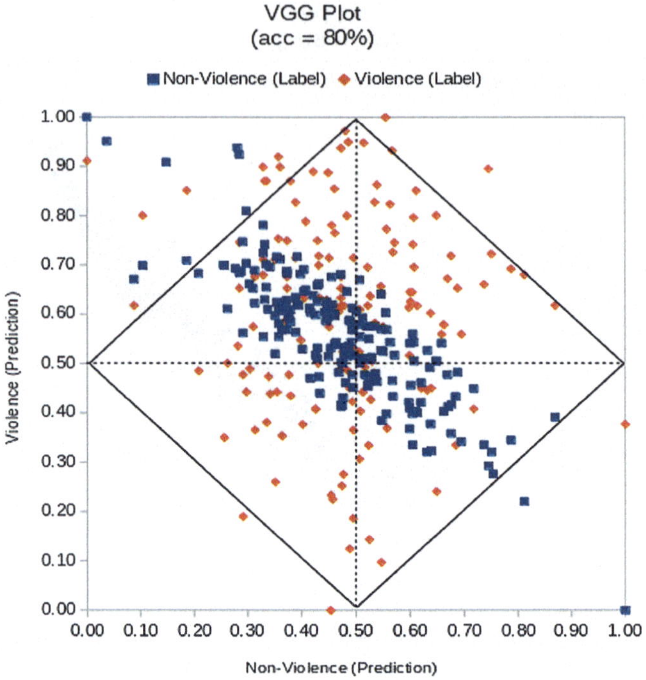

Fig. 4. VGG19 plot in the paraconsistent reticulate

This plot shows a concentration of non-violence labeled instances around the probability line but, mostly, within the uncertainty threshold; and violence labeled instances scattered throughout the reticulate but, also mostly, within the uncertainty threshold. Nevertheless, by comparing this plot with the one presented in Fig. 3, it can be suggest that a model's accuracy relates with its paraconsistent shape (this can be considered as oriented by the probability line and the other by the pertinence line). In this sense, a third find is that *(3) the APA shape may relate with accuracy.*

Notice also that, despite scattered, most of violence labeled instances are on the top quadrants while non-violence predictions are equally distributed between the two classes. Regarding the non-violence labeled distribution, the proportion of instances containing features related with violence is similar to instances containing features related with non-violence. In other words, non-violence instances can encompass violent and non-violent features. For instance on audio, a shout may be violence as well non-violence according to the situation. This may suggest that the identified features are not suitable for these classes, reinforcing finding (2). The violence labeled distribution suggests the model is being able in not attributing the wrong class for violence instances yet with both ambiguity and complement, therefore, the success of this model is on identifying the lack of non-violence features in violence instances. For another instance, it may be a shout on violence and non-violence instances yet all violence audios have shouts. This, in turn, corroborates with finding (1), biased dataset.

4.3 The Explosion Analysis

After discussing CNN6 and VGG19 architecture through the paraconsistent lens, it is possible to provide an explanation for the explosion that happened on violence and non-violence classification. In short, every instance owning a certain feature (*e.g.* shouts) is labeled as a certain class (*e.g.* violence) but any class label (*e.g.* violence and non-violence) may own that same feature. In this sense, it is not possible to say that model is capturing a class (*e.g.* violence) but that the lack of that feature is capturing a class (*e.g.* non-violence). Therefore, only a distinctive set of features was identified resulting on an untrustworthy model despite its accuracy. Considering that violence detection are part of surveillance systems, an untrustworthy model is undesirable; leading to the finding is that *(4) APA enables assessing the quality of a model despite its performance.*

5 Conclusions

This paper presented an attempt of using Annotated Paraconsistent Analysis (APA) for supporting eXplained Artificial Intelligence (XAI) on Neural-Networks. For the study case, it was presented a situation where a binary classification model was able to properly recognize one label but not the other with the form of the principle of explosion ($p \wedge \neg p$). By plotting the linear output max-min scaled data into the paraconsistent reticulate, it was possible to properly understand that the selected architectures were being able to predict. This result is early support for using APA for Neural-Network XAI, corroborating with this paper hypothesis.

For future works, the proposed approach shall be further explored and refined for application on several scenarios. Also, it may be explored how this approach may be used for enhancing the a network model by find deficit areas, refining the feature set or partitioning the dataset for enhance the network performance. Finally, it may be considered as a potential tool of SE4AI, specially focused on quality issues, by aid on understanding if the model meets the expected behavior (either functional or non-functional).

Acknowledgments. This work has been supported by FCT – Fundação para a Ciência e Tecnologia within the R&D Units Project Scope: UIDB/00319/2020.

References

1. Rodrigues, M., Novais, P., Santos, M.: Future challenges in intelligent tutoring systems - a framework, recent research developments in learning technologies. In: Méndez Villas, A., Gonzalez Pereira, B., Mesa González, J., Mesa González, J.A. (eds.)Proceedings of the 3rd International Conference on multimedia and Information & Communication Technologies in Education (m-ICTE2005), pp 929–934. Formatex (2005). ISBN 609-5994-5

2. DeRemer, F., Kron, H.H.: Programming-in-the-large versus programming-in-the-small. IEEE Trans. Softw. Eng. **SE-2**(2), 80–86 (1976). https://doi.org/10.1109/TSE.1976.233534
3. Gharibi, G., Walunj, V., Rella, S., Lee, Y.: ModelKB: towards automated management of the modeling lifecycle in deep learning. In: Proceedings of the 7th International Workshop on Realizing Artificial Intelligence Synergies in Software Engineering (RAISE 2019), pp. 28–34. IEEE Press (2019). https://doi.org/10.1109/RAISE.2019.00013
4. Durães, D., Carneiro, D., Bajo, J., Novais, P.: Using computer peripheral devices to measure attentiveness, trends in practical applications of scalable multi-agent systems, the PAAMS collection. In: de la Prieta, F., et al. (eds.) Advances in Intelligent Systems and Computing, pp. 147–155. Springer (2016), ISBN 9783319401584, ISSN 2194-5357. https://doi.org/10.1007/978-3-319-40159-1-12
5. Csiszár, O., Csiszár, G., Dombi, J.: Interpretable neural networks based on continuous-valued logic and multicriteria decision operators. Knowl. Syst. **199**, 105972 (2020). https://doi.org/10.1016/j.knosys.2020.105972
6. Durães, D., Bajo, J., Novais, P.: Characterize a human-robot interaction: robot personal assistance. In: Costa, A., Julian, V., Novais, P. (eds.) Personal Assistants: Emerging Computational Technologies. Intelligent Systems Reference Library, vol 132. Springer, Cham (2018). https://doi.org/10.1007/978-3-319-62530-0-8
7. Graham, P., Tanaka, K., Weber, Z.: Paraconsistent logic. In: Zalta, E.N. (ed.) The Stanford Encyclopedia of Philosophy (2018). https://plato.stanford.edu/archives/sum2018/entries/logic-paraconsistent/
8. Soliman, M.M., Kamal, M.H., El-Massih Nashed, M.A., Mostafa, Y.M., Chawky, B.S., Khattab, D.: Violence recognition from videos using deep learning techniques. In: 2019 Ninth International Conference on Intelligent Computing and Information Systems (ICICIS), Cairo, Egypt, pp. 80–85 (2019). https://doi.org/10.1109/ICICIS46948.2019.9014714
9. Abe, J.M., Akama, S., Nakamatsu, K.: Introduction to annotated logics: foundations for paracomplete and paraconsistent reasoning, vol. 88. Springer (2015). https://doi.org/10.1007/978-3-319-17912-4
10. Andrade, F., Neves, J., Novais, P., Machado, J., Abelha, A.: Legal security and credibility in agent based virtual enterprises. In: Camarinha-Matos, L., Afsarmanesh, H., Ortiz, A., (eds.) Collaborative Networks and their Breeding Environments. Springer-Verlag, pp. 501–512 (2005). https://doi.org/10.1007/0-387-29360-4-53. ISBN 0-387-28259-9
11. World Health Organization: WHO guidelines on ethical issues in public health surveillance. World Health Organization (2017). https://apps.who.int/iris/handle/10665/255721
12. Poppe, R.: A survey on vision-based human action recognition. Image Vis. Comput. **28**(6), 976–990 (2010). https://doi.org/10.1016/j.imavis.2009.11.014
13. Sun, Q., Liu, H.: Learning spatio-temporal co-occurrence correlograms for efficient human action classification. In: 2013 IEEE International Conference on Image Processing (ICIP 2013), Melbourne, 15–18 September, pp. 1–5. Research Collection School of Computing and Information Systems (2013). https://doi.org/10.1109/ICIP.2013.6738663
14. Liu, J., Luo, J., Shah, M.: Recognizing realistic actions from videos "in the wild". In: 2009 IEEE Conference on Computer Vision and Pattern Recognition, Miami, FL, USA, pp. 1996–2003 (2009). https://doi.org/10.1109/CVPR.2009.5206744

15. Mateen, M., Wen, J., Song, S., Huang, Z.: Fundus image classification using VGG-19 architecture with PCA and SVD. Symmetry **11**(1), 1 (2019). https://doi.org/10.3390/sym11010001
16. Zhao, Z. et al.: ML-CNN: a novel deep learning based disease named entity recognition architecture. In: 2016 IEEE International Conference on Bioinformatics and Biomedicine (BIBM), Shenzhen, China, p. 794 (2016). https://doi.org/10.1109/BIBM.2016.7822625

Computational Architecture of IoT Data Analytics for Connected Home Based on Deep Learning

Carlos Andres Castañeda Osorio[1] (✉), Luis Fernando Castillo Ossa[1,2], and Gustavo Adolfo Isaza Echeverry[1]

[1] Departamento de Sistemas e Informática, Universidad de Caldas, Manizales, Caldas, Colombia
{carlos.castaneda,luis.castillo,gustavo.isaza}@ucaldas.edu.co, lfcastilloos@unal.edu.co

[2] Departamento de Ingeniería Industrial, Universidad Nacional de Colombia Sede, Manizales, Colombia

Abstract. The internet of things (IoT) is a computing paradigm that expands every day along with the number of devices connected to the network including the home environment, that is why transmit information safely and be able to use all the computational capacity of the devices that compose it to analyze the generated data and find relevant information is one of the great challenges that it is tried to solve under the computational architecture proposed in the present article. The architecture proposed was tested in a laboratory environment with kitchen products information and the data obtained demonstrated that the architecture can solve the safe transmission using MQTT, storage in non-relational database and analysis of information with deep learning.

Keywords: Internet of Things (IoT) · Deep learning · Fog computing · IoT architecture · Machine learning · TensorFlow · MQTT

1 Introduction

The Internet of Things (*IoT*) is defined as the infrastructure that makes objects remotely accessible and connects them to each other, currently this paradigm addresses more and more devices and in 2010, 12,500 million devices were already connected to the IoT, a number about twice the world's population at the time (6.8 billion) [1]. This work addresses an architecture that allows the devices of an IoT infrastructure to transmit their information safely and can analyze the sensed data, combining the use of the fog computing paradigm [4] with the computational capacity offered by cloud infrastructure. These devices generate large amounts of data [5], which is why their transmission and subsequent appropriate analysis becomes more relevant to find important information for each field of study where it is applied.

At the household level, it is increasingly evident that devices are acquired with the ability to connect to the home network. These devices sense relevant data from

J. M. Corchado and S. Trabelsi (Eds.): SSCTIC 2021, LNNS 253, pp. 150–159, 2022.
https://doi.org/10.1007/978-3-030-78901-5_14

the environment where they are run and normally help to solve, facilitate, or automate common tasks within the connected home [2].

This work is structured as follows: in Sect. 2 the general concepts underlying the theories and technologies for its approach are described, in Sect. 3 the proposed computational architecture is explained with its complete description, in Sect. 4 there are the results of the validation test carried out and finally in Sect. 5 the conclusions of the work.

2 IoT, MQTT, Deep Learning and Fog Computing

The Internet of Things refers to addressable objects and their virtual representations in a structure like the Internet. Such objects can link to information about them, or it can transmit real-time sensor data on their state or other useful properties associated with the object [1].

With this emerging technology, companies are examining the opportunities where data transmission will create new markets to inspire positive change or improve existing services [5].

The continued growth of Internet of Things (IoT) applications is generating large amounts of data. Every day 2.5 quintillion bytes of data are created and by 2020, it is estimated that each person will create 1.7 MB per second [6]. The transmission of these volumes of information can be carried out under different technologies such as radio frequency identification (RFID), WLAN (IEEE 802.11), WPAN (IEEE 802.15) and WMAN (IEEE 802.16) for communications at the lower level. Regardless of the specific radio technology used to deploy the Machine-to-Machine (M2M) network, all end devices must make their data available to the industrial Internet. Unlike the web, which uses a single standard HTTP messaging protocol, the IoT cannot rely on a single protocol for all its needs. Consequently, there are hundreds of messaging protocols available to choose from for various types of IoT system requirements [11].

The protocol chosen according to the requirements and advantages analysis is MQTT. This is one of the oldest M2M communication protocols, introduced in 1999. It is a publish/subscribe messaging protocol designed for lightweight M2M communication on restricted networks. The MQTT client publishes messages to an MQTT agent, which are subscribed by other clients or can be held for future subscription. Each message is posted to an address, known as a topic. Customers can subscribe to multiple topics and receive every message posted on each topic. MQTT is a binary protocol and typically requires a fixed 2-byte header with small message payloads of up to 256 MB. It uses TCP as the transport protocol and TLS/SSL for security. Therefore, the communication between the client and the broker is a oriented connection. Another great feature of MQTT is its three levels of quality of service (QoS) for reliable message delivery [11].

The ease of generating and transmitting data is evidenced by the fact that in recent years, hand in hand with the increase in connected devices, people are using more and more devices in their home that have the capacity of sensing and generating relevant data for itself and for devices in its environment, likewise it is intended that all these devices provide support to tasks and that their function scales to decision-making based on data and learning that can be have to analyze them.

When saying that devices are going to learn based on data, it is necessary to define that learning is a multifaceted phenomenon. Learning processes include the acquisition of new declarative knowledge, the development of motor and cognitive skills through instruction or practice, the organization of new knowledge into general and effective representations, and the discovery of new facts and theories through of observation and experimentation. Since the dawn of the computer age, researchers have struggled to implement such capabilities in computers. Solving this problem has been, and continues to be, the most challenging and fascinating long-range goal in artificial intelligence (AI). The study and computational modeling of learning processes in their multiple manifestations is the subject of importance of machine learning [2].

Machine learning powers many aspects of modern society, from web searches to content filtering on social networks and recommendations on e-commerce websites, and is increasingly present in consumer products such as cameras and smart phones. Machine learning systems are used to identify objects in images, transcribe speech into text, match news, publications or products with user interests, and select relevant search results. Increasingly, these applications make use of a class of techniques called deep learning [3].

Deep learning allows computational models that are composed of multiple layers of processing to learn representations of data with multiple levels of abstraction. These methods have dramatically improved the state of the art in speech recognition, visual object recognition, object detection, and many other domains such as drug discovery and genomics. Deep learning discovers the intricate structure in large data sets by using the backpropagation algorithm to indicate how a machine should change its internal parameters that are used to compute the representation in each layer and the previous one. Deep convolutional networks have brought advances in image and video processing, while recurrent neural networks (RNNs) have focused on sequential data such as text and speech [3].

Section 3 describes in greater detail the devices used together with the proposed solution.

3 Proposed Architecture

The proposed architecture is the product of an investigation carried out in a macro-project in partnership between the university and the industry, the details of the institutions involved in the project are described in Sect. 6.

An indispensable part of an IoT architecture are the devices that sense data from their environment and share them through the network, in the smart kitchen it is possible to use many devices with different purposes that sense different parts and actions that are carried out in the kitchen. Table 1 describes the devices used in the architecture validation test, the device used, the current version and the function it fulfills within the IoT data analytics architecture are named.

The devices chosen for the validation test meet the following characteristics:

- Sensing relevant information about the smart kitchen environment.
- Transmit the sensed data to whoever corresponds in the architecture making use of the MQTT communication protocol.

Table 1. Physical devices, model and functionality within the IoT architecture.

Device	Version	Function within architecture
Raspberry PI 3	Model B	• *Broker MQTT* of internal devices • Publisher MQTT to the cloud
Node	MCU ESP8266	• Reading of data from the sensors • *Publisher MQTT* to the Raspberry
Magnetic door-window sensor	MC-38	• Sensing the opening or closing of kitchen drawers
Electric current sensor	ACS712	• Sensing the on or off of electrical appliances
Pyroelectric motion sensor	HC-SR501	• Sensing motion detection in the kitchen

- Encrypt/Decrypt the data transmitted and received in each case that applies.

According to the choices made, the following architecture is proposed that achieves the transmission of data in a secure way, an aspect that is essential in an IoT architecture.

Cloud services were previously validated in the main platforms (Amazon Web Services [8], Google Cloud Platform [9] and Microsoft Azure [10]) and all of them support, under different names, the architecture proposed with required services.

In the development of this project and its validation test, the Amazon Web Services (AWS) cloud platform was chosen given the result of the benchmarking carried out, in which the economic costs of the platforms and services provided were evaluated, in the Fig. 1 the architecture can be evidenced with the Amazon services chosen for each task.

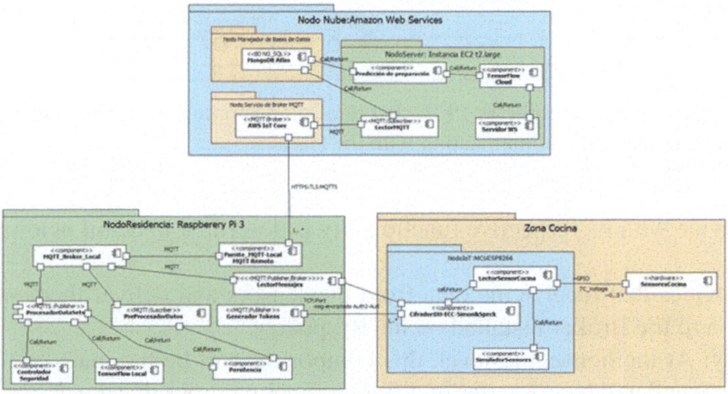

Fig. 1. Architecture with AWS services for the transmission and analysis of data from IoT devices.

The architecture is divided into two main components:

1. **Home components**: The first part of the architecture is made up of the elements that go directly into the home's local network, Figs. 2 and 3 show in detail the two key parts of this component of the architecture.

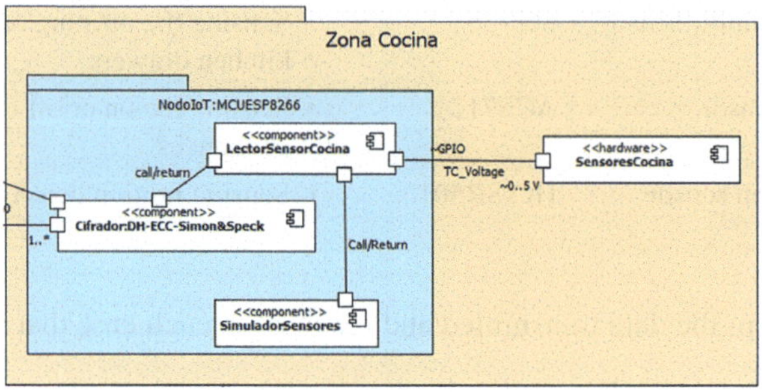

Fig. 2. Home component, kitchen area that contains N number of sensors from those described in Table 1.

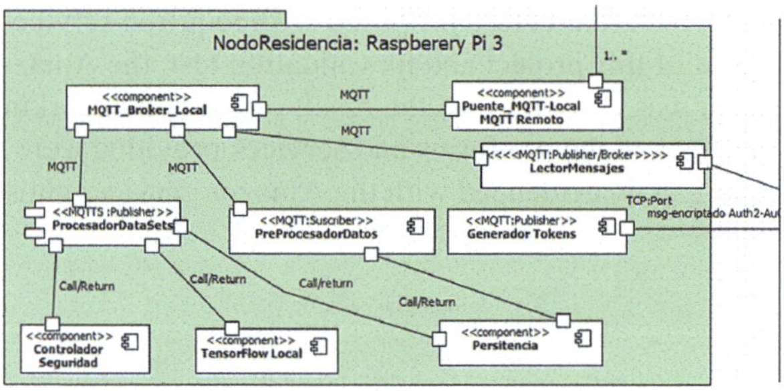

Fig. 3. Home component, Raspberry PI in charge of orchestrating the home sensors, analyzing part of the information and communicating with the cloud.

Each device with its respective functionality is described in detail below:

- Node: Devices that receive the data from the sensors and deliver them with double encryption to the Broker installed on the Raspberry.
- Raspberry: At the household level, this component will be the main one within the architecture and will have a series of different functionalities that are described below:
  - MQTT Broker: Receive requests from all home sensors through the MQTT protocol with double encryption factor. The MQTT broker Installed is Fly.
  - FOG Computing: Analysis of images taken through the Raspberry camera, classifying them into categories of food type previously defined in the trained model. Use of the TensorFlow data analytics framework.

- Publisher MQTT to the Cloud: In charge of publishing the Amazon Web Services MQTT Broker in the AWS IoT Core service.

- **Components of the Cloud**: The second part of the architecture is made up of cloud services that allow information to be received, stored and finally analyzed. Figure 4 describes the proposed cloud node in detail.

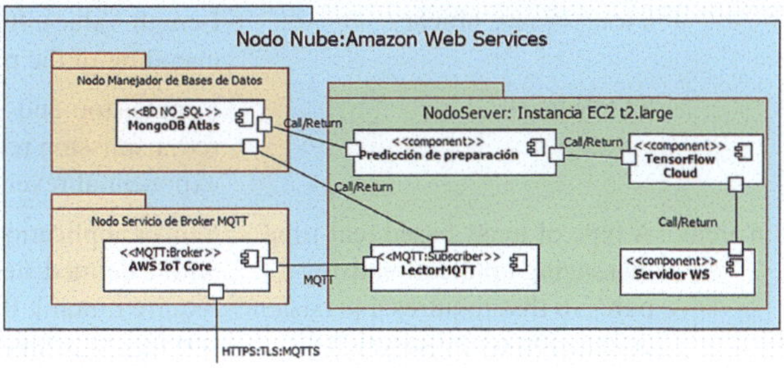

Fig. 4. Cloud node, in charge of receiving the information from the raspberry, storing it and later analyzing it.

The services and their respective use within the architecture are described below:

- AWS IoT Core: MQTT broker that allows the reception of messages by the home and their delivery to the server. All the operation of this broker is based on the publish/subscribe model.
- MongoDB Atlas: MongoDB service that allows you to have a cluster of non-relational databases. The non-relational model is better adjusted to what will later be the analysis of the data. This service is contracted directly with MongoDB, however, it is executed on the Amazon Web Services Cloud service.
- VPS Linux: Private virtual server service in the Amazon Web Services cloud, the server will fulfill the following functions:

  - Subscriber MQTT: The server will be subscribed to the messages received in the AWS IoT MQTT broker.
  - Storage in the database: After having the messages on the server, it will proceed to store them in the MongoDB database.
  - Data Analysis: Through the TensorFlow framework, the server will have the task of executing scripts that take the information from the database and perform a classification task about the task that is being executed in the kitchen.

In Table 2 you can find a comparison of the recommended protocols for the transmission of information in IoT devices, from this comparison and the consulted bibliography the decision is made to use MQTT within the architecture:

In summary, in the proposed architecture, starting with the housework, the raspberry PI will be in charge of receiving the results of the measurements of the sensors installed

Table 2. Comparative table of transmission protocols.

Properties	MQTT	CoAP
Transport layer	Mainly TCP	Mainly UDP, TCP can be used but it is not usual
Supported architecture	Based on REST request/response	Resource Observer (Publish-Subscribe)
Payload header structure	2-byte header. Payload treated as a long binary object	4-byte header + TLV (Type Length Value). It supports encoding of the payload as XML
Reliability	3 QoS levels	Simple stop-and-wait retransmission reliability with exponential reverse
Typical application areas	A type of mass-based real-time messaging application using pub/sub that requires a persistent connection to the server. The Message Broker is responsible for linking the sensors with the rest of the Web	Sensor applications (possibly with a defined sleep cycle) that require running RESTful web services to achieve direct connectivity to the Web (or optionally to some G/W) with methods similar to HTTP and URI using one or both request resources/response or observation (variant of pub/sub mode) in real time. Ideal for applications that require easy integration with the HTTP-based Web. Suitable for applications such as smart energy, building automation, etc.

in the kitchen through the MQTT protocol, being the broker of the household devices Afterwards, fog computing is carried out on the same device, in this case study in order to classify the images captured from the camera installed in the raspberry PI to the food found in the mise en place; After classifying them and grouping the data received by the sensors, it publishes the AWS IoT Core service located in the Amazon Web Services cloud in an MQTT topic already specified by the architecture.

4 Results

4.1 Data Transmitted via MQTT, Received in the Cloud and Stored in MongoDB Atlas

The first task of the cloud node is to receive the data in AWS both in the platform's administration console and in the VPS server proposed for this purpose, Figs. 5, 6 and 7 show the way in which the data They are transmitted from a home device and received in the cloud through the MQTT protocol using the AWS Cloud SDK with the python programming language.

Fig. 5. Data transmission from a local home environment to the cloud service.

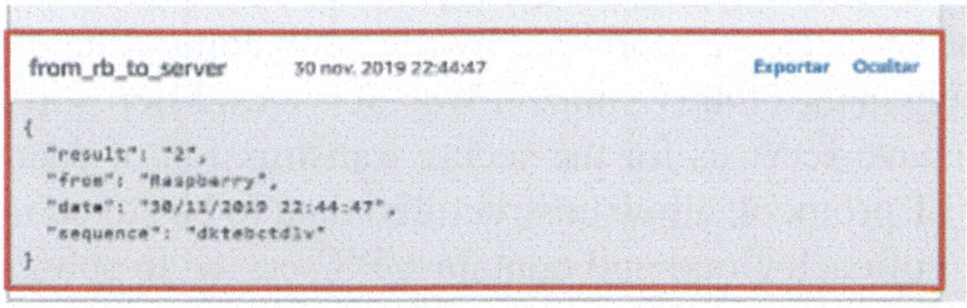

Fig. 6. Data received in the AWS cloud management console.

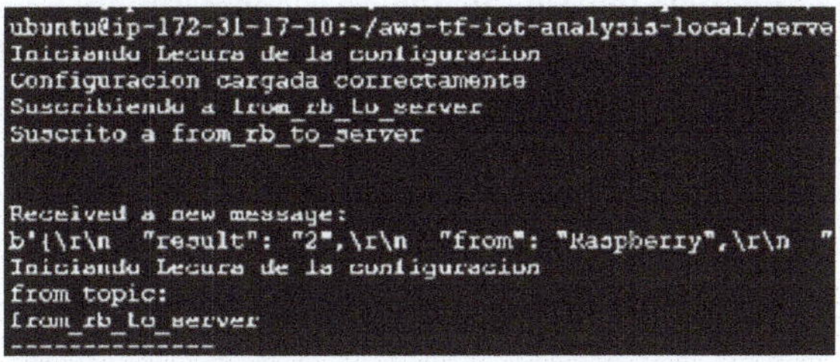

Fig. 7. Data received in the VPS server with Linux operating system subscribed through the MQTT protocol to the topic "from_rb_to_server"

After receiving the data, the VPS server proceeds to store them in the chosen non-relational database, in this case mongoDB atlas, in Fig. 8 you can see the transmitted data already stored using the client of the MongoDB Compass database engine.

The task after the data storage is the analysis of these, this work is not within the scope of this article.

When starting the analysis tests on the Raspberry and monitoring the use of resources of this, it is concluded.

Fig. 8. Data stored in MongoDB Atlas viewed through MongoDB compass

5 Conclusions and Future Work

The main conclusions of the proposed architecture and the work carried out are presented below:

- The architecture proposed in this article has as its main characteristics, the secure transmission of data, through the MQTT protocol, unstructured data storage, and VPS server, these services are supported by the main computing platforms that allow have the processing in the cloud (Amazon Web Services, Microsoft Azure and Google Cloud Platform) have services for the secure transmission of data from IoT devices through the MQTT protocol, allow unstructured data to be stored under conventional databases or own of each cloud and contain VPS service to subscribe to the MQTT topic, receive and analyze the information. The architecture was worked on from the generalization and was tested in Amazon Web Services, concluding that its conception lends itself to working in the different cloud providers with their specific services.
- In this work it was possible to verify that there is compatibility between the different implementations of Tensorflow in the same version, both on IoT devices and on servers, the results change with respect to execution time due to the difference in processing.
- As future work, it is proposed to test the architecture on the other cloud platforms and make performance and cost comparisons between them. In addition, it is proposed that the architecture be validated with different IoT devices with a different business core.

Acknowledgments. This project is part of my master's thesis in computational engineering at the University of Caldas and has contributed to the inter-institutional project between the university of caldas, the national university of Colombia and the company *Mabe* with *hermes* code 36715 entitled "Computational prototype for the fusion and analysis of large volumes of data in IoT environments (Internet of Things) from Machine Learning techniques and secure architectures between sensors, to characterize the behavior and interaction of users in a Connected Home ecosystem".

References

1. Stolpe, M.: The Internet of Things: opportunities and challenges for distributed data analysis. ACM SIGKDD Explor. Newsl. **18**(1), 15–34 (2016)

2. Carbonell, J.G., Michalski, R.S., Mitchell, T.M.: An overview of machine learning. In: Machine Learning, pp. 3–23. Morgan Kaufmann (1983)
3. LeCun, Y., Bengio, Y., Hinton, G.: Deep learning. Nature **521**(7553), 436–444 (2015)
4. Bonomi, F., Milito, R., Zhu, J., Addepalli, S.: Fog computing and its role in the internet of things. In: Proceedings of the First Edition of the MCC Workshop on Mobile Cloud Computing, pp. 13–16, ACM (2012)
5. Stergiou, C., Psannis, K.E., Kim, B.G., Gupta, B.: Secure integration of IoT and cloud computing. Future+ Gener. Comput. Syst. **78**, 964–975 (2018)
6. Renart, E.G., Veith, A.D.S., Balouek-Thomert, D., de Assuncao, M.D., Lefèvre, L., Parashar, M.: distributed operator placement for IoT data analytics across edge and cloud resources (2019)
7. Yuan, X., He, P., Zhu, Q., Li, X.: Adversarial examples: attacks and defenses for deep learning. IEEE Trans. Neural Netw. Learn. Syst. **30**, 2805–2824 (2019)
8. Jackson, K.R., et al.: Performance analysis of high performance computing applications on the amazon web services cloud. In: 2010 IEEE Second International Conference on Cloud Computing Technology and Science, pp. 159–168. IEEE November 2010
9. Krishnan, S.P.T., Gonzalez, J.L.U.: Building Your Next Big Thing with Google Cloud Platform: A Guide for Developers and Enterprise Architects. Apress (2015)
10. Barga, R., Fontama, V., Tok, W.H., Cabrera-Cordon, L.: Predictive Analytics with Microsoft Azure Machine Learning. Apress, Berkely (2015)
11. Naik, N.: Choice of effective messaging protocols for IoT systems: MQTT, CoAP, AMQP and HTTP. In: 2017 IEEE International Systems Engineering Symposium (ISSE), pp. 1–7. IEEE October 2017

Distributed Programming and Applications

Datacentric Analysis to Reduce Pedestrians Accidents: A Case Study in Colombia

Michael Puentes[1(✉)], Diana Novoa[2], John Manuel Delgado Nivia[3], Carlos Jaime Barrios Hernández[1], Oscar Carrillo[4], and Frédéric Le Mouël[5]

[1] Universidad Industrial de Santander, Bucaramanga, Colombia
michael.puentes@correo.uis.edu.co, cbarrios@uis.edu.co
[2] Universidad Nacional de la Plata, La Plata, Argentina
[3] Unidades Tecnológicas de Santander, Bucaramanga, Colombia
territoriointeligente@uts.edu.co
[4] Univ Lyon, CPE, INSA Lyon, Inria, CITI, EA3720, 69621 Villeurbanne, France
oscar.carrillo@cpe.fr
[5] Univ Lyon, INSA Lyon, Inria, CITI, EA3720, 69621 Villeurbanne, France
frederic.le-mouel@insa-lyon.fr

Abstract. Since 2012, in a case-study in Bucaramanga-Colombia, 179 pedestrians died in car accidents, and another 2873 pedestrians were injured. Each day, at least one passerby is involved in a tragedy. Knowing the causes to decrease accidents is crucial, and using system-dynamics to reproduce the collisions' events is critical to prevent further accidents. This work implements simulations to save lives by reducing the city's accidental rate and suggesting new safety policies to implement. Simulation's inputs are video recordings in some areas of the city. Deep Learning analysis of the images results in the segmentation of the different objects in the scene, and an interaction model identifies the primary reasons which prevail in the pedestrians or vehicles' behaviours. The first and most efficient safety policy to implement - validated by our simulations - would be to build speed bumps in specific places before the crossings reducing the accident rate by 80%.

Keywords: Data-centric · Traffic violation · Dynamic system

1 Introduction

This work aims to reduce accidents in a city by knowing pedestrian behaviour in a city's urban area.

Yang et al. [28] contributed notably to the field by separating two kinds of pedestrians - those obeying the law and those having opportunistic behaviours. Under this assumption, Yang made a questionnaire evaluating Chinese citizens' behaviour related to the pedestrian cross path. In his survey, some variables are

J. M. Corchado and S. Trabelsi (Eds.): SSCTIC 2021, LNNS 253, pp. 163–174, 2022.
https://doi.org/10.1007/978-3-030-78901-5_15

essential in the model construction for micro-simulation, like age and gender. This characteristic is relevant according to the perception of the pedestrians in the case study in Colombia [21,26]. Indeed, pedestrians clearly show in our dataset an obedience issue to the traffic authorities. Another related work by Chen Chai et al. [6] evaluates pedestrians' behaviour also by gender and age, but adds extra information from fuzzy logic-based observations about if the subject is a child. This work question the indicators that influence pedestrians' behaviour. Aaron et al. [1] compare real situations where different variables related to the environment were evaluated, knowing the reality will always be different. This work defines the variables of a micro-simulation which have to be part of the causal model to refine the modelled reality. Camara et al. [3] implement a decision tree to determine the pedestrians' vehicle interaction. This implementation, looking for the design of new policies for pedestrians in Bucaramanga - Colombia, seeks for less critical accidents in 2019 over 200 pedestrians involved in an accident[1]. They identify the need to find the correct variables that can reduce pedestrians accidents. For instance, Holland et al. [10] highlight gender as an essential factor in the behavior of a pedestrian to decide a crossing. Other works consider pedestrian and vehicles flows [4,19], even if the use of specific PTV-VISSIM and VISWALK software modules[2] can recognize pedestrian events and drivers' behaviour as individuals.

These different works well-illustrate the importance of determining the exact criteria influencing pedestrians and drivers to take an action at a crossing intersection. In Sect. 2, we go through several causes found in related works. Then, in Sect. 3, we describe the methodology used in our work to build the dataset and the simulation. In Sect. 4, we detail our model and simulation results. We finally conclude and give future works.

2 Related Work

Several research works try to identify the variables intervening in pedestrians' jaywalking producing accidents in a city - either from pedestrian behaviour or external causes. To find these causes, we establish different scenarios with real-life information and people perception. To validate these variables in our micro-simulation model, we refine them by comparing their value with video-recordings in a spiral process of refined simulation by Jordan (see Fig. 1) [22]. This validation is repeatedly processed in an endless task - as citizens behaviours are not trivial and only a near prediction is possible. It allows identifying new potential variables to be in the model of the citizens' perception. Furthermore, to compare the results with existing models [5].

To minimize the accidents that involved citizens and vehicles, it is necessary to find the reasons that affect pedestrians' opportunistic decisions when the traffic light allows the pass of the vehicles or scenarios where do not exist signals.

[1] https://www.datos.gov.co/Transporte/Accidentes-de-Transito-en-Bucaramanga-ocurridos-de/7cci-nqqb.

[2] http://vision-traffic.ptvgroup.com/es/productos/ptv-vissim/.

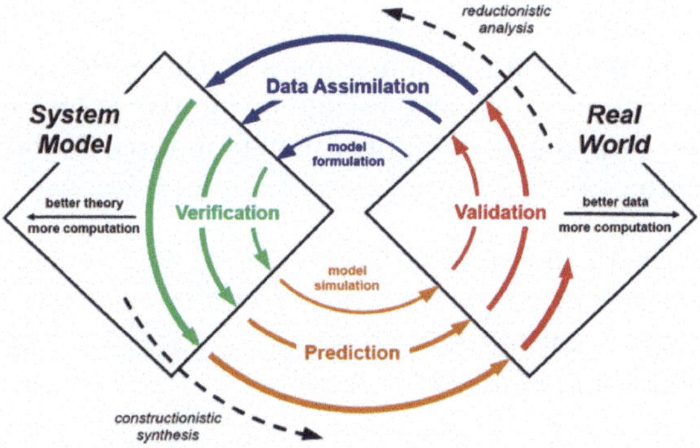

Fig. 1. Inference spiral of system science (Jordan 2015)

Pedestrians must have priority in the way, and the vehicle has to stop in that case. The empirical analysis of Sanghamitra et al. [8] lists the crossing decisions of passersby in a cross-side of the street based on the time gap until the next car. Even if all the variables could not be detected by this research tool, it is possible to recognize similar walkers' behaviours. Another well-known variable taking place in pedestrian behaviour is the "social force". It occurs when the pedestrians are guided by another citizen, without knowing if the principal citizen decision is correct, but at least having a partial vision of the path. Different passersby models can be found in the literature: magnetic force, social force, and Benefit-Cost Cellular. Teknomo does a review of microscopic simulations of pedestrians, detailing every pedestrian as an individual [25]. In his work, different variables are necessary to the mathematical modelling, but different causes could be part of a pedestrian accident. In a simulation, several causes can be considered in the model:

- Time gap between car and pedestrian [8,28]
- Social force [9,25]
- Environment (weather, pollution, noise) [13]
- Vehicle factors [13]
- Human factors (driver skills, fatigue, alcohol, drugs, too quick glance) [7,13]
- Road Conditions (corner, visibility, straight, wet, dry) [12]

These causes allow identifying different scenarios where a pedestrian have a specific behaviour. Pau et al. [20] select scenarios at different times of the day: with many pedestrians in the streets (peak hours) or when almost no pedestrians walk through the street. Rasouli et al. [24] determine the people who cross-traffic line and the street and who made a signal with the hands, showing a petition of the stop to the driver. This article's methodology looks at different environments: night, day, rain, snow - in the same place when crowded or not. This work is not related to implementing objects in the video sequence, but when the behaviour of pedestrian changes according to luminosity, it is a parameter we need to consider. Kouabenan et al. [12] analyze 55 reports of pedestrians' accidents, randomly selected from a police report on the IvoryCoast. In their article, they analyze the characteristics and circumstances of the accidents. Those previous works conclude their research by suggesting alternative solutions, simplifying or new public policies, campaigns, or improvements in a particular spot of the city. Proposed actions are:

- Accident prevention campaigns [12,18]
- Road safety policies [13,27]
- Improved lighting conditions [27,29]
- Vehicle conditions campaigns [27]

3 Methodology

Our research work aims to create a micro-simulation model that considers factors that interfere in the citizen's behaviour (pedestrian or driver). Later on, the simulation is refined with real-life video data from cameras installed in different spots in five cities in Colombia. This work is then validated with a case study on the city of Bucaramanga by using the paradigm of system dynamics; the model evolves by applying the following four steps:

1. Identify and analyse the factors influencing on the event
2. Modify the parameters of the simulation model
3. Execute the simulation with the modified model
4. Observe and differentiate between the real-life behaviour on the observed event and the simulation outcomes.

The process previously described is a cycle that will improve the model as soon as we add more variables according to the phenomenon's observations (see Fig. 2). If it is possible to find the factors that significantly impact the pedestrians' accidents in the city, these causes can become the targeted part of the new public policies to reduce future tragedies.

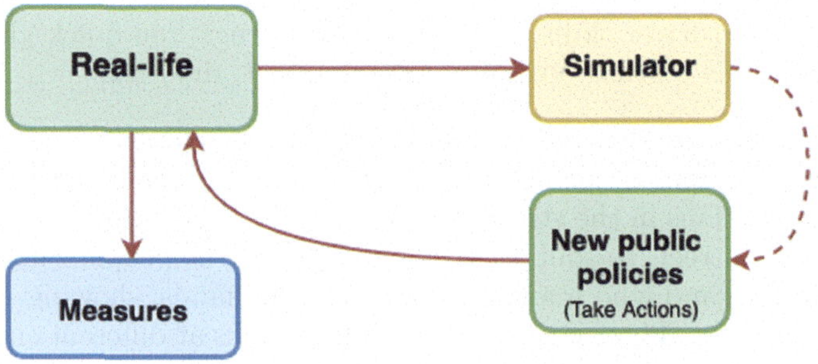

Fig. 2. General methodology implemented

To validate our model and apply our methodology, we collaborate with the Santander department's government in Colombia. One of the projects - funded by the government - aims to analyse the citizens' behaviour and improve civics in the city. To this end, 900 cameras were deployed in five cities of the department. Different projects are proposed to use these cameras: mobility, public spaces and harmony.

This research focuses on pedestrians to further a better mobility and reduce the rate of accidents in the city. Thus we select, from the deployed cameras, the spots where more accidents in particular conditions occur. An image of one of the top five selected spots where collisions happen very often is shown in Fig. 3. This particular is in a relevant neighbourhood collecting different variables according to the related works (see Sect. 2): no traffic lights, the second area in the city with more accidents, presenting architecture heterogeneity: one church, two universities, one park, and numerous residential areas nearby.

Fig. 3. Camera selected in Bucaramanga-Colombia

The initial simulation model starts from the previous research carried out on the evaluated phenomenon, bearing in mind that all the models can vary because we created the model for the particular case of Bucaramanga. The real-life factors can be analysed and measured thanks to the cameras of the city. To measure these factors, we analyse hours of video on the Briefcam[3] software. This software has tools to measure causes and find objects via deep learning on video-recordings and accelerate the process of identifying pedestrians. Then, with the video-recordings and the software, it is possible to get the data detailed in Table 1. This report is necessary to have extra information from particular behaviours from the video and the perception of the people who live in Bucaramanga.

[3] https://www.briefcam.com/.

Table 1. Variables describing the pedestrians' behaviours

Variable	Type	Observation
Date of Video	Date	
Range hour	e.g.: 21:00–22:00 hrs	
Pedestrians who cross the street	Numeric	Categorized by gender
Not safe events	Numeric	Those are events which involves an accident
Pedestrians against the law	Numeric	
Pedestrians who cross by the zebra	Numeric	
Pedestrians velocity average	Numeric	

4 Model and Simulation Results

According to the related works reviewed in Sect. 2, some causes directly entailing pedestrians' accidents are observable. Otherwise, for non-trivial causes, we estimate the necessity of micro-simulation [2]: seeing the pedestrian as an individual, identifying the general causes seen in the observed simulation, and measurable in the cameras installed for this project (see Fig. 4). In previous researches, many simulations were performed [5,13,18,24,27], and the proposed solutions produced new policies, which only have a sinusoidal behaviour, according to Mendez [18]. Therefore, it is necessary to implement a micro-simulation to identify and analyse the pedestrians' causes of accident as an individual [25,28] and determine the particular factors that will reduce the accident rate.

With micro-simulation, factors can directly be compared to measurable variables thanks to video recordings from the initial causes evaluated. In particular, this work assesses human factors in the micro-simulation for pedestrians as individuals. The numeric data is not enough. Hence we use the citizens' perception to have extra information not visible in the videos.

In some accidents, pedestrians endanger their own lives, and thus they interfere with traffic. In this simulation, we aim to look for a safer pedestrian crossing. Therefore, we use the Viswalk simulation tool[4] to find the causes of accidents in a micro-simulation of walkers in a specific sector of Bucaramanga (Colombia), using the same method represented in Fig. 2.

Firstly, we will analyse the priority, which in Colombia is for vehicles instead of pedestrians. To know the causes that a jaywalker could have, we interviewed students and other passers-by about how good pedestrians they were. The results are shown in Tables 2 and 3. This information was used to feed the micro-simulation, for example, and one of the more exciting information, *69% of people think that the pedestrian is disrespected*, even in a zebra crossing, the car or motorbike has priority. This is important information, when the simulation has this rule of priority, the accidents in the micro-simulation appear (see Fig. 5).

[4] https://www.ptvgroup.com.

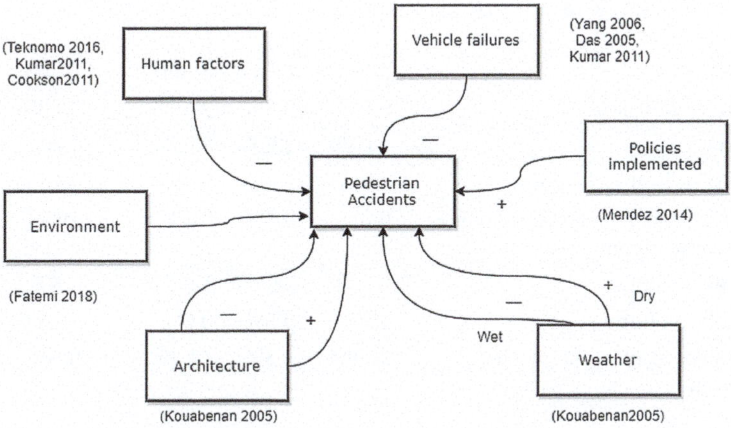

Fig. 4. Causal diagram for pedestrian accidents

Fig. 5. Viswalk micro-simulation in San Francisco neighborhood

The measured information for several video recordings, from the spot shown in Fig. 3, are presented in Table 4. The date and time were selected for working days, and the values shown in the table are the mean values during the week.

Table 2. Pedestrians' behavior at intersections

Question	Always	Often	Sometimes	Rarely	Never
Do you walk on the zebra crossing when you cross the road?	42.9%	47.6%	9.5%	0.0%	0.0%
Do you look at the state of traffic lights when you cross the road?	95.2%	4.8%	0.0%	0.0%	0.0%

Table 3. Probabilities of pedestrians' signal non-compliance under specific situations

Question	Always	Often	Sometimes	Rarely	Never
In general conditions	0.0%	9.5%	23.8%	33.3%	33.3%
In a hurry	14.3%	23.8%	38.1%	19%	4.8%
Long duration of red light	14.3%	14.3%	9.5%	33.3%	28.6%
Presence of other pedestrians who violate traffic signal	9.5%	9.5%	9.5%	19%	52.4%
Low traffic volume	33.3%	38.1%	14.3%	4.8%	9.5%
High traffic volume	19%	9.5%	0.0%	23.8%	47.6%
Police officer is on duty at the intersection	23.8%	4.8%	14.3%	9.5%	47.6%

The information was gathered from the videos thanks to the deep learning tool on Briefcam. We use this information as input for the simulations. Additionally, the same spot's heat-map is shown in Fig. 6. It reveals the frequent zones that pedestrians go, and that we will use to narrow down the solutions to implement.

With the additional information about the people's perception and the numeric data gathered from the videos, we have the micro-simulation input data. It is not possible to strictly imitate when walkers and vehicles appear, but they are defined by the number of objects that appear per minute. One frequent improvement in the pedestrians' care is the implementation of a bump. As a counterpart of that implementation, the queue of cars on the street produces traffic jams in the zone, altering the order and regular circulation. By changing the number of vehicles per minute, the simulation shows that maximum three vehicles stayed queuing in a row in an hour of simulation. One of the additional

Fig. 6. Heat map showing where walkers go through

Table 4. The quantifiable information from the video recording

Variable	Type	Observation
Date of Video	22/06/2019	
Range hour	e.g.: 9:00–10:00 hrs	
Pedestrians who cross the street	316	70% man and 30% woman
Not safe events	0	Not registered in the video
Pedestrians against the law	2	
Vehicles in the video	668	
Cars and motorbikes vehicles velocity average	60 km/hr	

factors of accidents in a row is a pedestrian who walks seeing his cellphone, this increases the probability of accidents to 88% for adults [11]. This number is possible to reproduce for all the pedestrians in the micro-simulation, then accidents appear.

Overall, the road speed reducer decreases the speed to 2 km/h from the 40 km/h mandatory limit, and even from 60–70 /h in real-life events, according to the video records. From the results of the simulation, this small but significant change can save over 80% of the people related in an accident - especially with the 88% people presenting distraction or bravery conditions (Fig. 7).

Fig. 7. Bump implemented in the simulation

5 Conclusions

The proposed model helps in determining the fittest variables that are more important in pedestrians' unsafe crossings in fast-growing cities. In other areas of a city or other cities having a similar environment, the model can be transferred to reduce the area's accidental rate.

Implementing a slight change in architecture - such as the speed bump deployment proposed in this article - produces a significant number of people's lives saved. 80% of accident decrease can be achieved in our study.

6 Future Work

The method proposed is a good first step to use infrastructure (cameras) and information (video-recordings) to build a smart city in the Bucaramanga case-study [21]. Our method analyzes different causes identified in the video recordings, other parameters such as the weather, building architectures, disabled people identification, vehicle conditions could be integrated and extend the method to global system dynamics. This pedestrian behaviour analysis can also provide urban services such as traffic optimization [15,16], smart parking [17], taxi recommendation [23], crisis management [14].

Acknowledgements. This work is supported by the Government and the Unidades Tecnologicas de Santander (project 879/2017). Thanks to the SC3-UIS Lab, the Colifri association, the CITI Lab at INSA Lyon and the CATAI workgroup - where this project was already discussed and received feedbacks.

References

1. Aarón, M.A., Gómez, C.A., Fontalvo, J., Gómez, A.J.: Análisis de la Movilidad Vehicular en el Departamento de La Guajira usando Simulación. El Caso de Riohacha y Maicao. Información tecnológica **30**(1), 321–332 (2019). https://doi.org/10.4067/s0718-07642019000100321
2. Álvarez-Pomar, L.: Modelo de inteligencia colectiva de los sistemas peatonales. Universidad Distrital Francisco Jose de Caldas **23**(45), 5–24 (2016)
3. Camara, F., et al.: Predicting pedestrian road-crossing assertiveness for autonomous vehicle control. In: IEEE Conference on Intelligent Transportation Systems, Proceedings, ITSC November 2018, pp. 2098–2103 (2018). https://doi.org/10.1109/ITSC.2018.8569282
4. Campisi, T., Tesoriere, G., Canale, A.: The pedestrian micro-simulation applied to the river Neretva: the case study of the Mostar "old bridge". In: AIP Conference Proceedings, vol. 2040, November 2018. https://doi.org/10.1063/1.5079193
5. Cantillo, V., Arellana, J., Rolong, M.: Modelling pedestrian crossing behaviour in urban roads: a latent variable approach. Transp. Res. Part F Traffic Psychol. Behav. **32**, 56–67 (2015). https://doi.org/10.1016/j.trf.2015.04.008. https://linkinghub.elsevier.com/retrieve/pii/S1369847815000716

6. Chai, C., Shi, X., Wong, Y.D., Er, M.J., Gwee, E.T.M.: Fuzzy logic-based observation and evaluation of pedestrians' behavioral patterns by age and gender. Transp. Res. Part F Traffic Psychol. Behav. **40**, 104–118 (2016)
7. Cookson, R., Richards, D., Cuerden, R.: The characteristics of pedestrian road traffic accidents and the resulting injuries (2011)
8. Das, S., Manski, C.F., Manuszak, M.D.: Walk or wait? An empirical analysis of street crossing decisions. J. Appl. Econometrics **20**(4), 529–548 (2005). https://doi.org/10.1002/jae.791
9. Helbing, D., Molnár, P.: Social force model for pedestrian dynamics. Phys. Rev. E **51**(5), 4282–4286 (1995). https://doi.org/10.1103/PhysRevE.51.4282
10. Holland, C., Hill, R.: Gender differences in factors predicting unsafe crossing decisions in adult pedestrians across the lifespan: a simulation study. Accid. Anal. Prev. **42**(4), 1097–1106 (2010)
11. Koopmans, J.M., Friedman, L., Kwon, S., Sheehan, K.: Urban crash-related child pedestrian injury incidence and characteristics associated with injury severity. Accid. Anal. Prev. **77**, 127–136 (2015). https://doi.org/10.1016/j.aap.2015.02.005
12. Kouabenan, D.R., Guyot, J.M.: Study of the causes of pedestrian accidents by severity. J. Psychol. Afr. **14**(2) (2005). https://doi.org/10.4314/jpa.v14i2.30620
13. Kumar, N., Umadevi, G.: Application of System Dynamic Simulation Modeling in Road Safety. Trasnportation Research Board (2011)
14. Le Mouël, F., Hernández, C.B., Carrillo, O., Pedraza, G.: Decentralized, robust and efficient services for an autonomous and real-time urban crisis management (2017). https://arxiv.org/abs/1703.04519
15. Lèbre, M.A., Le Mouël, F., Ménard, E.: On the importance of real data for microscopic urban vehicular mobility trace. In: Proceedings of the 14th International Conference on ITS Telecommunications (ITST 2015), Copenhagen, Denmark, pp. 22–26, December 2015. https://doi.org/10.1109/ITST.2015.7377394
16. Lèbre, M.A., Le Mouël, F., Ménard, E.: Partial and local knowledge for global efficiency of urban vehicular traffic. In: Proceedings of the IEEE 82nd Vehicular Technology Conference (VTC 2015-Fall), Boston, MA, USA, pp. 1–5, September 2015. https://doi.org/10.1109/VTCFall.2015.7391065
17. Lin, T., Rivano, H., Le Mouël, F.: A survey of smart parking solutions. IEEE Trans. Intell. Transp. Syst. **18**(12), 3229–3253 (2017). https://doi.org/10.1109/TITS.2017.2685143
18. Méndez-Giraldo, G., Álvarez-Pomar, L.: Dynamic model to analyze pedestrian traffic policies in Bogota. Dyna 81(186), 276 (2014). https://doi.org/10.15446/dyna.v81n186.45219
19. Oskarbski, J., Gumińska, L.: The application of microscopic models in the study of pedestrian traffic. In: MATEC Web of Conferences, vol. 231, pp. 1–7 (2018). https://doi.org/10.1051/matecconf/201823103003
20. Pau, G., Campisi, T., Canale, A., Severino, A., Collotta, M., Tesoriere, G.: Smart pedestrian crossing management at traffic light junctions through a fuzzy-based approach. Future Internet **10**(2) (2018). https://doi.org/10.3390/fi10020015
21. Puentes, M., Arroyo, I., Carrillo, O., Barrios, C.J., Le Mouël, F.: Towards smart-city implementation for crisis management in fast-growing and unplanned cities: the colombian scenario. Ingeniería y Ciencia **16**(32), 151–169 (2020). https://doi.org/10.17230/ingciencia.16.32.7
22. Puentes, M., Novoa, D., Nivia, J.D., Hernndez, C.B., Carrillo, O., Le Mouël, F.: Pedestrian behaviour modeling and simulation from real time data information. In: 2nd Workshop CATAI - SmartData for Citizen Wellness, Bogotá, Colombia, October 2019. http://www.catai.fr/catai2019/

23. Qian, S., Cao, J., Le Mouël, F., Sahel, I., Li, M.: SCRAM: a sharing considered route assignment mechanism for fair taxi route recommendations. In: Proceedings of the 21st ACM SIGKDD Conference on Knowledge Discovery and Data Mining (KDD 2015), Sydney, Australia, pp. 955–964, August 2015. https://doi.org/10.1145/2783258.2783261
24. Rasouli, A., Kotseruba, I., Tsotsos, J.K.: Understanding pedestrian behavior in complex traffic scenes. IEEE Trans. Intell. Veh. **3**(1), 61–70 (2018). https://doi.org/10.1109/tiv.2017.2788193. http://ieeexplore.ieee.org/document/8241847/
25. Teknomo, K., Takeyama, Y., Inamura, H.: Review on microscopic pedestrian simulation model (March), 1–2 (2016). http://arxiv.org/abs/1609.01808
26. Torres, G., al.: La ciudad - región sostenible como proyecto: desafíos actuales. Visiones cruzadas y perspectivas. Universidad Nacional de Colombia, Sede Bogotá, Bogotá, October 2019. http://bdigital.unal.edu.co/74466/
27. Ulfarsson, G.F., Kim, S., Booth, K.M.: Analyzing fault in pedestrian-motor vehicle crashes in North Carolina. Accid. Anal. Prev. **42**(6), 1805–1813 (2010)
28. Yang, J., Deng, W., Wang, J., Li, Q., Wang, Z.: Modeling pedestrians' road crossing behavior in traffic system micro-simulation in China. Transp. Res. Part A Policy Pract. **40**(3), 280–290 (2006). https://doi.org/10.1016/j.tra.2005.08.001
29. Zhang, G., Yau, K.K., Zhang, X.: Analyzing fault and severity in pedestrian-motor vehicle accidents in China. Accid. Anal. Prev. **73**, 141–150 (2014)

Enhancing SmartKADASTER 3D City Model with Stratified Information in Supporting Smart City Enablement

Nur Zurairah Abdul Halim[1(✉)], Hairi Karim[2], Siew Chengxi Bernad[3], Chan Keat Lim[1], and Azhari Mohamed[1]

[1] Department of Survey and Mapping Malaysia (JUPEM), 50578 Kuala Lumpur, Malaysia
nurzurairah@jupem.gov.my

[2] 3D GIS Laboratory, Universiti Teknologi Malaysia, 81310 Johor Bahru, Malaysia

[3] Geotech Solutions Sdn Bhd, Kuala Lumpur, Malaysia

Abstract. SmartKADASTER Interactive Portal (SKiP) was developed as a 3D visual and analysis platform for JUPEM clients to better understand the dynamic and spatial relationships between cadastral parcels and their objects surroundings. However, SKiP was incomplete as it lacks strata title parcels information. A study was conducted to determine a suitable conceptual method for incorporating stratified information into the current SmartKADASTER environment. This paper will highlight the related strata survey practice in Malaysia and the advancement of SmartKADASTER 3D city modelling. Considerations and methods for using Strata XML and CityGML datasets are also discussed. Furthermore, some insights into using Strata XML to complement the SmartKADASTER city model with stratified information and 3D models are highlighted. The recommendations proposed at the end of this paper are hoped to contribute to the body of knowledge for 3D city model development and be adopted, particularly in improving the SmartKADASTER system for smart city enablement.

Keywords: Strata XML · SmartKADASTER · City modelling · CityGML

1 Introduction

Malaysia is experiencing an increase of 139% of population growth. To date, there are more than 32.7 mil population (including 3 mil non-citizens) in Malaysia compared to 13.7 mil in 1980 [1]. By 2050, 68% of Malaysians are expected to live in urban areas. Consequently, systematic and proper handling of space, right and ownership are necessary to provide an equitable relationship between people and their property, providing strong evidence for urban dwellers' tenure security and well-being. These are among the challenges to be handled with smart and sustainable manners for city to expand [2]. This paper aims at sharing the experience of incorporating 3D stratified information into the SmartKADASTER city model developed to support smart city decision making based on 3D geospatial analysis. According to previous researchers [3], 3D city models are

J. M. Corchado and S. Trabelsi (Eds.): SSCTIC 2021, LNNS 253, pp. 175–186, 2022.
https://doi.org/10.1007/978-3-030-78901-5_16

being used in the smart city paradigm, typically serving as a framework for the urban environment.

Naturally, location, land-use patterns, distances, and interactions are the essential factors of geospatial approach, as key supporting smart city implementation [3]. The Department of Survey and Mapping of Malaysia (JUPEM) is responsible for Malaysia's land administration's geospatial component, which is the fundamental dataset of Malaysia's Spatial Data Infrastructure (SDI). As of to date, Malaysia has more than 8 million land parcels stored in the National Digital Cadastral Database (NDCDB). While cadastral survey data can be acquired in 3D, it is stored planimetrically (x,y), thus inhibits further 3D-based analysis. 2D information cannot serve urban complexed situations, especially when different types of land use and properties were placed in a complicated 3D scenario. As a result, the SmartKADASTER Interactive Portal (SKiP) was developed as a 3D visual and analysis tool for the clients to better understand the complex and spatial relationships of the land or building parcels with its surroundings [4].

Although the essence of SKiP (Phase 1) was for 3D-based analysis, the cadastral data's foundation was mainly from the 2D survey accurate NDCDB. 3D cadastral information is available in Malaysia but through Strata Title, referring to the building parcels in a strata scheme for aboveground properties. A Strata Title Plan shows an accurate scale model of the strata parcels and their area size and building storey height. It provides the spatial dimensions of assets and properties containing legal interests, legal boundaries, and legal attributes. The Strata Title's legal boundaries are defined as median lines inside physical structures such as walls and ceilings. The management of strata and stratum titles in Malaysia are under the purview of the respective state's JUPEM and Land and Mine Office (PTG). Both authorities differently handle strata title preparations. PTG manages strata ownership and registration with *eTanah* (*eLand*). At the same time, JUPEM handles the spatial component and Strata Title Plan preparation through *eKadaster*. Both systems are still 2D in nature.

To enable 3D spatial analysis in SKiP, other 3D geodata such as building footprints, digital terrain models and meshed city model was applied and incorporated into SKiP. The approach enabled various 3D-based spatial analysis, such as water rising simulation and shadow analysis, for application such as forecasting an area's property market value. Such 3D spatial analysis has directly helped users to democratise SKiP data and was found fit for its purpose at providing ease of doing spatial analysis [4]. However, SKiP is still lacking of 3D cadastral information. A complete 2D and 3D cadastre information in SKiP would enable a holistic spatial analysis result that has linkages to the object's Rights, Restrictions and Responsibility (RRR). Therefore, a study was carried out to determine the conceptual approach to be applied for incorporating stratified 3D cadastral information into SmartKADASTER Phase 2. Another source for 3D cadastral information in Malaysia is the Stratum Title, but it is not included in the study.

The remaining sections of this paper cover the rest of the study, including the related strata survey practice in Malaysia in Sect. 2. SmartKADASTER 3D City Modelling development is described in Sect. 3, followed by leveraging Strata XML and SmartKADASTER 3D database. Lastly, Sect. 5 on recommendations and conclusion.

2 Related Works on Strata Survey Practice

Existing cadastre in Malaysia is managed in a 2D environment, except for aboveground stratified properties that are 2D + 1D (Storey Height), which can also be known as the volumetric parcel or 2.5D. Stratified titles typically include high-rise properties such as condominiums, apartments, and landed properties gated and guarded [5], with underground properties like basement or bunkers. Before a strata submission, a Licensed Land Surveyor (LLS) must confirm the subdivided buildings are situated wholly within the boundaries of the cadastral lot, and the subdivision would not contravene any restriction in interest to which the land comprised in that lot is for the time being subject. The LLS is then required to carry out the as-built survey of the strata parcel. The strata parcel's spatial properties are described as 'x', 'y' and 'z'. However, the 'z' value does not refer to a specific or established vertical datum or the terrain but rather the strata parcel's geometry. The 'z' refers to the parcel's relative height measured from its ceiling to the floor surface, providing volumetric of the parcel [6].

A strata parcel's legal boundaries are defined in the building elements' median location, most of the time as equal parts of a wall and ceilings or other physical structure. Although 3D data are acquired on-site, the legalistic cadastre system and land law still use 2D geometric description. The legal and land law expression for land and property tenure have not been prepared to register in a 3D situation. The strata title plan is the by-law document prepared and embedded into the paper-based Strata Title to prove ownership. A sample of a Strata Title Plan is shown in Fig. 1.

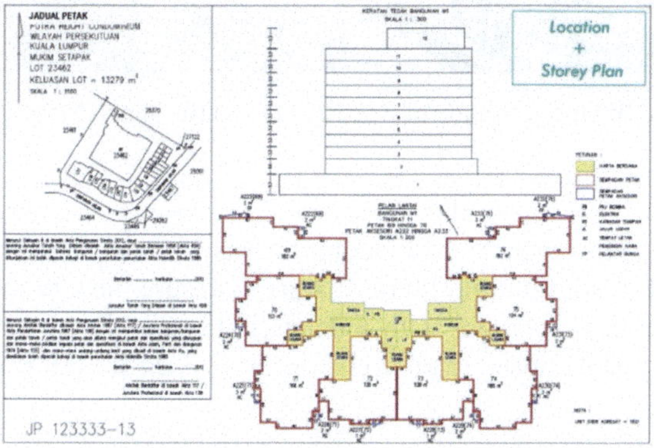

Fig. 1. A sample of a Strata Plan showing the strata unit parcel and its corresponding storey plan.

The Strata Title is a unique title given to parcel units owner of a residential or commercial multi-storey building that shares common facilities such as security, parking, and common governing space [6]. Sometimes it could be owned by more than a person. With the Strata XML data's availability, the strata parcels can now be rendered and visualised in a 3D scene. A 3D model could further illustrate the 'x', 'y' and 'z' coordinates of each parcel unit, thus, become more significant as evidence on ownership rights for the vertical space. A strata Unique Parcel Identifier (UPI) is created to help identify every strata parcel and link it to the corresponding 2D cadastral lot. A detailed explanation of the strata UPI can be referred to [7].

3 SmartKADASTER 3D City Modelling Toward Smart City

3.1 The Motivation

One of the unique features of the SmartKADASTER project is the 3D city model. A 3D city model is a mixture of both geometric and non-geometric data. 3D reconstruction and data integration are usually performed to build the city model [8]. The city model of the Phase 1 SmartKADASTER project was constructed and displayed as a 3D mesh with Skyline's proprietary format in SKiP, as shown in Fig. 2.

Fig. 2. 3D mesh sample of SmartKADASTER phase 1 city model

Naturally, NDCDB was the basis for cadastral survey information in SmartKADASTER. However, it was incomplete as it lacks strata title parcels information. Because of this limitation, Phase 2 aims to ensure the city model for a Greater Kuala Lumpur complies (Selangor state and a part of Negeri Sembilan) with a universal exchange standard and database-ready to incorporate cadastre information.

3.2 CityGML Schema

CityGML is an open data model and XML-based format for storing and exchanging common 3D features in virtual 3D city and landscape models [9]. The development of CityGML is to reach a standard definition of the basic entities (such as buildings, roads, rivers, bridges, vegetation and city furniture), attributes, and relations of a 3D city model. It also defines different standard levels of detail (LoDs) for the 3D objects, which allows the representation of objects for various applications and purposes, such as simulations, urban data mining, facility management, and thematic inquiries. Thus, the schema is a crucial feature of 3D city modelling in terms of cost-effective, long-term model maintenance, enabling the reuse of the same data in multiple application areas, such as the 3D model of Helsinki, which supports more than ten applications as of 2017. Consequently, CityGML2.0 is utilised in SmartKADASTER to support existing strata XML files (converted into Strata-CityGML compliences) and for the same purpose of visualisation and management within SKiP environment.

3.3 3D Database

In Phase 2 SmartKADASTER, PostgreSQL is used to handle the attribute table for 2D and 3D data, while PostGIS is the spatial extender to serve geometry and coordinate system of vector data. The 3D CityGML schema in PostgreSQL database is based on the CityObject relationship – 3DCityDB class (one of the elements supported in Object-based Spatial Database, OSD). An OSD is a spatial database that stores the location as objects (e.g. cities, rivers), which exist independently of their locations. While CityObject in this case capable of representing an object in several predefined details (LoDs) and linking with respective LoD attributes. The attributes storing with objects within an OSD provide better presentation result, improved security, searching query, manipulation (update/add/delete) capabilities in a more efficient way.

3DCityDB is used to provide SQL functions to create a CityGML (CityObject) schema for PostgreSQL. The selection is based on previous researchers' recommendations, such as [11], where 3DCityDB offers more compact and allows fewer tables to automatically create a spatial relational database schema. The database is essentially supporting the CityGML version 2.0 and other tools such as the Data Publishing Web Feature Service and the Importer/Exporter tool. With CityGML Importer/Exporter, CityGML features and geometries are the first to be read and imported, neglecting all XLink reference information, but temporarily stored in the database. To complete the entire CityGML import process, the XLink relationship information stored in the database are being re-resolved and written to the respective CityGML data tables. At the same time, CityGML datasets are validated for 3DCityDB syntax error and geometric-topological accuracy with CityDoctor. Apart from software, information and communication technology (ICT) also plays important roles in measuring smartness level of smart city indication [10]. As a consequence, for the SmartKADASTER 3D citymodel, a database approach is preferred over file-based management to allow functions such as ease of queries and data or model updating.

3.4 3D UPI Supporting Multiple Representation Details (LoDs)

As the study focuses on the Building Module of CityGML and the migration process has completed, CityObject generates the Building Table. The minimum attributes of the 3D CityGML LoD and 3D UPI models are shown in Fig. 3. User queries (SKiP) are performed in this database for LoD information and its geometry. The 3D UPI allows users to quickly access a list of CityObjects by querying the selected geometry and attributes. For the SmartKADASTER 3D model, it was decided that the 3D UPI is the extension of the 2D UPI of the cadastre lot too, but "D" is introduced to represent the respective LoD type. For example, 1601400141393(S)x(B)xD1 for LoD1, …D2 for LoD2, …D3 for LoD3, …D4 for LoD4. Attributes can be managed directly using the PostgreSQL database. However, the geometry for each 3D CityGML LoD still demands manual intervention. Nevertheless, PostgreSQL allows any obsolete model to be replaced with a new model based on the 3D UPI while preserving attributes.

The LoDs model was constructed using SketchUp software from the various sources of surveyed point clouds tied up to the ground control points (GCP) and 2D cadastre marks. Figure 4 show a building sample of LoD1–3 according to CityGML version 2.0 schema. For Phase 2, the minimum number of building in LoD2 and LoD3 is 1500 buildings, while LoD1 is for the whole project area, 1,430 km^2. All these models will be tied up using 3D UPI ID with D extension, stored in Postgres database and visualise/query using SKiP.

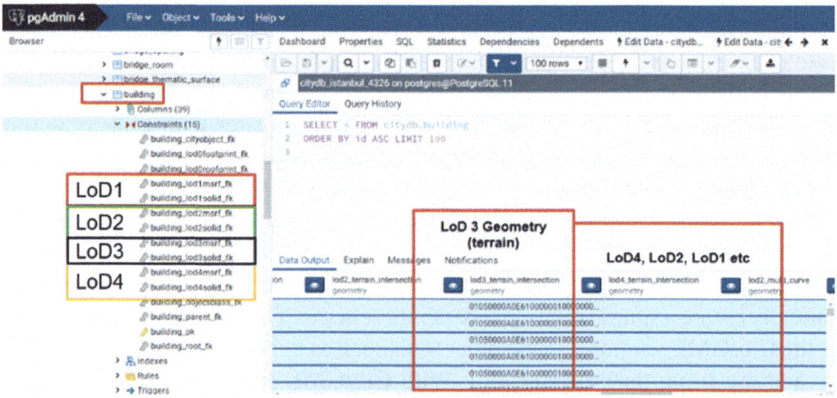

Fig. 3. Building table for respective LoD in building module of CityObject table.

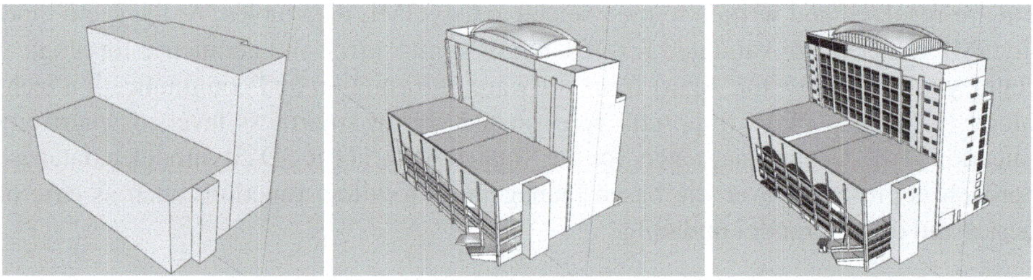

Fig. 4. L-R: Samples of LoD 1, 2 and 3 of *Wisma Perbadanan Kemajuan Pertanian Selangor*.

4 Leveraging Strata XML in 3D SmartKADASTER Databases

4.1 Differences Between Strata XML and City Modelling

During the project implementation, it was found that there are some significant differences in comparing the dataset schema of strata and the new SmartKADASTER city model using CityGML. Six main differences are highlighted; details information, 3D UPI, geometry, attributes, focus users and storage management.

Level of Details. Strata XML only have a single schema for a building. The schema includes metadata, UPI ID, geometry, classes (e.g. owners, shared facilities etc.), attributes (e.g. name of unit) and others whereas CityGML with 5 LoDs (4 of them are for 3D).

Geometry. The geometry of the strata file focuses on the middle wall of the building structure (boundary). However, CityGML is based on each wall's surface coordinates (exterior in general and interior for LoD4). In other words, the Strata concept is defined by boundary and rights (hidden) while CityGMLis based on actual representation by eyes.

3D UPI Strata vs 3D UPI City Model. Both strata and city models have UPI 3D depicting each other's characteristics. For example, "*Tingkat*" and Accessories are available in Strata XML and LoD categories are presented for city models. However, both UPIs have a continuation of the 2D UPI, which refers to the cadastral lots as found in the NDCDB. The 2D UPI in NDCDB is equivalent to 3D UPI for LoD 0 City Model, while for Strata XML refers to the UPI codes set by JUPEM on strata properties.

Attributes. In terms of attributes, existing Strata XML have rich information linked with the RRR information available in PTG (*eTanah*). While in CityGML LoD, most of the attributes incorporated and linked in the database are in fact shared with other datasets. There are very minimum attributes migrated during the construction of the 3D model and database migration since model extractions are from point clouds.

Focus Users and Professional Domains. Strata is a file data with specific purposes and may interest users where linkage of geometry and RRR ownership is their primary concern. Meanwhile, CityGML is introduced to offer different standard LoDs and cost-effectiveness, so more comprehensive users and cases can utilise the 3D data.

Storage Management (File-Based and Database). The main differences of both schema implementation lie on storage management; strata in file-based XML format and CityGML in the PostgreSQL database. In general, file-based data provides less security, less efficient in updating and searching and multi-user viewing/updating data. In Strata XML format, all strata files refer to a specific development area with plenty of files, IDs, and information of the 3D space. Accessing a specific file in multi-user mode without secure account security will be a devastating drawback compared to database query. Similarly, proprietary CityGML format is also in XML type file-based but more specialised as in Geographic Markup Language (.gml format). However, with advanced and recent development in import/export tool such as 3DCityDB, has allowed migration of CityGML file-based into PostgreSQL database. Accessing 3D city model from database gives any 3D online platform a booster in supporting massive datasets, especially city level and even a country. Fast information retrieval, updating work and security could be enhanced drastically using database environment.

4.2 3D Model Presentation of Strata + CityGML for SmartKADASTER

Essentially, the purpose of modelling 3D cadastre objects highlighted by [11] is intended to provide boundary certainty for 3D cadastre objects. Therefore, with the notion that a 3D visualisation is a form of geovisualisation, the 3D model presentation is the key to disseminating, visualising and utilising the 3D model to support decision-making especially as part of smart city module. Ideally, the rich Strata XML information should be retained. However, because of the differences between the strata and city model schema, the presentation of the 3D model may require a different grouping of data. Nevertheless, both strata and city model models can still be integrated with 3D UPI. An actual 3D model of shop lots reconstructed using SketchUp is shown in Fig. 5. Instead, as a long block, the LoD models for party wall buildings are reconstructed into individual models based on exact building footprints with 3D UPI with respective 2D Cadastral lot (2D UPI). The 3D UPI ID later shall be linked to Strata UPI ID for any information on space ownership and RRR without opening a new 3D viewer for the visualise Strata model. The overall concept of integrating CityGML model in SmartKADASTER Phase 2 with existing Strata XML, database and single viewer (SKiP) is illustrated in Fig. 6. Previous researchers [12] had successfully integrated PostgreSQL database to support the smart city concept in their work with 4D web application.

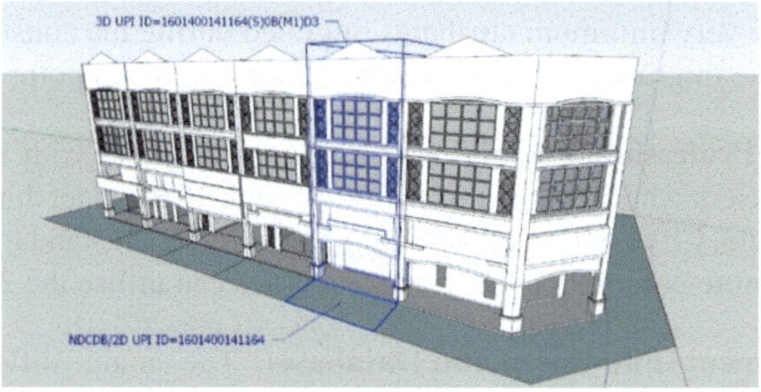

Fig. 5. Example of proposed 3D UPI (LoD3, D3) based on NDCDB 2D UPI.

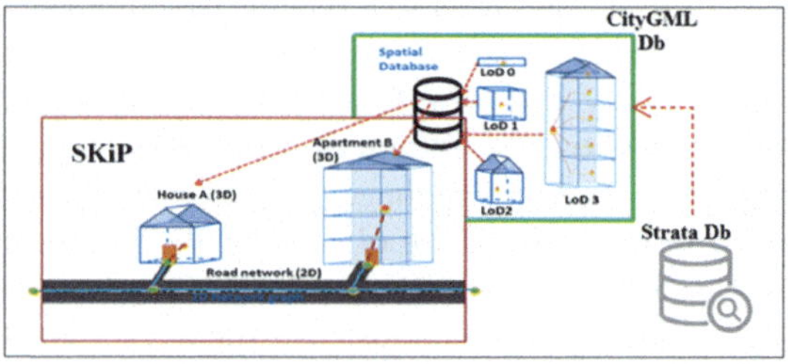

Fig. 6. 3D visualisation concept and data workflow (edited from [12]).

4.3 Development of Strata XML to CityGML Converter

Strata XML as part of the legalised data and primary output of Malaysia strata legal framework, comprises information but not limited to attribute information such as types of building usage, types of parcels, strata project information, the height and altitude of the parcel; as well as spatial information such as surveying information that includes bearing and distances of the corner of the associated parcel. The XML has a customised schema that explains how elements represent the real-world entity, e.g. *Petak* is equivalent to Building Parcel. The underlying concept that pushes the idea of future-cadastre in 3D smart city management is to enable customised schema in the city modelling domain, a subset of Smart City Management. Enabling customised schema in city modelling requires the use of Application Domain Extension (ADE) in CityGML. On the one hand, the application is city modelling, while on the other hand, it comprises specific strata information beyond the CityGML standard scope. Figure 7 depicted how strata XML data be utilised for a city model that contains cadastre information.

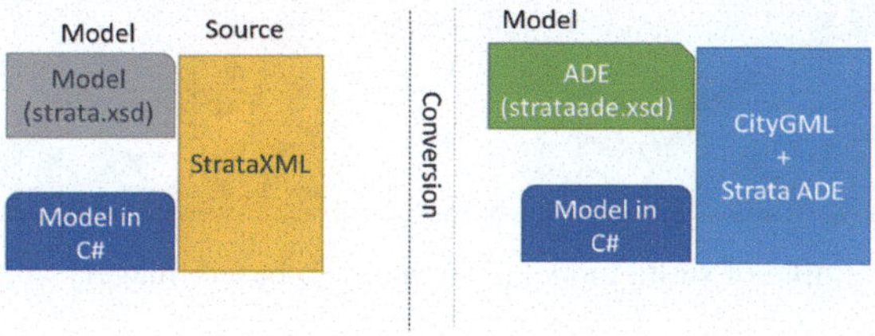

Fig. 7. The logical model of both input and output to enable cadastre information in smart city.

The conceptual workflow is then further realised into a physical workflow where each input and output model is created in C#, complied to the formatting of Schema Definition (XSD). The XSD (strata.xsd) was used to develop the model, while the new XSD (strataade.xsd) was developed and further developed into C# model. Figure 7 also shows the workflow that is physicalised using C#. The program was then developed using C#, whereby it converts StrataXML to CityGML with custom attributes and naming using the ADE. The ADE was validated using XML validator. With the ADE the CityGML was produced and further visualised in FME Safe software. The CityGML is additionally imported into database as required on the application side.

4.4 Visualisation Platform

Some potential web-based visualisation platforms such as Skyline (commercial) and Cesium (open source) are in the exploration stage. [13] managed in integrating 2D (current land registration systems) and 3D geospatial data (3D properties) via utilising Cesium JS as a 3D geospatial platform. Nevertheless, the project aim is to utilise strata information in the same visualisation platform as the SmartKADASTER city model in SKiP. The 3D visualisation platform's functionalities should at least meet what previous researchers [14, 15] have outlined. It is unnecessary to have the same database as CityGML; it could be a different database name/schema but supported in the same

PostgreSQL database and schema to link via the same 3D UPI standard. Figure 8 shows an example of the strata model (3D) converted from Strata XML, while Fig. 9 shows another converted sample with existing 2D NDCDB cadastre system.

To enable smart, on-demand and online purchasing, JUPEM's eBIZ application is being integrated with the visualisation platform of SKiP. Primary Users can enjoy SKiP services and make online payments by prepaid or e-wallets and instantly download their purchased digital data.

Fig. 8. Example of strata model (3D) in a web-based imagery application.

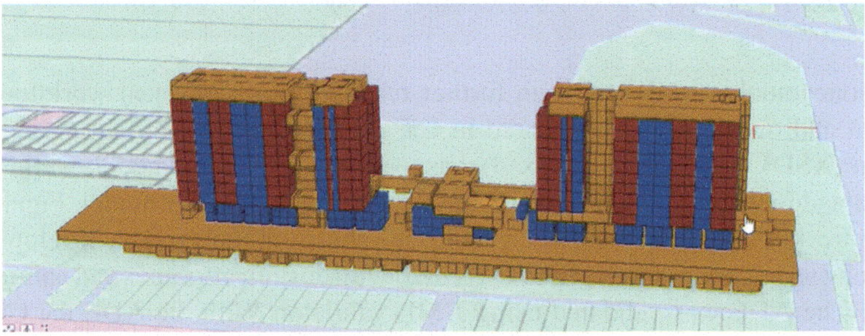

Fig. 9. Example of strata model (3D) with CityGML schema overlaid with 2D NDCDB lot.

4.5 3D UPI Linkages Between Both Models

The schema structure and differences are "not connected" to one another based on geometrical details (LoD). Otherwise, another new schema of data structure is required to incorporate both geometric features. However, there is a way to connect them, that is by using 3D UPI ID. Basically, all strata parcels are associated with every LoD type. Although it can be from different database sources, it can be linked and displayed in a single viewer in SKiP. This relationship between tables was made using the specific 3D UPI in each object class as a connection continuity in the 3D database.

5 Concluding Remarks

5.1 Recommendations and Future Plans

Several recommendations based on the findings of the study are highlighted for future work consideration, particularly incorporating stratified information into city database:

i. 2D cadastre data should still be employed for 3D visualisation in a 2D/3D hybrid cadastre approach.
ii. 3D models are typically resulting in higher storage when converting into CityGML.
iii. Traditionally, strata's value is based on parcel area (length x wide) without volume (height) calculation. This new information should be added into Strata GML database, and help owners understand their parcel's vertical dimensions.
iv. Seamless 3D visualisation between Strata GML and the CityModel is crucial to increase user experience. Generally, a strata parcel usually consists of several utilities like column, beam, and sewerage pipe that can be considered unusable areas; could be updated. These exclusive use areas cannot be determined in the Sale and Purchase Agreement, Property Valuation Report and Cadastral Map.
v. Concerning the above, combination of LoD and stratified information, has potential, including indoor navigation.
vi. The 3D buildings can be seamlessly integrated with other datasets such as terrain, roads, traffic, sewerage, sensors, landscapes, and others to complete and understand the city with a common reference datum.
vii. The 3D city model of SmartKADASTER comprises all spectrums of LoDs (from LoD 0 to 4). Under the new CityGML 3.0, the LODs 0/1/2/3 remain, and the interior of objects that are typically modelled and described as LoD 4 is removed. Instead, they are expressed and integrated with the LODs 0/1/2/3.
viii. Underground legal object (utilities etc) is best represented in city model CityGML and integrated with owner's information and RRR from cadastral law.
ix. The developed Strata converter can act as a new medium for validating Strata XML provided by LLS to improve as-built document for strata title reference.

5.2 Conclusion

This paper has highlighted the background, the aim and motivation to have a 3D city model database for SmartKADASTER (CityGML) and 3D Strata for smart city-based 3D model. The description on CityGML, Strata file, and database were also explained in this paper for a better understanding and idea of integration. A strata-to-CityGML converter is developed from strata information into CityGML-based schema for standardisation of information retrieval in SKiP. These two data sources will provide significant improvement in spatial analysis and smart city data management, especially for Phase 2 area of interest. Thus, issues related to the land or building's RRR can be verified seamlessly. In addition, several insights on leveraging the Strata XML to complete the stratified information within the SmartKADASTER environment were also touched in this paper that can be further studied in the future. Recommendations proposed in this paper is hoped to contribute to the 3D city model development body of knowledge and be adopted especially for smart city management in Malaysia.

References

1. DOSM: Current Population Estimates, Malaysia. DOSM (2020). https://www.dosm.gov.my/v1/index.php?r=column/cthemeByCat&cat=155&bul_id=OVByWjg5YkQ3MWFZRTN5bDJiaEVhZz09&menu_id=L0pheU43NWJwRWVSZklWdzQ4TlhUUT09. Accessed 9 Feb 2021
2. Joshi, S., Saxena, S., Godbole, T.: Developing smart cities: an integrated framework. Procedia Comput. Sci. **93**, 902–909 (2016)
3. Li, W., Batty, M., Goodchild, M.F.: Real-time GIS for smart cities. Int. J. Geogr. Inf. Sci. **34**(2), 311–324 (2020). https://doi.org/10.1080/13658816.2019.1673397
4. Halim, N., Chan, K.: SmartKADASTER Interactive Portal (SKiP), is it fit for purpose? In: IOP Conference Series: Earth and Environmental Science, vol. 540, no. 1, p. 012025, IOP Publishing (2020)
5. Hanafi, F.M., Hassan, M.I.: The integration of 3D spatial and non–spatial component for strata management. Int. Arch. Photogrammetry, Remote Sens. Spat. Inf. Sci. **42**(4), W16 (2019)
6. Choon, T.L., Abdul Rahman, A.: An overview of strata and stratum objects for Malaysian 3D cadaster. In: Advances Toward 3D GIS. Johor: UTM, pp. 327–352 (2018)
7. Zulkifli, N.A., Rahman, A.A., Van Oosterom, P.: Developing 2D and 3D cadastral registration system based on LADM: illustrated with Malaysian cases. In: Proceedings of the 5 th FIG Land Administration Domain Model Workshop, pp. 24–25 (2013)
8. Julin, A., et al.: Characterizing 3D city modeling projects: towards a harmonized interoperable system. ISPRS Int. J. Geo-Inf. **7**(2), 55 (2018). https://www.mdpi.com/2220-9964/7/2/55
9. Siew, C.B., Kumar, P.: CitySAC: a query-able cityGML compression system. Smart Cities **2**(1), 106–117 (2019)
10. Meijer, A., Bolívar, M.P.R.: Governing the smart city: a review of the literature on smart urban governance. Int. Rev. Adm. Sci. **82**(2), 392–408 (2016)
11. van Oosterom, P.: Research and development in 3D cadastres. Comput. Environ. Urban Syst. **40**, 1–6 (2013)
12. Karim, H., Rahman, A.A., Jamali, A.: Unified topological framework for retrieving 2D and 3D multi-scale spatial information. Int. Arch. Photogrammetry, Remote Sens. Spatial Inf. Sci. **42**(4), W9 (2018)
13. Aditya, T., Laksono, D., Susanta, F.F., Istarno, I., Diyono, D., Ariyanto, D.: Visualization of 3D survey data for strata titles. ISPRS Int. J. Geo-Inf. **9**(5), 310 (2020)
14. Shojaei, D., Kalantari, M., Bishop, I.D., Rajabifard, A., Aien, A.: Visualization requirements for 3D cadastral systems. Comput. Environ. Urban Syst. **41**, 39–54 (2013)
15. Shojaei, D., Rajabifard, A., Kalantari, M., Bishop, I.D., Aien, A.: Design and development of a web-based 3D cadastral visualisation prototype. Int. J. Digit. Earth **8**(7), 538–557 (2015)

A Novel Model for Detection and Classification Coronavirus (COVID-19) Based on Chest X-Ray Images Using CNN-CapsNet

Dahdouh Yousra(✉), Anouar Boudhir Abdelhakim, and Ben Ahmed Mohamed

List Laboratory, FSTT UAE Tangier, Tangier, Morocco
{aboudhir,mbenahmed}@uae.ac.ma

Abstract. The coronavirus disease 2019 (COVID-19) first emerged in Wuhan China, and spread across the globe with unprecedented effect. It has became a major healthcare challenge threatening health of billions of humans. Due to absence of therapeutic drugs or vaccines for all, discovering this virus in the early stages will help in diagnosis, evaluation and fast recovery using one and most commonly of the key screening approaches being radiological imaging --Chest X-Ray--.

With the advances in Artificial Intelligence algorithms and especially Deep Learning models, and within order to help the radiologists to analyze the vast amount of Chest X-Ray images, which can be crucial for diagnosis and detection of COVID-19. In this paper, we proposed an effective Chest X-Rays image classification model named CNN-CapsNet. Our main idea is to make full use of the merits of these two models: CNN and CapsNet. First, we used a CNN without fully connected layers are used as an initial feature maps extractor. In detail, we have used a pre-trained deep CNN model --VGG19-- that was fully trained on the ImageNet dataset is selected as a feature extractor. Then, we have fed the initial feature maps into a newly designed CapsNet to obtain the final classification result.

The performance of our model was evaluated with the four metrics: Accuracy, Sensitivity, Precision and F1 Score. The result is based on the data available in the repositories of Kaggle and Mendeley. Where the experimental results demonstrate that the proposed method can lead to a competitive classification performance compared with the state-of-the-art methods, as achieved high accuracy of 94%.

Keywords: Coronavirus · COVID-19 · Chest X-Ray · Transfer learning · Deep learning · CNN · VGG19 · CapsNet · Classification

1 Introduction

The new coronavirus 2019 (also known as Covid-19 or SARSCoV-2) is an infectious disease was identified in Wuhan, China [1] in December 2019, and spread globally which is considered on February 11, 2020 [3], by the World Health Organization W.H.O. as a pandemic [2]. Seriousness of this virus is due to its nature of expanding with very fast

J. M. Corchado and S. Trabelsi (Eds.): SSCTIC 2021, LNNS 253, pp. 187–199, 2022.
https://doi.org/10.1007/978-3-030-78901-5_17

pace, where cause howed fever, cough, fatigue, and myalgias in human body during early phases [1], in addition variety of infections related to respiratory system. In the absence of effective treatment or specific therapeutic vaccines the COVID-19 outbreak forms a major concern for the health community [4]; it has become one of the major health care challenges worldwide. Rather, it has become the greatest challenge for humanity in the twenty-first century. Therefore, remote control of the disease, including diagnosis, early quarantine, and follow-up, is imperative. The clinical studies have shown people infected by COVID19 may suffer from pneumonia because the virus spreads to the lungs. Diagnosis is typically associated with both the symptoms of pneumonia and Chest X-Ray tests [6], where using X-Ray imaging techniques is a faster, easier, cheaper and the most common available.

Recently, Artificial Intelligence (AI) using Deep Learning technology has demonstrated great success in the field of medicine for diagnosis purpose [7], due to its accuracy for auto-detection and classification for lung diseases [8, 9], cardiology [10, 11], and brain surgery [12, 13]. As became the core of advanced Computer-Aided Diagnosis (CAD) systems in many medical applications to assist physicians to automatically detect and diagnostic diseases fast and reliable.

All of Deep Learning models, especially the Convolutional Neural Network (CNN), are more applicable for medical image classification specially X-Ray images, it is considered one of the most successful algorithms that have been proved its ability to diagnosis medical images with high accuracy for detecting COVID-19 in recent months. For example, in [14] constructed a CNN architecture based on various ImageNet pre-trained models to extract the high-level features. Those features were fed into a Support Vector Machine SVM as a machine learning classifier in order to detect the COVID-19 cases using Chest X-Ray images. However, CNNs typically require large amount of training data and cannot properly handle input transformations. To overcome the limitations of CNN networks, new structures called CapsNet [15] have recently been proposed, are robust to rotation and affine transformation, and require far less training data, which is the case for processing X-Ray images datasets to detecting COVID-19. The CapsNet uses a group of neurons as a capsule to replace a neuron in the traditional neural network.

To further improve the accuracy of the X-Ray image classification and motivated by the powerful ability of feature learning of deep CNN and the property of equivariance of CapsNet for detection of COVID-19. In this paper, we proposed a new architecture named CNN-CapsNet is composed of two parts. First, we have used a pre-trained convolutional network architecture "VGG19" thanks to the good results that VGG19 shows in different classification problems [16], is fully trained on the ImageNet [17] dataset, and its intermediate convolutional layer is used as an initial feature maps extractor. Then, the initial feature maps are fed into a newly designed CapsNet to classification X-Ray images. Experimental results show that the proposed architecture achieves a more competitive accuracy compared with state-of-the-art methods.

The remainder of this paper is organized as follows. The related works are presented in Sect. 2. Section 3 introduces the theory of CNN, CapsNet and Transfer Learning first, and then describes the proposed method in detail. Section 4 gives our experiment results, including the introduction to the dataset used, and the experiment settings. We have presented the experimental evaluation of the proposed model in the same section.

Finally, Sect. 5 concludes the current study and describes some possibilities for future works.

2 Related Works

The COVID-19 coronavirus is one of the newest viruses on the earth which was announced in late December 2019. Caused a global epidemic problem that could spread quickly from the individual to the individual in the community. Since the vaccine is not yet developed, the right measure to reduce the epidemic is to making a quick diagnostic detect patients at an early stage and to immediately isolate infected patients from the healthy population.

The coronavirus (Covid-19) [18] attracts the attention of many researchers to do further investigation about the symptoms of this viral disease, where X-Ray imaging is one of the most important ways relatively cost effective and commonly done for lung infection detection and is useful for COVID-19 detection [19]. With the rapid development of computer technology, digital image processing technology has been widely applied in the medical field using new technologies and Artificial Intelligence (AI) technologies, such as deep learning for reduce the growing burden on radiologists especially in case the COVID-19 to screen mild cases, triage new infections, and monitor disease advancements. In this section, we describe some remarkable works that have directly influenced the development of our work:

In [20] Narin et al. proposed a deep convolutional neural network based automatic prediction model of COVID-19 with the help of pre-trained transfer models using CXR images. In this research, authors used ResNet50, InceptionV3 and Inception-ResNetV2 pre-trained models to obtain a higher prediction accuracy for a subset of X-Ray dataset.

In another research, Sethy et al. [15] proposed a deep learning-based model to identify coronavirus infections using CXR images. Deep features from CXR images have been extracted and support vect or machine (SVM) classifier is used to measure accuracy, false positive rate, F1 score, Matthew's Correlation Coefficient (MCC) and kappa. It is found that ResNet50 in combination with SVM is statistically superior when compared to other models.

Further, in [21] Bukhari et al. They used ResNet-50 CNN architectures on 278 CXR images, partitioned under 3 groups as normal, pneumonia and COVID-19. This approach has promising results and indicated substantial differentiation of pulmonary changes caused by COVID-19 from the other types of pneumonia.

For help radiologists to automatically identify Covid19 were proposed. In [22] Hemdan et al. they introduced a new deep learning framework; namely COVIDX-Net, allows classifying Covid19 X-Ray images into positive and negative Covid19. Authors used seven DCNN architectures (VGG19, DenseNet121, ResNetV2, InceptionV3, Inception ResNetV2, Xception, and MobileNetV2). They also used a dataset including 50 X-Ray images split into two categories normal and Covid19 positive cases. The obtained results depicted that VGG19 and DenseNet201 architectures have good performances with an F1 score of 89% and 91% for normal and covid19.

Following this context, this article proposes to contribute for early diagnosis of COVID-19 using the state of-the-art deep learning architectures, assisted with transfer learning and CapsNet a new model show superior performance for detection of COVID-19 using Chest X-Ray images. We have used a pre-trained deep CNN model --VGG19-- that was fully trained on the ImageNet dataset is selected as a feature extractor. Then, we have fed the initial feature maps into a newly designed CapsNet to obtain the final classification result.

3 Method

In this section, we provide a brief introduction to CNNs, CapsNets and Transfer Learning will be made first and then the proposed model will be detailed.

3.1 Convolutional Neural Network (CNN)

The convolutional neural network (CNN) [23] is a type of feed-forward artificial neural network based on shared weights [24], which is biologically inspired by the organization of the animal visual cortex. Additionally, CNNs have been the most popular framework in recent years have wide applications in image and video recognition, recommender systems and natural language processing, and widely applied for medical imaging analysis.

As shown in Fig. 1, A CNN consists of an input layer, output layer, as well as multiple hidden layers. The hidden layers of a CNN typically consist of convolutional layers, pooling layers, fully connected layers and normalization layers (ReLU).

The convolutional layer is the core building block of a CNN, which outputs feature maps by computing a dot product between the local region in the input feature maps and a filter. Each of the feature maps is followed by a nonlinear function for approximating arbitrarily complex functions and squashing the output of the neural network to be within certain bounds, such as the rectified linear unit (ReLU) nonlinearity, which is commonly used because of its computational efficiency. The pooling layer performs a down sampling operation to feature maps by computing the maximum or average value on a sub-region. Usually, the fully connected layers follow several stacked convolutional and pooling layers and the last fully connected layer, for example the softmax layer computing the scores for each class.

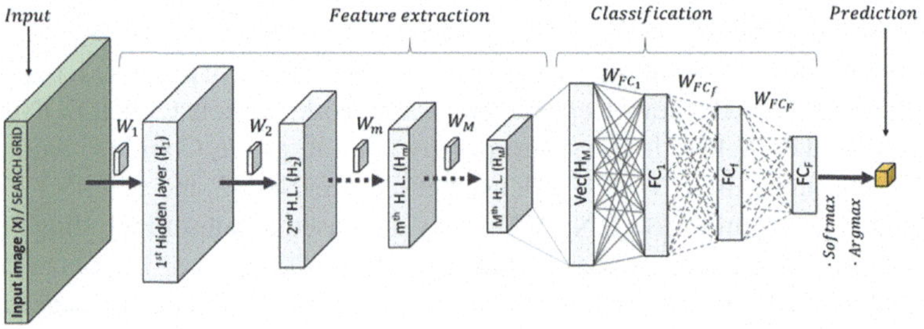

Fig. 1. A figure caption is the convolutional neural network architecture.

CNN has been incorporated into medical imaging analysis and clinical diagnostic. In this paper, we thus proposed CNN network architecture without fully connected layers for extract an initial features map.

3.2 Capsule Networks (CapsNets)

Capsule Networks (CapsNets) represent a completely novel type of deep learning architectures have been introduced by Sabour et al. (2017) to improve and attempt to overcome the limits and drawbacks of CNNs, such as lacking the explicit notion of an entity where not take into account many spatial relations between simpler objects, and losing valuable information about the position of some entity that the network tries to recognize during max-pooling. She is characterized by its strength due to two main ideas:

1. Dynamic routing by-agreement instead of max pooling, and;
2. Squashing, where scalar output feature detectors of CNNs are replaced with vector output capsules.

The CapsNet architecture contains three types of layers: the convolutional layer, the primary capsule layer and the classification capsule layer [16] (Fig. 2).

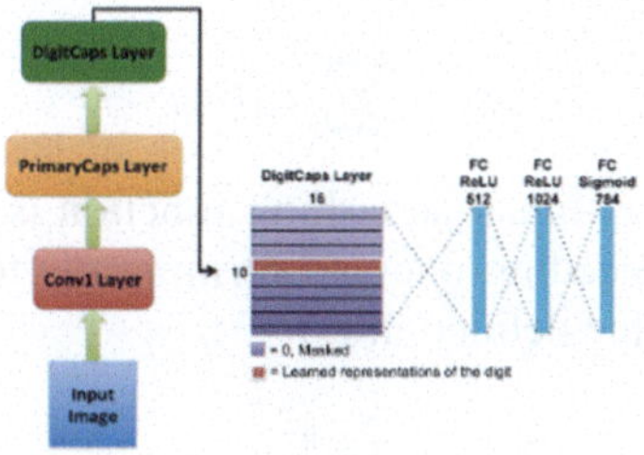

Fig. 2. CapsNet architecture.

A capsule is composed of a group of neurons [16], whose parameters can represent various properties of a specific type of entity that is presented in an image, such as position, size, and orientation. Each capsule is responsible for determining a single component in the object, and all capsules jointly determine the overall structure of the object. Figure 3 illustrates the way that CapsNet routes the information from one layer to another layer by a dynamic routing mechanism [16]. Which means capsules in lower levels predict the outcome of capsules in higher levels and higher-level capsules are activated only if these predictions agree.

Considering *Ui* as the output of lower-level capsule i, its prediction for higher level capsule j is computed as:

$$\widehat{U}_{j|i} = W_{ij} U_i \tag{1}$$

Where *Wij* is the weighting matrix that needs to be learned in the backward pass. Each capsule tries to predict the output of higher-level capsules, and if this prediction

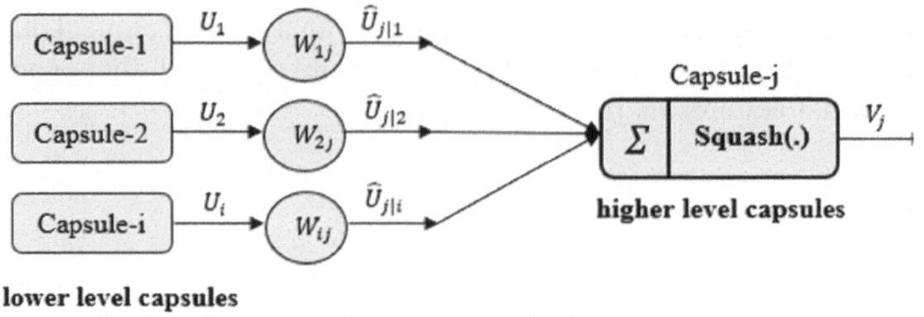

Fig. 3. Connections between the lower level and higher-level capsules.

conforms to the actual output of higher-level capsules, the coupling coefficient between these two capsules increases. Based on the degree of conformation, coupling coefficients *cij* are calculated using the following softmax function:

$$c_{ij} = \frac{\exp(b_{ij})}{\Sigma_k \exp(b_{ik})} \tag{2}$$

Where *bij* is the log probability that whether lower-level capsule i should be coupled with higher level capsule j and it is initially set to 0 at the beginning of the routing by agreement process. Therefore, the input vector to the higher-level capsule j is calculated as follows:

$$s_j = \sum\nolimits_{\mathrm{i}} c_{ij} \hat{U}_{j|i} \tag{3}$$

Finally, the following non-linear squashing function is used to prevent the output vectors of capsules from exceeding one and forming the final output of each capsule based on its initial vector value defined in Eq. (3).

$$v_j = \frac{\|s_j\|^2}{1 + \|s_j\|^2} \frac{s_j}{\|s_j\|} \tag{4}$$

Where *sj* is the input vector to capsule j and *vj* is the output. The log probabilities should be updated in the routing process based on the agreement between *vj* and $\widehat{U}_{j|i}$ using the fact that if the two vectors agree, they will have a large inner product. Therefore, agreement *aij* for updating log probabilities *bij* and coupling coefficients *cij* is calculated as follows:

$$a_{ij} = \widehat{U}_{j|i} v_j \tag{5}$$

Each capsule k in the last layer is associated with a loss function l_k, the loss function l_k, is computed as follows:

$$l_k = T_k \max\left(0, m^+ - v_k\right)^2 + \lambda(1 - T_k \max\left(0, v_k - m^-\right)^2 \tag{6}$$

Where T_k is 1 whenever class k is actually present, and is 0 otherwise. Terms m +, m −, and λ are hyper parameters to be indicated before the learning process. The total loss is simply the sum of the loss of all output capsules of the last layer.

3.3 Transfer Learning

Transfer learning or knowledge transfer is a method of reusing a model trained from large dataset such as ImageNet [25] to solve another problem or task. This technique can be helpful when the available data is small and saves time labeling large amount of data and also building a more generalized model. There are many ways to use the pre-trained models. First, it can be used to classify images from new dataset. Second way, the pre-trained model is used for image pre-processing and feature extraction. Third way, first few layers of the model trained are frozen during training and the last layers are retrained according to the one's dataset. The best state-of-the-art architecture models used as transfer learning are (VGG16) [26], (VGG19) [26] Residual Networks [27].

In our work we have used a pre-trained convolutional network architecture "VGG19" thanks to the good results that VGG19 shows in different problems. The VGG19 model [28] has 19 layers with weights, formed by 16 convolutions and 3 fully-connected (fc) layers and its input is an image of size 224 × 224. The convolutional layers have a small kernel size 3 × 3 with 1 pixel of padding and stride. The network has 5 max-pooling layers with a kernel size of 2 × 2 and stride of 2 pixels. Rectified Linear Units (ReLUs) are used as the non-linear function. After the convolutional part there is a linear classifier with 3 fully-connected (fc) layers and dropout between them, first two fc layers have 4096 features while the last one has only 1000. The last fc layer is followed by a softmax layer with the same number of outputs which gives the probabilities of the input to be long to each of the 1000 classes of the ImageNet dataset (Fig. 4).

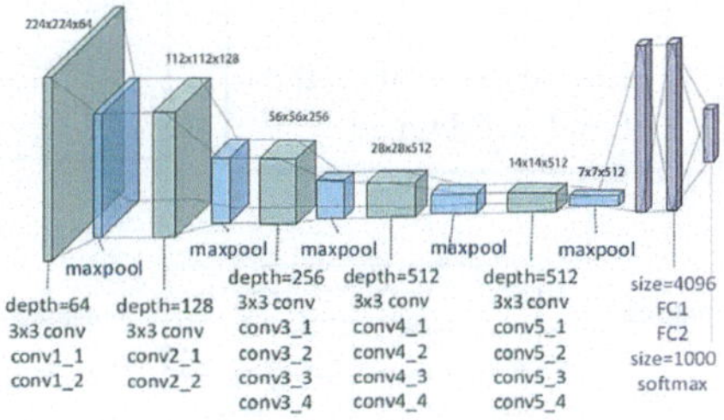

Fig. 4. The network architecture of VGG-19 model [28].

3.4 Proposed Model

In this section, we will highlight the key components of our proposed classification model -- CNN-CapsNet -- for detecting COVID-19 using Chest X-Rays image.

As illustrated in Fig. 5, the proposed model CNN-CapsNet can be devised into two parts: CNN and CapsNet. First, a Chest X-Ray image is fed into a CNN model, and the initial feature maps are extracted from the convolutional layers. Then, the initial feature maps are fed into CapsNet to obtain the final classification result. The main design principles of our model are as follows:

(1) Collecting the Chest X-Ray images for the dataset from COVID-19 Positive cases with Normal and Viral Pneumonia images.

(2) The image preprocessing, such as reading images, resizing images to 224 × 224 and apply One Hot Encoding on the labels of dataset.
(3) Splitting the dataset into two sets: a training set and a test set.
(4) Extracting prominent features from images using the CNN network pre-trained -VGG19-.
(5) Using dropout to reduce overfitting and local response normalization [25] to reduce error rates.
(6) Classifying the Chest X-Ray image using CapsNet.
(7) Choosing the proper learning rate and other hyper-parameter.
(8) Apply data augmentation methods using Keras ImageDataGenerator during training.
(9) Evaluating the performance of our model on the test data.

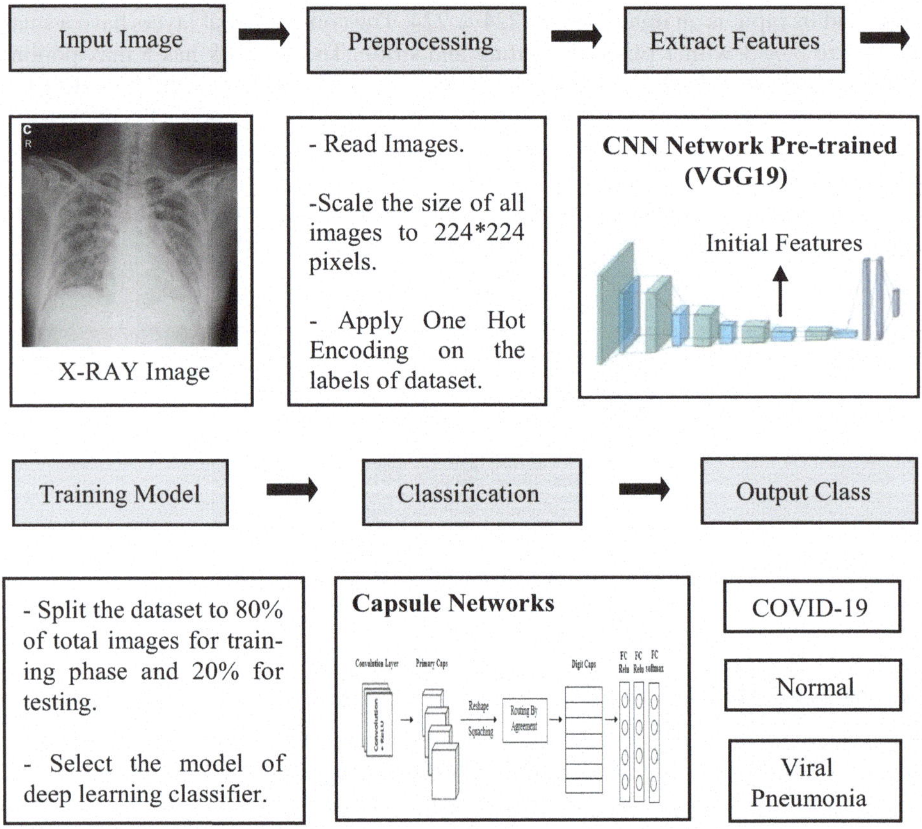

Fig. 5. The architecture of the proposed classification method.

Preprocessing: It is a common practice to perform several simple preprocessing steps before attempting to generate features from data. In this work, the all images were resized to 224 × 224 pixels to make the computation and training faster. Then the whole dataset is normalized using zero mean and unit standard deviation.

Feature Extractor Training Based on CNN Networks Pre-trained (VGG19): In our case, the size of the input image is 224 × 224 pixels, the image will be passed to a convolutional neural network. We used the CNN model (VGG-19) fully trained on the ImageNet dataset as an initial feature map extractors, considering their popularity in different classification problems. We extracted features from an intermediate layer with VGG19 "block4" as the layer of initial feature maps where the size is 14 × 14 × 512, if the input image size is 224 × 224 pixels.

Classification Based on CapsNet: Once the initial feature maps are extracted from the VGG19, it will be passed to a CapsNet classifier for predicting a label for the input patterns. The CapsNet architecture included three layers: one convolutional layer, one Primary Caps layer and one DigitCaps layer.

A 5 × 5 convolution kernel with a stride of 1, and a ReLU activation function is used in the convolution layer. The number of output feature maps is set as 256. The dimension of the capsules in the PrimaryCaps and DigitCaps layers are the vital parameters of the CapsNet. The last layer is a softmax layer that outputs the probability of each class being present.

Once the CapsNet classifier has been well trained, it performs the recognition task and makes new decisions on images.

4 Experiments and Evaluation

4.1 Dataset Description

Since COVID-19 is a new disease, there is no appropriate sized dataset available that can be used for this research work. Therefore, we had to create a dataset by collecting chest X-Ray images from two different open-source image databases: Kaggle [29] and Mendeley [30]. Figure 6 shows few example Chest X-Ray images of all the three cases, normal, viral pneumonia and COVID-19.

The Chest X-Ray images of healthy persons, patients suffering from viral pneumonia have been obtained from Kaggle repository [29]. It contains 1341 Chest X-Ray images of normal cases persons, 1345 images of patients suffering from viral pneumonia. And were collected 1280 X-Ray images of COVID-19 infected patient's cases from Mendeley repository [30]. In this work, we collected a total of 3966 images from these two sources. We then resized all the images to the dimension of 224 × 224 pixels to fit our model.

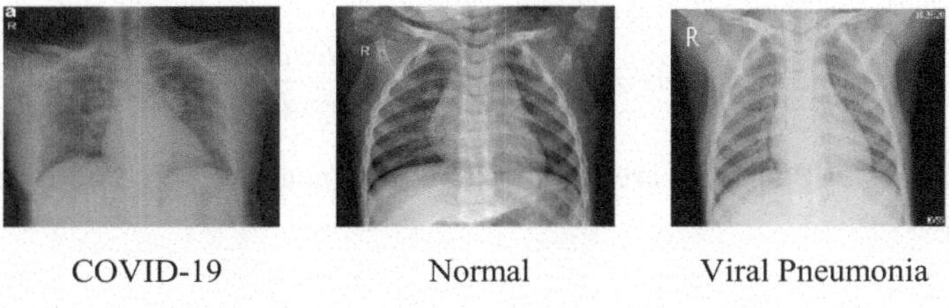

Fig. 6. A sample of X-ray images dataset for COVID-19 patients, normal cases, and viral pneumonia cases.

4.2 Data Augmentation

Data augmentation is an AI method for increasing the size and the diversity of training sets. Data augmentation methods are commonly used in deep learning to address class imbalance problems, reduce overfitting problem, and improve convergence, which ultimately contributes to better results.

In this work, due to the lack of a larger number of available samples, data augmentation using Keras ImageDataGenerator during training. The transformations that employed include random rotation of the images, horizontal flips, scaling, zooming. Data augmentation improves the generalization and enhance the learning capability of the model [31].

4.3 Evaluation Metrics

In order to evaluate of our model we use the most common performance metrics such as accuracy, sensitivity/recall, precision, F1- score [32]. And they are presented from Eq. (7) to Eq. (10). Accuracy of a method determines how correct the values are predicted. It is simply measures how often the classifier makes the correct prediction. It's the ratio between the number of correct predictions and the total number of predictions.

$$Accuracy = \frac{TP + TN}{TP + TN + FP + FN} \tag{7}$$

Recall or Sensitivity or True Positive Rate shows how many of the correct results are discovered. It's the ratio between the true positive values of prediction and the summation of predicted true positive values and predicted false negative values.

$$Recall/\text{Sensitivity} = \frac{TP}{TP + FN} \tag{8}$$

Precision determines the reproducibility of the measurement or how many of the predictions are correct. It's the number of correct positive results divided by the number of positive results predicted by the classifier.

$$Precision = \frac{TP}{TP + FP} \tag{9}$$

F1-score uses a combination of precision and recall to calculate a balanced average result.

$$F1Score = 2 \times \frac{(Recall \times Precision)}{(Recall + Precision)} \tag{10}$$

Where TP, TN, FP and FN are true positive, true negative, false positive, and false negative respectively.

Table 1. Results of our proposed model.

Method	F1-score	Precision	Accuracy	Optimizer
	0.51	0.49	0.51	Adam
CNN-CapsNeT	0.66	0.69	0.67	RMSprop
	0.83	0.87	0.83	RMSprop
	0.94	**0.95**	**0.94**	**SGD**

4.4 Results

In order to train and evaluate our model, we randomly splitting data using technique Train_Test_Split. Where 80% of the images are used for training and 20% are used for testing. We trained our model conducting several experiments for different optimizers such RMSprop [33], Adam [33], and SGD; learning rates and other hyper-parameters; the following table shows the results obtained (Table 1):

We have found the best result for our model with the accuracy = 94% with the following hyper-parameters: Dropout: 0.25, Optimizer: Stochastic Gradient Descent (SGD) with momentum, Learning rate: 1e-5, Batch size: 64, Epochs: 120.

5 Conclusion and Future Work

The emergence and spread of COVID-19 is rapidly becoming a global concern as it has highlighted the need for an early, rapid, sensitive, and computer-aided (CAD) diagnostic tool to help diagnose COVID-19 quickly and accurately at low cost. With the aim of reducing the risk of infection spreading and relieving pressure on the healthcare system.

In this paper, we developed a novel system based with deep learning and transfer learning using Chest X-ray images to predict COVID-19 patients automatically and help doctors to make decisions in clinical practice. The proposed architecture consists of two parts: CNN and CapsNet. The CNN part is transferring the original Chest X-ray images to the original feature maps. In addition, the CapsNet part converts the original feature maps into various levels of capsules to obtain the final classification result.

Through various experiments, the results demonstrate the effectiveness of our model --CNN-CapsNet-- where he achieved achieves high performance about 94% of accuracy. And there is still room be improved if we provide more data using larger datasets, in addition, exploit the potentials of capsules and achieve state-of-the-art in the future work, and investigate effects of having more capsule layers on the classification accuracy.

References

1. Huang, C., Wang, Y., Li, X., et al.: Clinical features of patients infected with 2019 novel coronavirus in Wuhan, China. Lancet 2020. **8**(5), 475–481 (2020). https://doi.org/10.1016/S2213-2600(20)30079-5

2. Yang, W., Cao, Q., Qin, L., et al.: Clinical characteristics and imaging manifestations of the 2019 novel coronavirus disease (COVID-19): a multi-center study in Wenzhou city, Zhejiang, China. J. Infect **80**(4), 388-393. (2020). https://doi.org/10.1016/j.jinf.2020.02.016
3. Sohrabi, C., et al.: World Health Organization declares global emergency: a review of the 2019 novel coronavirus (COVID-19). Int. J. Surg. **76**, 71–76 (2020)
4. Rothe, C., et al.: Transmission of 2019-nCoV infection from an asymptomatic contact in Germany. N. Engl. J. Med. **382**(2020), 970–971 (2020). https://doi.org/10.1056/nejmc2001468
5. Harel, D. (ed.): First-Order Dynamic Logic. LNCS, vol. 68. Springer, Heidelberg (1979). https://doi.org/10.1007/3-540-09237-4
6. Franquet, T.: Imaging of pneumonia: trends and algorithms. Eur. Respir. J. **18**(1), 196–208 (2001)
7. Bakator, M., Radosav, D.: Deep Learning and medical diagnosis: a review of literature. Published: 17 August 2018
8. Liang, C.H., Liu, Y.C., Wu, M.T., Garcia-Castro, F., Alberich-Bayarri A., Wu, F.Z.: Identifying pulmonary nodules or masses on chest radiography using deep learning: external validation and strategies to improve clinical practice. Clin. Radiol. **75**(1), 38–45 (2020)
9. Hussein, S., Kandel, P., Bolan, C.W., Wallace, M.B., Bagci, U.: Lung and pancreatic tumor characterization in the deep learning era: novel supervised and unsupervised learning approaches. IEEE Trans. Med. Imaging **38**(8), 1777–1787 (2019)
10. Karar, M.E., El-Khafif, S.H., El-Brawany, M.A.: Automated diagnosis of heart sounds using rule based classification tree. J. Med. Syst. **41**(4), 60 (2017)
11. Karar, M.E., Merk, D.R., Chalopin, C., Walther, T., Falk, V., Burgert, O.: Aortic valve prosthesis tracking for transapical aortic valve implantation (in Eng). Int. J. Comput. Assist. Radiol. Surg. **6**(5), 583–590 (2011)
12. Ghassemi, N., Shoeibi, A., Rouhani, M.: Deep neural network with generative adversarial networks pretraining for brain tumor classification based on MR images. Biomed. Signal Process. Control. **57**, 101678 (2020)
13. Rathore, H., Al-Ali, A.K., Mohamed, A., Du, X., Guizani, M.: A novel deep learning strategy for classifying different attack patterns for deep brain implants. IEEE Access **7**, 24154–24164 (2019)
14. Sethy, P.K., Behera, S.K.: Detection of coronavirus disease (COVID-19) based on deep features (2020)
15. Sabour, S., Frosst, N., Hinton, G.E.: Dynamic routing between capsules. In: Proceedings of the 31st Conference on Neural Information Processing Systems (NIPS), Long Beach, CA, USA, 4–9 December 2017, pp. 3859–3869 (2017)
16. Anil kumar, B., Rajesh Kumar, P.: Classification of MR brain tumors with deep plain and residual feed forward CNNs through transfer learning (2019)
17. Russakovsky, A., Deng, J., Su, H.: ImageNet large scale visual recognition challenge. Int. J. Comput. Vis. **115**, 211–252 (2015)
18. Hageman, J.R.: The coronavirus disease 2019 (COVID-19). Pediatr. Ann. (2020)
19. Wong, H.Y.F., et al.: Frequency and distribution of chest radiographic findings in covid19 positive patients. (inEN) (2020). https://doi.org/10.1148/radiol.2020201160
20. Narin, A., Kaya, C., Pamuk, Z.: Automatic detection of coronavirus disease (covid-19) using x-ray images and deep convolutional neural networks. arXiv preprint arXiv:2003.10849 (2020)
21. Bukhari, S.U.K., Bukhari, S.S.K., Syed, A., Shah, S.S.H.: The diagnostic evaluation of convolutional neural network (CNN) for the assessment of chest x-ray of patients infected with covid-19. medRxiv (2020)
22. Hemdan, E.E.D., Shouman, M.A., Karar, M.E.: COVIDX-Net: a framework of deep learning classifiers to diagnose COVID-19 in x-ray images. arXiv preprint arXiv:2003.11055 (2020)

23. Krizhevsky, A., Sutskever, I., Hinton, G.: Imagenet classification with deep convolutional neural networks. In: NIPS (2012)
24. LeCun, Y.: LeNet-5, convolutional neural networks, p. 20 2015. Available online: http://yann.lecun.com/exdb/lenet. Accessed 17 Aug 2018
25. Deng, J., Dong, W., Socher, R., Li, L.-J., Li, K., Fei Fei, L.: ImageNet: a large-scale hierarchical image database. In: 2009 IEEE Conference on Computer Vision and Pattern Recognition, Miami, FL, USA (2009)
26. Simonyan, K., Zisserman, A.: Very deep convolutional networks for large-scale image recognition. arXiv Prepr. arXiv:1409.1556 (2014)
27. He, K., Zhang, X., Ren, S., Sun, J.: Deep residual learning for image recognition. arXiv preprint arXiv:1512.03385 (2015)
28. Zheng, Y., Yang, C., Merkulov, A.: Breast cancer screening using convolutional neural network. In: Conference Paper (2018)
29. Covid-19 Radiography Database. https://www.kaggle.com/tawsifurrahman/covid19-radiography-database.
30. Curated Dataset for COVID-19 Posterior-Anterior Chest Radiography Images (X-Rays). https://data.mendeley.com/datasets/9xkhgts2s6/1.
31. Perez, L., Wang, J.: The effectiveness of data augmentation in image classification using deep learning. arXiv preprint arXiv:1712.04621 (2017)
32. Goutte, C., Gaussier, E.: A probabilistic interpretation of precision, recall and f-score, with implication for evaluation. In: Losada, D.E., Fernández-Luna, J.M. (eds.) ECIR 2005. LNCS, vol. 3408, pp. 345–359. Springer, Heidelberg (2005). https://doi.org/10.1007/978-3-540-31865-1_25
33. Ruder, S.: An overview of gradient descent optimization algorithms∗. arXiv:1609.04747v2 [cs. LG] 15 Jun 2017

Distributed Platform for the Extraction and Analysis of Information

Francisco Pinto-Santos[1(✉)], Niloufar Shoeibi[1], Alberto Rivas[1,2], Guillermo Hernández[1,2], Pablo Chamoso[1,2], and Fernando De La Prieta[1]

[1] BISITE Research Group, University of Salamanca, Salamanca, Spain
{franpintosantos,Niloufar.shoeibi,rivis,guillehg,chamoso,fer}@usal.es
[2] Air Institute, IoT Digital Innovation Hub, Salamanca, Spain

Abstract. Information analysis has become a key tool today. Most large companies use generic or custom-developed applications that allow them to extract knowledge from data and translate that knowledge into greater benefit. However, in the field of information ingesting and processing, there are not many generic tools in terms of purpose and scalability to process larger amounts of information or perform the processing tasks faster. In this article, we present a tool designed to perform all kinds of personalized searches, and later, on the information retrieved from the Internet apply different transformations and analysis. The platform that supports the tool is based on a distributed architecture capable of adapting to the automatically available computing resources and guaranteeing optimal performance for these resources, allowing it to scale to various machines with relative easiness. However, in the area of information intake and processing, there are not many generic tools that are purposeful and scalable enough to process larger amounts of information or perform the processing tasks faster. In this article, we present a tool designed to perform all kinds of personalized searches, and later, on the information retrieved from the Internet apply different transformations and analysis. The platform that supports the tool is based on a distributed architecture capable of adapting to automatically available computing resources and ensuring optimal performance for those resources, allowing it to scale to multiple machines with relative ease. The system has been designed, deployed and evaluated successfully, and is presented throughout this document.

Keywords: Big data · Distributed systems · Data analysis · Text analysis · Data ingest · Data mining

1 Introduction

Data has become an integral part of the daily work of most large companies in the most developed countries. Over the last five years, thanks to advances in computing, both at the hardware level (processing and storage capacity) and at the software level (optimisation in techniques for processing data in a distributed

J. M. Corchado and S. Trabelsi (Eds.): SSCTIC 2021, LNNS 253, pp. 200–210, 2022.
https://doi.org/10.1007/978-3-030-78901-5_18

manner -Big Data-), analysing large volumes of information has proven to provide key knowledge for any business. It can therefore be said that data analysis provides a great capacity to considerably improve the profit of companies [8].

But information analysis is not only relevant for companies, it can provide knowledge to any type of user for any purpose (investment, research, etc.). In this sense, the present work is presented, which aims to provide a computer tool capable of extracting information requested by users on any subject.

More specifically, the information will be extracted from a set of relevant sources such as search engines or reputable specialized media, which will then be analyzed with artificial intelligence methodologies to extract the most relevant aspects of the user's search [7]. For example, methodologies are applied to analyse the feeling of the information related to the user's search, relationships with similar terms are analysed, or monetary amounts are analysed if they exist.

The main novelty of the proposed system is the possibility of deploying the system in a simple and innovative software architecture that allows the analysis of large volumes of data, always adapting to the computer resources provided. This will allow any user to access all the functionality of the system, which will be able to obtain all the results of the analysis of their search in a dynamic control panel [2].

The paper is organized as follows. Section 2 reviews related literature. The proposed system is presented in Sect. 3. Then, Sect. 4 presents the evaluation of the presented system. Finally, the conclusions drawn from the results are presented in Sect. 5.

2 Related Work

In the era of Big Data, the Internet has become the main source of data, both in terms of usage and volume and type of information. Although it is true that many companies obtain added value by analyzing their own data, on many occasions, complementing that information with information from the Internet can help to enrich the results [6,9].

Over the last few years, several approaches have appeared in the literature that propose methodologies for the extraction and analysis of information from the Internet, such as [3,11] or [12]. The main source is social networks, such as Twitter, where trends are made by analyzing people's or companies' profiles according to their publications, such as the studies presented in [5] and in [10].

However, there are also studies such as the one presented in [1], where the results of the main search engines are used to profile people and obtain all the results associated with an individual more quickly than a human would do. This type of tool is especially useful, for example, for journalistic research tasks.

Unfortunately, no scientific work has been found that analyzes news hosted on the Internet in order to extract information on a particular topic, although it is true that tools such as Google Trends [4], provide tools that facilitate their analysis. For this reason, this paper focuses on this aspect, trying to automate the analyses that can be applied to news related to the searches carried out by users.

3 Proposed System

The objective of the system is to create an architecture for the intake of data from the Internet. Adapted to the characteristics of the current techniques of data extraction from web pages (crawling and scraping techniques).

To test the functioning of this architecture, a use case is proposed based on the ingesting, processing of news, for its later visualization.

The proposed architecture is based on ETL (Extract Transform Load) and presents characteristics such as flexibility in terms of data scheme, adaptability and scalability. In addition, the architecture is distributed and allows replication of one or more of the processing stages in the ETL.

The main innovation provided by this artwork is the foundation on two pillars that are Apache Kafka (or simply Kafka) and Celery, which allow to generate packages of tasks to be consumed on demand by workers distributed in several machines.

3.1 General Overview

Among this section, the most relevant aspects of architecture will be explained. After that, each part of the system proposed as a case study of architecture will be explained in depth.

As the system is presented, there are 5 different processing stages, which are independent and only linked by the input and output data format.

Due to this, it was decided to use an architecture in the form of ETL, interpreting the obtaining of news as the data loading phase, to which successive transformations are applied, after which they are formatted to be stored.

The innovation proposed by this ETL-based architecture lies in two pillars: Kafka and Celery.

For this purpose, a data flow is designed through Kafka, with which the different elements of the system communicate.

By means of this communication through Kafka, some typical problems of the software life cycle are solved and the system acquires a series of relevant characteristics.

The first is the minimum coupling, since each module is independent from the others and all types of coupling are eliminated. Although the modules communicate by messages, communication between them is "passive", since each module will read messages from an Kafka topic, perform a processing and write the results back to another Apache Kafka topic.

From the communication through Kafka, also derives the high cohesion, because each module has a functionality assigned and is self-contained, making these modules robust, reusable and facilitating the understanding of the code.

As a consequence of the two previous features, the system is also equipped with easy maintenance and replacement.

In relation to the inclusion of Kafka, finally, the capacity of reuse of the processing results is derived, since if it is desired to obtain the results from a stage it is enough to only subscribe to the topic of Kafka where the module overturns the results to obtain them, without necessity to modify configurations nor to make developments or modifications on the present system.

Once the communications are established through Kafka, the second pillar of this architecture is used in each module, which is Celery by RabbitMQ to generate asynchronous tasks within each module to be consumed later by Celery workers.

The way of proceeding inside each module to process a request is the one that can be seen in the Fig. 1, which in each module is developed in the following way: The entry point to the module is a while loop, in which the system remains blocked reading messages from a Kafka topic.

Once a certain number of messages have been read, which is configurable (and is set to 10 in almost all modules), a request packet is generated. With that request packet, a Celery task is generated in a specific queue.

On one or more computers, there are Celery workers, connected and authenticated with RabbitMQ, who when they have processing space within the configuration parameters provided locally on each computer, consume a Celery task.

Once in each worker in a local way, the task is executed, and the results are dumped in another Kafka topic different from the input one.

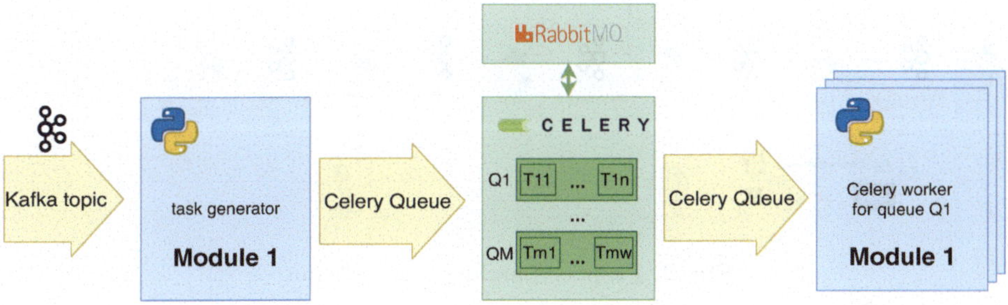

Fig. 1. Process of generating and executing tasks using Kafka, Celery and RabbitMQ

Thanks to this mechanism of generation-consumption of tasks, a series of favorable characteristics for the system are achieved.

Among these characteristics for the system, there is scalability, since this architecture makes it possible to grow in number of machines in a simple and affordable way. This is possible due to the presence in the system of the transparency of replication, since it is fully replicable, or partially replicable only one or some of the modules through Celery's workers.

As a result of replicating the system, the availability of the system increases, because in case of error all the machines have to fall down so that the system stops working. The result is a robust and flexible system because each machine running a Celery worker is the same in terms of the functions it performs, but it executes and consumes tasks as far as its computing capabilities allow, making it an adaptive system.

Finally, it should be noted that it avoids bottlenecks, because if one processing stage is slower than the others, slowing down the system, it can be replicated in several machines.

3.2 News Items Ingest and Visualization Components Overview

Once the theoretical part of the architecture has been presented, each of its components is presented, which can be seen in Fig. 2.

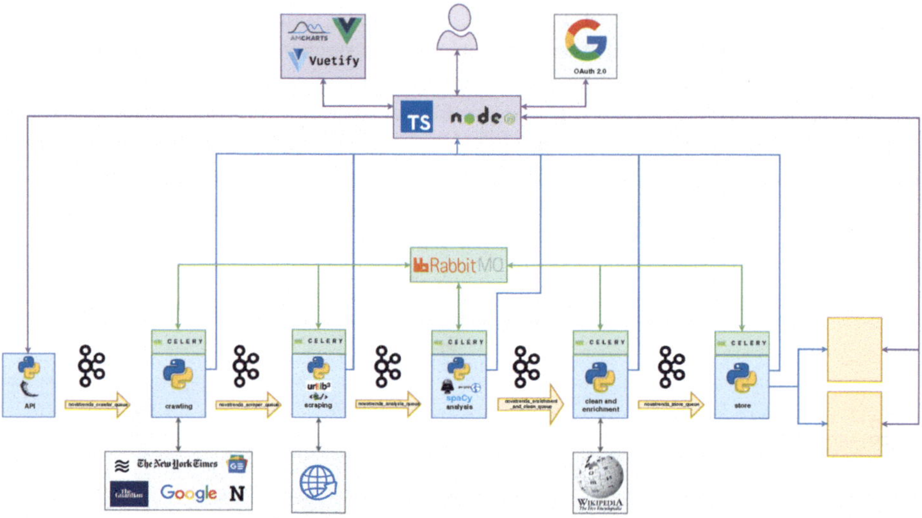

Fig. 2. Proposed architecture schema

API Module. This module is the most differentiated from the rest of the system's modules involved in the ETL architecture, since it serves as an input interface for news search and processing requests.

The implementation of this module has been done in Python, and serves a REST API, through Flask and Flask-RESTPlus, which are the libraries used in this programming language, most used for this purpose.

The API contains a single endpoint, which takes the input request in HTTP format, which contains the search keywords. It then constructs a message in JSON request format (containing the received keywords) and dumps it into a Kafka topic, to which the crawling module subscribes.

Crawling Module. This module is implemented in Python, and houses the responsibility for news ingest for further processing.

Regarding the data sources on which the ingest is made, the following news search APIs were selected, to obtain news in the English language: NewsAPI, Currents API, Google News API, Google Search API, The Guardian API and finally, The New York Times API.

In each task, once obtained the news, it overturns each one to a topic of Kafka designated of exit so that they are read by the module of scraping.

It is important to highlight that this module is the one that limits the number of news items retrieved per request, since the intake is done through public API with usage limitation. Because of this, in each request it recovers between 600 and 700 news items.

Scraping Module. This module is implemented in Python, and has the responsibility to take a news and download its content in HTML, and then parse it to obtain the maximum amount of content of a news.

In the scraping tasks that are generated to be consumed by Celery workers, they arrive among other data of the news from which to obtain the content, the URL of this one. With which, the Python requests library is used to do the HTTP GET request and to obtain the HTML document of the news. After that, the LXML library is used to parse the HTML content and obtain the content of the news itself.

To extract the content of the news, it was decided to develop a single scraper that takes the content of the elements type <p>, since in them lies most of the content of the news.

In each task, once the news content has been downloaded, it is individually dumped into a Kafka topic designated as output to be read by the analysis module.

Analysis Module. This module is implemented in Python, and houses the responsibility to take a news item (whose content has been downloaded and parsed) and perform a series of analyses on it to take.

The analysis tasks generated by this module consist mainly of two analyses:

- Entity Recognition Analysis: This analysis is better known as NER (Name Entity Recognition). It is a technique of NLP (Natural Language Processing), which tries to detect the names of countries, companies, people, geopolitical organizations, etc. In this analysis, the context is very important.
- Sentiment analysis: this is another NLP technique, used to obtain the degree of subjectivity and bias of a text, in order to know if it has positive, negative or neutral connotations.

In each task, once the news has been analyzed, it is poured individually into a Kafka topic designated as output to be read by the processing module.

Processing Module. This module is implemented in Python, and is responsible for taking the NER analysis results of a news item and performing a series of normalizations and enrichments on them, as well as cleaning them up in case they are not correct or can be normalized.

In this module, the tasks generated, specifically perform three functions:

- Standardization of investments: It consists of converting the text fragments that have been detected as investments in the NER analysis, into a numerical quantity and a currency.
- Standardization of locations: This consists of converting location names (countries, cities, regions, etc.) that have been detected as locations in the NER analysis, into a standardized ISO format, such as ISO 3166-1 alpha-2.
- Enrichment of entities: consists of collecting more information about the entities, specifically descriptive texts, alternative names, images, etc.

In each task, once the news has been processed, it is individually dumped into a Kafka topic designated as output to be read by the storage module.

Storage Module. This module is implemented in Python, and holds the responsibility of taking the information of a news and the results of the analysis, to store them.

In MongoDB, all non-entity information is stored, i.e.:

- News: all the information obtained except the content of this one, since it is not legal to store the content as such.
- Result of the sentiment analysis associated with a news item.
- Results of the locations obtained through NER analysis, associated with a news item.
- Results of the investments obtained through NER analysis, associated with a news item.

Regarding the Neo4J part, as mentioned above, a request network is generated in which information regarding NER analysis of people, companies and geopolitical organizations is stored, in the form of a large network associated with a search.

4 Evaluation

To show a couple of study examples, two system searches will be used. A first search on quantum computers and a second on GAIA-X.

To do this, first Fig. 3 and Fig. 4 show the news statistics retrieved in the searches. Specifically, in the first one, 696 news have been obtained, and in the second one, 608 news have been obtained.

To measure the performance of the system, it uses the integral time in processing the search, of which it has the timestamp of beginning (moment in which they begin to look for news) until the end (moment in which it finishes to store the last news). Due to the fact that the news and processing results go through the different processing stages of the system in the form of a stream and it is not possible to measure how long each stage takes.

element group	value group
− category: general ×	
keywords	GAIA-X
completed status	100.00%
− category: status ×	
crawled news	696
downloaded news	696
analysed news	696
cleaned news	696
completed news	696
− category: time ×	
start date	19:18:15 04/06/2020
end date	21:21:09 04/06/2020

Fig. 3. GAIA-X search statistics

element group	value group
− category: general ×	
keywords	Quantum computing
completed status	100.00%
− category: status ×	
crawled news	608
downloaded news	608
analysed news	608
cleaned news	608
completed news	608
− category: time ×	
start date	08:34:52 04/06/2020
end date	09:11:03 04/06/2020

Fig. 4. Quantum computers search statistics

Because of this, measurements have been taken from 21 searches with the system working in distributed mode, resulting in an average time of 02:01:57 per search.

As for the improvement introduced by working with the distributed mode, when introducing a new machine, the time of integral computation of a search can be lowered up to 50%, because as it can be seen in Fig. 5, where the results of the processing of a search are shown using the system displayed in a single machine, the time from the beginning of a search to its end consists of almost 4 h, while in the examples shown above in Fig. 3 and Fig. 4, it can be seen that it is close to two hours.

Fig. 5. Example of search in non-distributed mode

Finally, we proceed to describe the results obtained after the processing stages, which are variable according to the search. Depending on the keywords you can get more or less news and with texts more or less rich in entities such as people, locations, etc. Due to this, from 21 searches with different keywords, among which some are very popular at present like "COVID" to others with less media impact like "GAIA-X", we have obtained on average the results shown in Table 1.

Before finishing, it's important to emphasize that these results, have to be placed in a development and test environment, because they are limited by the crawling module that restricts the number of news items obtained due to its public API. However, this system is capable of managing large amounts of information, which is demonstrated by the load tests, where two machines have

Table 1. Summary of results per search

Result type	Average results per search
News items	670.53
Locations	82.49
Investments	863.26
Entities	2540.31

been tested with 30 concurrent requests, which makes a total of news items higher in 20000 handled by the system, without appreciating major differences in performance.

5 Conclusions and Future Work

In conclusion, thanks to the application of ETLs together with Kafka, Celery and RabbitMQ, it is possible to implement a distributed ETL, which can be replicated totally or partially (only some stages of it) in a transparent way.

Thanks to these features, we can say that the system is focused on flexibility, scalability and adaptability to changes in data schemes, which makes it ideal for use cases focused on the extraction of data from the Internet through crawling and scraping techniques.

For future work, it is proposed to carry out tests between several machines of different capacities, to observe the adaptability of the system and how the performance varies in this situation.

Also, it is proposed to use other messaging and task management systems, to see how these affect the performance of the system.

Acknowledgments. This work has been supported by the project "XAI - XAI - Sistemas Inteligentes Auto Explicativos creados con Módulos de Mezcla de Expertos", ID SA082P20, financed by Junta Castilla y León, Consejería de Educación, and FEDER funds.

References

1. Chamoso, P., Bartolomé, Á., García-Retuerta, D., Prieto, J., De La Prieta, F.: Profile generation system using artificial intelligence for information recovery and analysis. J. Ambient Intell. Hum. Comput. **11**(11), 1–10 (2020)
2. Corchado, J.M., et al.: Deepint.net: a rapid deployment platform for smart territories. Sensors **21**(1), 236 (2021). https://doi.org/10.3390/s21010236. https://www.mdpi.com/1424-8220/21/1/236
3. Germann, J.E.: Approaching the largest 'API': extracting information from the internet with python. Code4Lib J. (39) (2018)
4. Jun, S.P., Yoo, H.S., Choi, S.: Ten years of research change using Google trends: from the perspective of big data utilizations and applications. Technol. Forecast. Soc. Change **130**, 69–87 (2018)

5. Liang, S., Zhang, X., Ren, Z., Kanoulas, E.: Dynamic embeddings for user profiling in Twitter. In: Proceedings of the 24th ACM SIGKDD International Conference on Knowledge Discovery & Data Mining, pp. 1764–1773 (2018)
6. Liu, P., Yi, S.P.: Investment decision-making and coordination of a three-stage supply chain considering data company in the big data era. Ann. Oper. Res. **270**(1–2), 255–271 (2018)
7. Magdy, A., et al.: Geotrend: spatial trending queries on real-time microblogs. In: Proceedings of the 24th ACM SIGSPATIAL International Conference on Advances in Geographic Information Systems. SIGSPACIAL 2016, Association for Computing Machinery, New York, NY, USA (2016). https://doi.org/10.1145/2996913.2996986
8. Rajaraman, V.: Big data analytics. Resonance **21**(8), 695–716 (2016). https://doi.org/10.1007/s12045-016-0376-7
9. Subramanian, N., Abdulrahman, M.D., Chan, H.K., Ning, K.: Big data analytics: service and manufacturing industries perspectives. In: Supply Chain Management in the Big Data Era, pp. 13–23. IGI Global (2017)
10. Verhoeven, B., Daelemans, W., Plank, B.: Twisty: a multilingual Twitter stylometry corpus for gender and personality profiling. In: Calzolari, N., et al. (eds.) Proceedings of the 10th Annual Conference on Language Resources and Evaluation (LREC 2016), pp. 1–6 (2016)
11. Viji Amutha Mary, A., Kumar, K.S., Sai, K.P.S.: An automatic approach to extracting geographic information from internet. J. Comput. Theor. Nanosci. **16**(8), 3216–3218 (2019)
12. Wu, F., Huang, X., Jiang, B.: A data-driven approach for extracting representative information from large datasets with mixed attributes. IEEE Trans. Eng. Manag. (2019)

RETRACTED CHAPTER: Intelligent Development of Smart Cities: Deepint.net Case Studies

Juan M. Corchado[1,3,4], Francisco Pinto-Santos[1](✉), Otman Aghmou[5], and Saber Trabelsi[2]

[1] BISITE Research Group, University of Salamanca, 37007 Salamanca, Spain
franpintosantos@usal.es

[2] Texas A&M University at Qatar, 23874 Doha, Qatar
saber.trabelsi@qatar.tamu.edu

[3] Air Institute, IoT Digital Innovation Hub, 37188 Salamanca, Spain

[4] Department of Electronics, Information and Communication, Faculty of Engineering, Osaka Institute of Technology, Osaka 535-8585, Japan

[5] Msheireb Downtown Doha, Doha, Qatar

Abstract. Deepint.net is a platform that has been designed to use Artificial Intelligence models for the analysis of datasets and real time data sources, all without programming. This platform can be customized to read any type of data from webs, files, databases, sensors, it can also stream data in real time if needed. This paper presents deepint.net and how it can be used to construct a smart city management platform in a highly attractive, user- friendly and intuitive visualization environment. Deepint.net guides the creation and configuration of algorithms that analyze data optimally. The platform makes it possible to create dashboards for better visualization experience, moreover, they can be easily integrated in any other online application. This paper presents an efficient cyberphysical platform for the smart management of smart cities. It is efficient because it facilitates the implementation of data acquisition and data management methods, as well as data representation and dashboard configuration. The platform allows for the use of any type of data source, ranging from the measurements of multifunctional IoT sensing devices to relational and non-relational databases. It is also smart because it incorporates a complete artificial intelligence suite for data analysis; it includes techniques for data classification, clustering, forecasting, optimization, visualization, etc. It is also compatible with the edge computing concept, allowing for the distribution of intelligence and the use of intelligent sensors. The concept of smart cities is evolving and adapting to new applications; before, the trend was to come up with solutions for megacity management, now the focus has shifted creating intelligent neighbourhoods, districts or territories. In this paper, the platform is presented, and its architecture and functionalities are described.

Keywords: Smart cities · Big data · Data analysis · Data visualization · Edge computing · Artificial intelligence

---

The original version of this chapter was retracted: The retraction note to this chapter is available at https://doi.org/10.1007/978-3-030-78901-5_31

J. M. Corchado and S. Trabelsi (Eds.): SSCTIC 2021, LNNS 253, pp. 211–225, 2022.
https://doi.org/10.1007/978-3-030-78901-5_19

1 Introduction

A smart city is an urban environment that implements disruptive technologies for greater effectiveness, flexibility and sustainability of city operations, improving the quality of life of the citizens. The concept of smart cities has emerged in response to the growing population densities in cities, as it has been estimated that in 2030 there will be 43 megacities; holding more than 10 million residents each. The population density will increase by 30% in most cities and 60% of world population will live in cities [2]. This trend is explained by the fact that highly populated areas foster sustainable economic growth [3].

Many smart city projects, which strive towards social equity, economic growth, sustainability and environmental friendliness, have failed to meet all the objectives they had initially set out for themselves. This is because achieving all of these values at once is rather unrealistic, and should be slowly and carefully worked towards. To prevent the failure of future smart city projects, the design should be based on a modular architecture, with each module targeted at a different service within a smart city, such as e-Health, Smart Grids, Transportation, Traffic Control, etc. The independence of the modules means that they can be added, changed or deleted at any time. This type of design makes smart city development flexible, preventing the loss of resources on unsuccessful projects.

The collection of data is fundamental for any smart city, as it offers valuable information when creating and delivering services to citizens. However, large amounts of raw data must undergo a series of processes before becoming useful and comprehensible information. For this reason, there has been much interest in AI-based data analysis and visualization methodologies over the last years. The objective of the platform proposed in this paper, called Deepint.net, is to facilitate data analysis through the application of numerous methods and algorithms [10], in a semi-automated and assisted manner. Moreover, dynamic dashboards can be created on the platform for smart cities, territories, neighbourhoods, etc.; the graphics make it easier for the user to understand the results of the analyses. The platform may be used by itself and/or in conjunction with other (existing) smart city tools/IoT systems, and it can be integrated easily thanks to the connectors it possesses [6, 9].

Deepint.net can be used in large cities or small territories, its configuration makes it suitable for the management of the data coming from any area, no matter its size. The platform automates all the processes involved in data management, ranging from data ingestion, to processing, analysis, dashboard creation. This platform is easy to use and does not require experts in artificial intelligence, edge computing or machine learning.

Deepint.net offers ease of use to city managers or governors, making it possible for them to perform data analyses and create models without the help of a data analysts. The platform aids the data analysis process at different levels: (i) it is capable of working on its own, given its high computing power, providing access to the most common methods and algorithms, this spares cities additional investment; (ii) it can work with data from multiple sources (files, repositories based on CKAN, relational and non-relational databases, streaming data, multi- functional IoT sensors, social networks, etc.); (iii) it automates data processing mechanisms, meaning that it is not necessary to do any programming (information fusion, data filtering, etc.); (iv) fast decision making thanks to interactive graphics that highlight important results; (v) the platform's dynamic data assimilation and neurosymbolic artificial intelligence systems assists the user or automatically select the correct combination of algorithms to be applied to the data. The workflow of the platform is illustrated in Fig. 1.

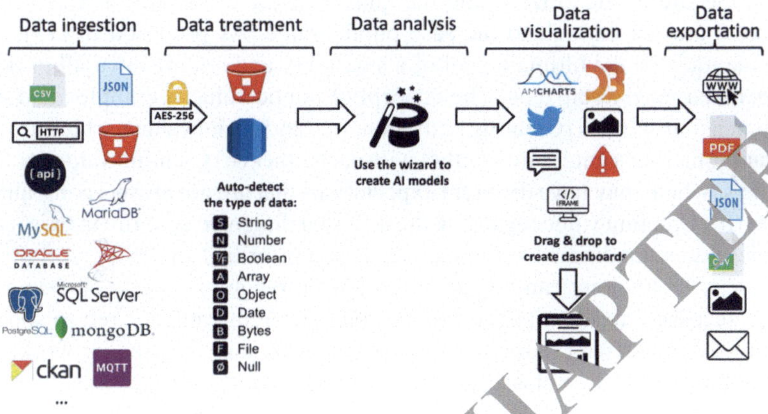

Fig. 1. Deepint.net data analysis flow (from [4])

A number of smart territory projects have already employed this platform, including in locations such as Caldas, Colombia; Panama City, Panama; Istanbul, Turkey; Barcelona and Tenerife in Spain. The state of the art is reviewed in Sect. 2. Section 3 approaches the use of Deepint.net for the management of smart city projects. Some of the dashboards created with this platform are presented in Sect. 4. Finally, the conclusions are drawn and future lines of research are discussed in Sect. 5.

2 Smart Cities

There are many different concepts, technologies and definitions surrounding the smart city paradigm. This has led to considerable confusion as regards to the functions, scope and level of technology implemented by cities that are branded as smart. According to the United Nations Educational, Scientific and Cultural Organization (UNESCO) "All the cities who claim the Smart City status are merely patchworks of opportunistic modernization, which is not always coherent and is sometimes juxtaposed without any real unity of function or meaning" [1]. Striving towards a global standard for smart cities, this section establishes the criteria according to which it will be possible to assess whether a city is smart or not [7, 8]. Thus, this section defines the technologies that should be implemented in any smart city architecture to enable the delivery of basic smart services.

Different definitions in state-of-the-art literature agree that a city is smart [13, 14] if it has the following three features: (i) Environmentally friendly; a city that not only avoids environmental degradation but also actively counteracts it, ensuring environmental conservation. (ii) Effective; refers to the ability of a city to successfully provide public and private services to all citizens, businesses, or non-profit organizations. (iii) Innovative; a city that employs cutting-edge developments and new technologies, continually striving to improve the services it offers to citizens.

These three features, which define a city as smart, can only be achieved through the expert use of new technologies, that converts them into tools of public value for improving the life of the citizens and the quality of the environment. However, these three aspects are not enough to increase public value. As proposed in [15], creating public value must be the ultimate goal of a smart city and requires that all projects and initiatives be targeted at citizens. The concept of public value is complex and includes several dimensions [16]: (i) Creating both economic and social values, which are difficult to unite and sometimes enter into conflict with each other (ii) Creating value for different stakeholders, which may have different expectations that are not always compatible with each other (iii) Creating value regarding the different dimensions of life in the city, which also implies understanding what the real needs and priorities are [8, 17].

Smart City technology can be applied to a wide range of aspects of daily city life. Thus, it is necessary to identify the services that are frequently offered to society in a structured way, classifying them according to their domain. In existing works such as the one published in [18], a classification according to domain is presented:

(i) Natural resources and energy: (a) smart grids: enhance the use of electricity grids by considering consumer habits and providing sustainable, affordable and secure distribution, and safe use [19, 20] (b) street lighting: offer features such as air pollution control or Wi-Fi connectivity that allow to incorporate software for reducing consumption based on a variety of criteria [21] (c) renewable energy sources: exploitation of natural resources that are regenerative or inexhaustible, such as heat, water, or air [22, 23] (d) waste management: collection, recycling, and deposit of waste using methods that prevent pollution [24, 25] (e) water management: analysis and management of the quantity and quality of the water used in agriculture, municipal, or industrial purposes [26] (f) food and agriculture: such as the use of wireless sensor networks for harvest management and knowledge of the conditions in which plants grow [27].

(ii) Transport and mobility: (a) city logistics: efficiently integrating business needs with traffic conditions and geographical and environmental variables [28] (b) mobility information: distribution and use of dynamically selected information, both prior to the completion of the journey and during the journey, with the aim of improving traffic and transport efficiency, as well as ensuring high quality travel experience [29] (c) mobility of people: use of different innovative and sustainable ways to provide transport to people in cities, such as the development of public transport modes and green- powered vehicles, all supported by advanced technologies and the proactive behavior of citizens [30] (d) services employing district information models: whichare domain-specific models that include data models of Building Information Models (BIM), Geographic Information Systems (GIS), and System Information Models (SIM) [31, 32].

(iii) Smart buildings: (a) facility management: cleaning and maintenance of urban facilities [33] (b) construction services: use of services such as power grids, lifts, fire safety systems, and telecommunications [34, 35] (c) housing quality: aspects related to the quality of life in residential buildings, such as comfort, lighting, heating, ventilation, and air conditioning. This category includes everything related to increasing the level of satisfaction of people in their home life [36].

(iv) Daily life: (a) entertainment: ways to stimulate tourism, provide information on leisure events, proposals for free time and nightlife [37] (b) hospitality: the ability of a city to accommodate students, tourists, and other non-residents by offering solutions that meet their needs [38] (c) pollution control: control of emissions and water waste from the use of different devices. Decision-making to improve the air, water, and environmental quality in general [39] (d) public security: protection of citizens and their belongings based on the active involvement of public organizations, the police, and even citizens themselves. Collection and monitoring of information for crime prevention [40, 41] (e) health: prevention, diagnosis, and treatment of diseases supported by information and communication technologies [42, 43] (f) welfare and social inclusion: improving the quality of life by stimulating social learning and participation. Certain groups of citizens require special attention, such as the elderly and persons with disabilities [44] (g) culture: dissemination of information on cultural activities and motivating citizens to get involved in them (h) management of public spaces: care, maintenance, and active management of public spaces to increase the attractiveness of the city and solutions that provide visitors with information on tourist attractions [45].

(v) Government: (a) e-governance: digitization of public administration through the management of documents and formalities using digital tools, in order to optimize work and provide fast and new services to citizens [46] (b) E-democracy: use of information and communication systems for the management of votes [47] (c) transparency: allowing citizens to access official documents in a simple way and decreasing the chances of abuse of authorities who may use the system for their own interests or withhold relevant information from authorities [48].

(vi) Economy and society: (a) innovation and entrepreneurship: measures to promote innovation systems and urban entrepreneurship, for example, by using incubators [49] (b) cultural heritage management: the use of digital systems can provide visitors to cultural heritage sites with new experiences. Asset information management systems are used for performing maintenance in historic buildings [11] (c) digital education: extensive use of methodologies and digital tools in schools [12, 13] (d) human capital management: policies that improve human capital investments and attract and retain talent, avoiding the flight of human capital [14, 15].

Although there are more classifications, the one presented above is the most complete one in terms of domains and subdomains. Given the great diversity of existing services and the continuous appearance of new ones, the proposed platform should be used as a basis for deploying new services, capable of supporting the heterogeneous set of technological solutions.

3 Building Smart City Control Systems

Developing a smart city is normally very expensive, time-consuming and takes a lot of careful planning. To facilitate this task there are multiple tools available for data analysis and dashboard construction, as shown in [4] and [5]. Deepint.net is a platform that aims to make smart city development less costly, facilitating all aspects of data management, processing and visualization. Deepint.net is deployed in a self- adapting cloud environment and enables non-expert users to employ artificial intelligence methodologies (random forest, neural networks, etc.) in data analysis.

Deepint.net has wizards that assist the user in the building of the data processing models, automating most of the process and selecting the correct set of methodologies for the user, in accordance with the data to be analysed. Moreover, the wizard indicates the user how to import the data to the platform, how to manage the data, apply a model and, lastly, extract and assess the results. One of the characteristics of the platform are dynamic and reusable dashboards that can be shared among users and exported in different formats to other smart city tools.

Regarding the type of support in which the original data is found, the following are allowed: i) direct sources: CSV or JSON files imported from local files, or URLs ii) derived sources: new data sources obtained from existing data sources iii) databases:; iv) Other services such as AWS S3, CKAN or MQTT, and others. The data that is uploaded on Deepint.net is stored in 'data sources'.

Data sources are fundamental because all the functionalities of the platform revolve around working with data. To export the data from a source to the platform, the user must define the source of the original data, and fill in the required information so that the data is imported. For example, if the original source is a database, the user must provide the host, the username, the password, the database name and the SQL query to be executed. In case the source is dynamic, i.e. new data is constantly being added to the database, the user must indicate how often they want the source to be updated with the new information on the database. Lastly, Deepint.net makes it possible for users to encrypt the data on the platform, which provides extra protection that other commercial tools do not offer. The users must be aware that if they select the encryption option, all functionalities will take longer to execute, as the data must be decrypted prior.

Thanks to Deepint.net, many different processes can be executed on the data contained in the sources. The platform automatically detects the type of data and the format (decimal numbers or dates), so the user does not have to worry about defining the data type. Nevertheless, at times the user may wish to manually introduce the data type or change it, when working on specific models or graphs or models. At this point in the flow, the user may consider creating a new data source from the data sources uploaded on the platform, by, for example, filtering the data according to different specificiations or merging two data sources together. The tool also offers the possibility of working with data sources from an API to edit them programmatically.

Deepint.net can be applied in both, classification and regression problems. The tool is equipped with supervised learning techniques and algorithms such as linear and logistic regressions, Decision Tree, Gradient Boosting, Random Forest, Extreme Gradient Boosting, Naive Bayes, Support Vector Machines and unsupervised learning methodologies, such as clustering methods (k-means or DBSCAN, etc.), association rule learning or dimensionality reduction techniques (PCA). Moreover, the platform uses Natural Language Processing is employed to retrieve similar text, as well as classify and cluster text.

The tool offers a wizard that allows to create dynamic and interactive graphics in a very easy way, in a few simple steps, as presented in Fig. 8. The first step involves creating the visualization. The person using the platform must introduce the data source, and then specify the conditions for the visualization (conditions can be nested with AND and OR operations). Then, the source is filtered according to these conditions and the obtained data subset is represented. Alternatively, subsets for visualization may be created at random in cases where pivot charts are used. This is because they are programmed in JavaScript, consuming the user's processing power. The second step involves selecting the chart on which the data is to be represented.

The graphics of the chart must be selected by the user, there are 30 different styles. Similarly, the style in which the data is presented on the chart can be configured, such as the title, legend, series, colors, etc. Once the visualizations have been configured, they can be placed on interactive dashboards. The user may decide on the layout of their dashboard through the drag and drop option. Moreover, in case the user wishes to eliminate any information that is no longer of interest, they may perform another round of filtering to their visualization. This feature is very helpful, as the dynamics of the urban environment may make it necessary to change the data being visualized. Regarding dashboards, it is also possible to add other elements, such as the results of the machine learning models, iframes, images or content through WYSIWYG editors and much more.

Compatibility is an essential feature of any smart city tool. Deepint.net has been designed to facilitate work with other smart city tools, making it very easy to export and exploit the results of the developments made on the platform. Deepint.net allows the export of all data sources to CSV or JSON files, as well as the results of artificial intelligence models or visualizations (such as static PNG images) to, for example, be able to include them in documents or reports. More importantly, sharing the developed dashboards is very straightforward among Deepint.net users and anyone who receives an access link.

4 Smart City Case Studies

This section reviews state-of-the-art smart city platforms for which Deepint.net has been used. There are different parameters by which the value of a city can be assessed according to a series of parameters, these are, potentially, the 10 most relevant parameters: governance, urban planning, public management, technology, environment, internationalprojection, socialcohesion,mobilityand transportation, human capital and economy.

Bearing in mind that the perfect city does not exist, we must strive towards that perfection with all the means at our disposal. Tokyo or New York, for example, should improve their social cohesion. Information and communication technologies must be harnessed to improve it, while maintaining a commitment to the environment.

By 2050, according to the OECD, unless drastic measures are taken, economic and demographic growth will have an unprecedented environmental and social impact. Given that the majority of the population will be concentrated in large cities, it is essential that cities are prepared for more than 2 billion additional inhabitants by 2050.

Moreover, cities must have mechanisms and tools to efficiently manage unexpected events, such as the emergence of a pandemic. Covid-19 has taught us that it is very important to react in time, with the best data, the best time estimation and the use of flexible tools providing decision support through the use of artificial intelligence. Deepint.net is a tool that is being used systematically in numerous projects in the field of smart cities to create decision support systems. In this regard, this section presents some of the products created with deepint.net in cities like Panama and Manizales, to manage the evolution of Covid-19 and to be able to take action on it. Moreover, the platform has been used to manage mobility data in the smart cities of Paris, Melbourne, Barcelona, Msheireb Downtown Doha, Istanbul or Tenerife in different mobility projects.

Panama Smart City: In Panama, the deepint.net platform has been used to build models that facilitate the acquisition of data related to Covid-19 infections and deaths, data from a scientific open data portal built by the Technological University of Panama and other information sources. Several intelligent systems have been created on Deepint.net to predict the evolution of the pandemic using expert mixture models that incorporate recurrent networks, case-based reasoning systems and mathematical pandemic prediction models. A dashboard created in the framework of this project to display the evolution of Covid-19 in the country, is presented here (Fig. 2).

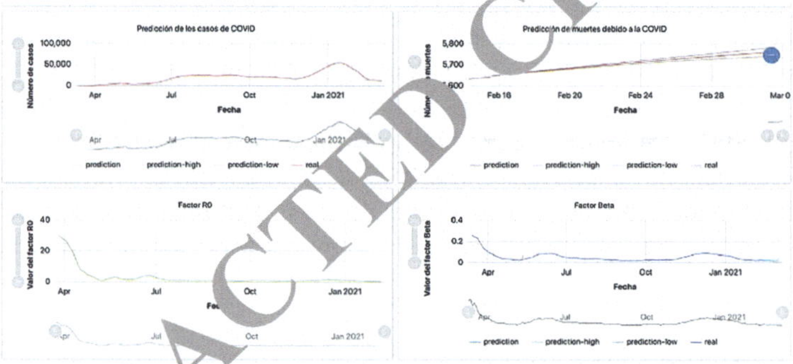

Fig. 2. Prediction of Covid-19 in Panama with Deepint.net

Manizales is a Colombian municipality, capital of the department of Caldas. The government of Colombia has funded the development of a project to build a platform that facilitates the prediction of the evolution of the pandemic. In this case, different models have been used, one of them being a hybrid artificial intelligence system that includes a SIR model, convolutional networks and knowledge about temporal mobility restrictions (Fig. 3).

Paris Smart City. Deepint.net enabled the creation of a system for bike hire management in Paris. The goal of the developed system is to predict areas with the higher bicycle density in real time, using historical data. The process is carried out using the Pareto optimal location algorithm. A smoothed plot was generated by overlapping Gaussian kernels weighted with the estimated availability probability. Figure 4 shows the Heat Map of a probability model for the availability of bikes in the city of Paris.

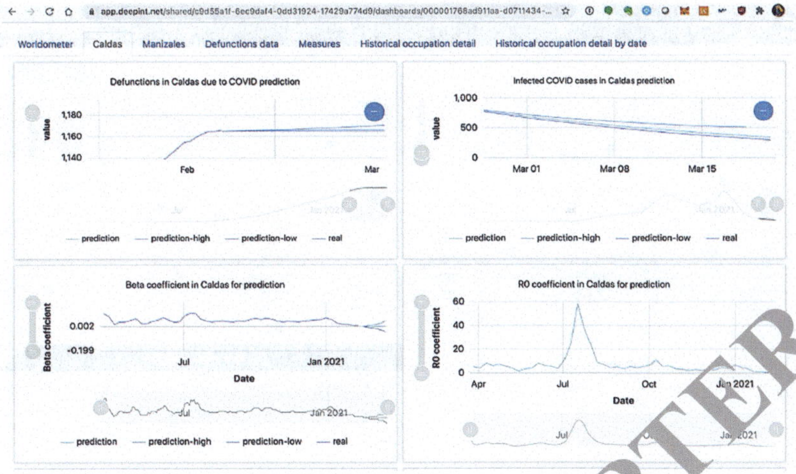

Fig. 3. Covid-19 prediction in Caldas with Deepint.net

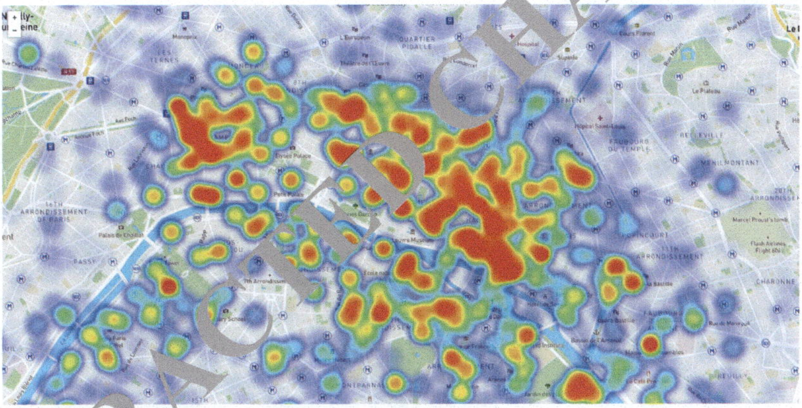

Fig. 4. Heat map of probability model for available bikes in the city of Paris developed with deepint.net (from [illegible])

Deepint.net has used Melbourne's large open data portal to determine the degree of people and vehicle traffic on the streets. Furthermore, Deepint.net has been used as a decision support system to manage the city's traffic efficiently (Fig. 5).

Deepint.net has also been used for open data management in Barcelona (Spain) to monitor both traffic and some user challenges, the degree of traffic on the streets and citizen mobility. The COAPS Project has allowed for the development of this management system for Barcelona (Fig. 6).

In the case of Tenerife, Spain, Deepint.net is used for metro monitoring. Nowadays, it is of vital importance to ensure safety, for which the Deepint.net tool is very valuable, making it possible to capture information about the occupation of the cars and to predict

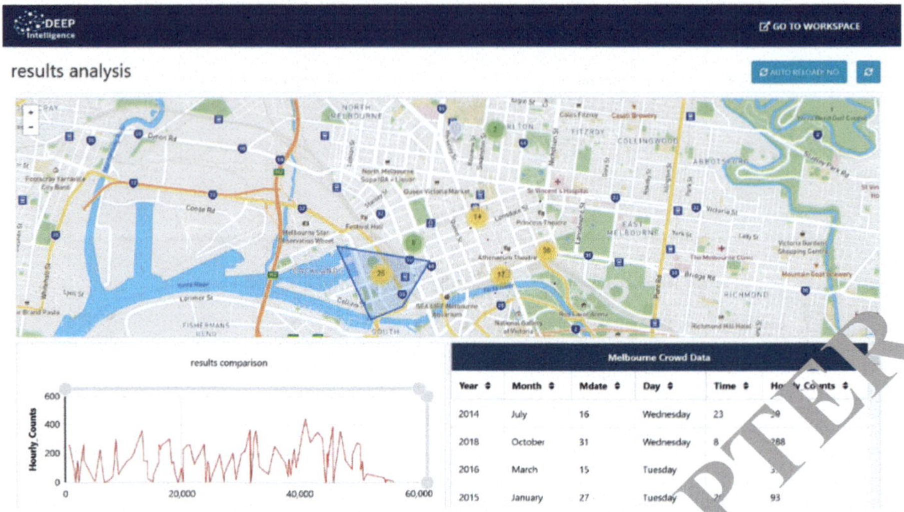

Fig. 5. System dashboard for real-time monitoring of crowds for Melbourne, created using deepint.net.

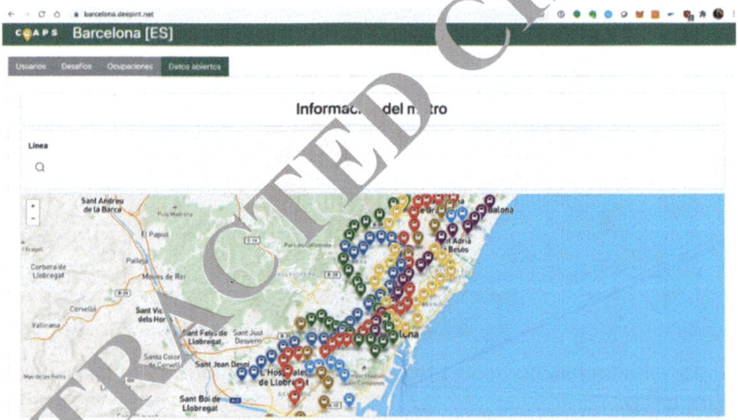

Fig. 6. A dashboard for the management of the smart city of Barcelona, created using deepint.net.

what may be their occupation in the future, so as to avoid risky situations and occupation outside the legal limits (Fig. 7).

Deepint.net has been used in several of the verticals of the Istanbul smart city project constructed in the framework of the project COAPS. Monitoring crowds, waste, natural gas, etc. in real time and taking smart decisions is a challenge, and Deepint.net is an extraordinary solution for the construction of a decision support system that uses real-time data, information and knowledge.

Msheireb Downtown Doha. Msheireb Downtown Doha is one of the most advanced smart city districts in the world. Located in the heart of Doha, Qatar, the city district. Its slogan is “Smart City with Soul”.

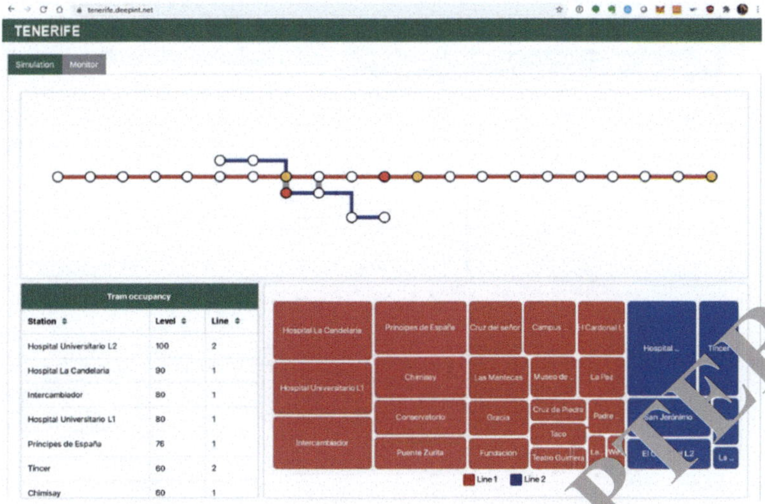

Fig. 7. Information on the occupation of the Tenerife su... using deepint.net.

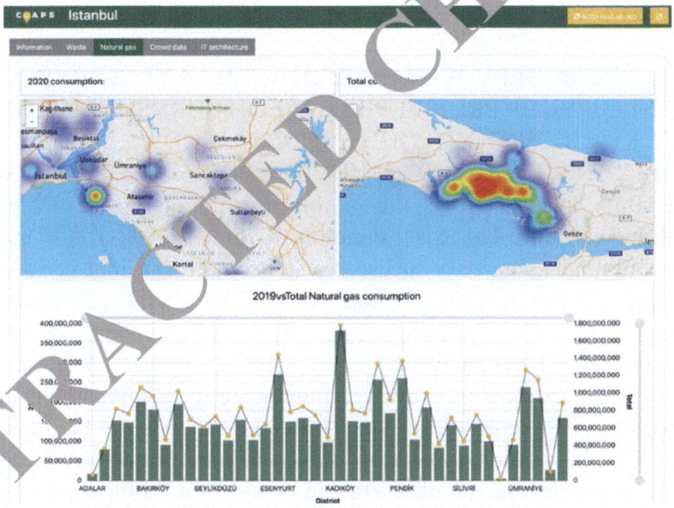

Fig. 8. A dashboard for the management of the natural gas data in Istanbul using deepint.net.

This city is presented as a model of a safe city, able to offer the best guarantees, even in the times of the pandemic. "Msheireb Downtown Doha represents the essence of Qatar through its position in the memory and heart of Qataris," said engineer Ali Al Kuwari, Acting CEO of Msheireb Properties, the developer behind the city district. This city is equipped with state-of-the-art technology, giving it access to the information it needs to ensure efficient, secure and sustainable development. Deepint.net has developed some of the data management and decision-making systems that can be accessed from within its control center (Fig. 9).

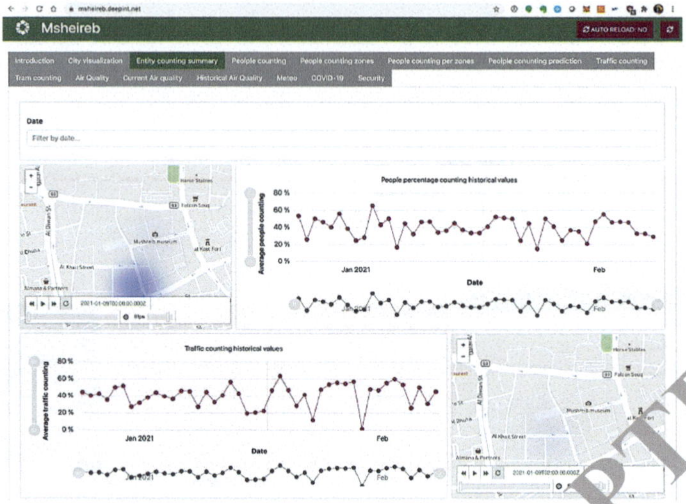

Fig. 9. A dashboard for Msheireb downtown Doha developed with deepint.net.

5 Conclusions

The smart city concept DLPV WR LPSURYH WKH OLYLQJ VWDQGDUGV LQ FLWLHV PDNLQJ WKHP PRUH FRPIRUWDEOH SURVSHURXV JUHHQ VDIH DQG KHDOWK\ + RZHYHU ZH QHHG WRROV WKDW.

ZLOO HQVXUH WKH REMHFWLYHV RI VPDUW FLW\SURMHFWV FDQ EH PHW VXFcessfully. In recent years, the combined use of disruptive technologies, such as IoT, Cloud and AI, have led to the creation of some of WRGD\ V most ZLGHO\ used tools. Deepint.net is an easy to use, versatile platform, that is very efficient in building decision support systems for smart cities. The platform facilitates the use of centralized intelligence and edge architectures, with intelligent nodes, allowing for both decentralized and centralized analyses.

Frequently, a large amount of a city's resources are spent on maintenance and management, especially if these processes are inefficient. Deepint.net facilitates these processes and provides the required tools, algorithms and computing power, leading to much lower spending. Furthermore, the platform can be used for multiple purposes, adapting perfectly to all the dimensions of urban life, such as traffic optimization, citizen opinion analysis from social networks, evaluating and counteracting pollution, etc. Moreover, territories of any size can use the platform, thanks to its scalability feature.

Thus, Deepint.net customizes plans to the needs of each client. Nevertheless, any administrator may create different roles for different employees so that each can access the Deepint.net features. Cities in which this platform would be used would not require experts in programming or data analysis. Any non-expert user is fit to operate the platform, perform the analyses, comprehend the results and extract conclusions.

References

1. Mortoja, M.G., Yigitcanlar, T.: Local drivers of anthropogenic climate change: quantifying the impact through a remote sensing approach in brisbane. Remote Sens. **12**(14), 2270 (2020)
2. Chamoso, P., González-Briones, A., Rodríguez, S., Corchado, J.M.: Tendencies of technologies and platforms in smart cities: a state-of-the-art review. Wirel. Commun. Mob. Comput. ID 3086854 (2018). https://doi.org/10.1155/2018/3086854
3. Butler, L., Yigitcanlar, T., Paz, A.: Smart urban mobility innovations: a comprehensive review and evaluation. IEEE Access **8**, 196034–196049 (2020)
4. Corchado, J.M., et al.: Deepint.net: a rapid deployment platform for smart territories. Sensors **21**(1), 236 (2021). https://doi.org/10.3390/s21010236
5. Yigitcanlar, T., Butler, L., Windle, E., Desouza, K.C., Mehmood, R., Corchado, J.M.: Can building artificially intelligent cities safeguard humanity from natural disasters, pandemics, and other catastrophes? An Urban scholar's perspective. Sensors **20**(10), 2988 (2020). https://doi.org/10.3390/s20102988
6. Chamoso, P., González-Briones, A., de la Prieta, F., Venyaga-Moorthy, K.G., Corchado, J.M.: Smart city as a distributed platform: toward a system for citizen-oriented management. Comput. Commun. **152**, 323–332 (2020)
7. Casado-Vara, R., Martín, Á., del Rey, S., Affes, J.P., Corchado, J.M.: IoT network slicing on virtual layers of homogeneous data for improved algorithm operation in smart buildings. Future Gener. Comput. Syst. **102**, 965–977 (2020)
8. González Bedia, M., Corchado, J.M.: A planning strategy based on variational calcu lus for deliberative agents. Comput. Inf. Syst. **9**, 2–13 (2002)
9. Wang, X., Li, T., Sun, S., Corchado, J.M.: A survey of recent advances in particle filters and remaining challenges for multitarget tracking. Sensors **17**(12), 2707 (2017)
10. González-Briones, A., Chamoso, P., Yoe, H., Corchado, J.M.: GreenVMAS: virtual organization based platform for heating greenhouses using waste energy from power plants. Sensors **18**(3), 861 (2018)
11. Ridel, B., Reuter, P., Laviole, J., Mellado, N., Couture, N., Granier, X.: The revealing flashlight: Interactive spatial augmented reality for detail exploration of cultural heritage artifacts. J. Comput. Cultural Heritage **7**(2), 6 (2014)
12. Kim, T., Cho, J.Y., Lee, B.G.: Evolution to smart learning in public education: a case study of Korean public education. In: Ley, T., Ruohonen, M., Laanpere, M., Tatnall, A. (eds.) OST 2012. IAICT, vol. 395, pp. 170–178. Springer, Heidelberg (2013). https://doi.org/10.1007/978-3-642-37285-8_18
13. Di Mascio, T., Vittorini, P., Gennari, R., Melonio, A., De La Prieta, F., Alrifai, M.: The learners' user classes in the TERENCE adaptive learning system. In: Proceedings of the 12th IEEE International Conference on Advanced Learning Technologies, ICALT 2012, pp. 572–576, Rome, Italy, July 2012
14. Kavaratzis, M.: From city marketing to city branding: towards a theoretical framework for developing city brands. Place Branding Public Diplomacy **1**(1), 58–73 (2004)
15. Chamoso, P., Rivas, A., Rodríguez, S., Bajo, J.: Relationship recommender system in a business and employment-oriented social network. Inf. Sci. **433–434**, 204–220 (2018)
16. Benington, J., Moore, M.H.: Public Value: Theory and Practice. Palgrave Macmillan, Basingstoke (2011)
17. Baccarne, B., Mechant, P., Schuurman, D.: Empowered Cities? An analysis of the structure and generated value of the smart city ghent. In: Smart City, Progress in IS, pp. 157–182. Springer International Publishing (2014). https://doi.org/10.1007/978-3-319-06160-3_8
18. Neirotti, P., De Marco, A., Cagliano, A.C., Mangano, G., Scorrano, F.: Current trends in smart city initiatives: some stylised facts. Cities **38**, 25–36 (2014)

19. Sousa, T., Morais, H., Soares, J., Vale, Z.: Day-ahead resource scheduling in smart grids considering vehicle-to-Grid and network constraints. Appl. Energy **96**, 183–193 (2012)
20. Gomes, L., Faria, P., Morais, H., Vale, Z., Ramos, C.: Distributed, agent-based intelligent system for demand response program simulation in smart grids. IEEE Intell. Syst. **29**(1), 56–65 (2014)
21. De Paz, J.F., Bajo, J., Rodríguez, S., Villarrubia, G., Corchado, J.M.: Intelligent system for lighting control in smart cities. Inf. Sci. **372**, 241–255 (2016)
22. Viitanen, J., Kingston, R.: Smart cities and green growth: Outsourcing democratic and environmental resilience to the global technology sector. Environ. Plann. A **46**(4), 803–819 (2014)
23. González-Briones, A., Prieto, J., De La Prieta, F., Herrera-Viedma, E., Corchado, J.M.: Energy optimization using a case-based reasoning strategy. Sensors **18**(3) (2018)
24. Chamoso, P., De Paz, J.F., Rodríguez, S., Bajo, J.: Smart cities simulation environment for intelligent algorithms evaluation. In: ISAmI 2016: Ambient Intelligence- Software and Applications–7th International Symposium on Ambient Intelligence (ISAmI 2016), vol. 476 of Advances in Intelligent Systems and Computing, pp. 177–187. Springer International Publishing (2016). https://doi.org/10.1007/978-3-319-40114-0_20
25. González-Briones, A., Chamoso, P., Yoe, H., Corchado, J.M.: GreenVMAS: virtual organization based platform for heating greenhouses using waste energy from power plants. Sensors **18**(3) (2018)
26. Barthel, S., Isendahl, C.: Urban gardens, agriculture, and water management: sources of resilience for long-term food security in cities. Ecol. Econ. **86**, 224–234 (2013)
27. Perera, C., Zaslavsky, A., Christen, P., Georgakopoulos, D.: Sensing as a service model for smart cities supported by internet of things. Trans. Emerg. Telecommun. Technol. **25**(1), 81–93 (2014)
28. Nowicka, K.: Smart city logistics on cloud computing model. Procedia-Soc. Behav. Sci. **151**, 266–281 (2014)
29. Hörold, S., Mayas, C., Krömker, H.: Towards paperless mobility information in public transport. In: Kurosu, M. (ed.) HCI 2015. LNCS, vol. 9169, pp. 340–349. Springer, Cham (2015). https://doi.org/10.1007/978-3-319-20901-2_31
30. Jeske, T.: Floating car data from smartphones: what google and Waze know about you and how hackers can control traffic. In: Proceedings of the BlackHat Europe, pp. 1–12, Amsterdam, Netherlands (2013)
31. Krylovskiy, A., Jahn, M., Patti, E.: Designing a smart city internet of things platform with microservice architecture. In: Proceedings of the 3rd International Conference on Future Internet of Things and Cloud (FiCloud 2015), pp. 25–30, Rome, Italy, August 2015
32. Liu, X., Wang, X., Wright, G., Cheng, J.C.P., Li, X., Liu, R.: A state-of-the-art review on the integration of Building Information Modeling (BIM) and Geographic Information System (GIS). ISPRS Int. J. Geo-Inf. **6**(2), 53 (2017)
33. Tobi, S.U.M., Amaratunga, D., Noor, N.M.: Social enterprise applications in an urbanfacilities management setting. Facilities **31**(5/6), 238–254 (2013)
34. Hernandez, L., Baladron, C., Aguiar, J.M., et al.: A survey on electric power demand forecasting: future trends in smart grids, microgrids and smart buildings. IEEE Commun. Surv. Tutorials **16**(3), 1460–1495 (2014)
35. Nguyen, T.A., Aiello, M.: Energy intelligent buildings based on user activity: a survey. Energy Build. **56**(1), 244–257 (2013)
36. Wilson, C., Hargreaves, T., Hauxwell-Baldwin, R.: Smart homes and their users: a systematic analysis and key challenges. Personal Ubiquitous Comput. **19**(2), 463–476 (2014)
37. Gretzel, U., Sigala, M., Xiang, Z., Koo, C.: Smart tourism: foundations and developments. Electron. Mark. **25**(3), 179–188 (2015)

38. Koo, C., Ricci, F., Cobanoglu, C., Okumus, F.: Special issue on smart, connected hospitality and tourism. Inf. Syst. Front. **19**(4), 699–703 (2017)
39. Kumar, P., Morawska, L., Martani, C., et al.: The rise of low-cost sensing for managing air pollution in cities. Environ. Int. **75**, 199–205 (2015)
40. Khan, Z., Pervez, Z., Ghafoor, A.: Towards cloud based smart cities data security and privacy management. In: Proceedings of the 7th IEEE/ACM International Conference on Utility and Cloud Computing, UCC 2014, pp. 806–811, London, UK, December 2014
41. González-Briones, A., Villarrubia, G., De Paz, J.F., Corchado, J.M.: A multi-agent system for the classification of gender and age from images. Comput. Vis. Image Underst. (2018). (in Press)
42. Solanas, A., Patsakis, C., Conti, M., et al.: Smart health: a context-aware health paradigm within smart cities. IEEE Commun. Mag. **52**(8), 74–81 (2014)
43. Black, A.D., Car, J., Pagliari, C., et al.: The impact of ehealth on the quality and safety of health care: a systematic overview. PLoS Med. **8**(1), e1000387 (2011)
44. Hussain, A., Wenbi, R., Da Silva, A.L., Nadher, M., Mudhish, M.: Health and emergency care platform for the elderly and disabled people in the Smart City. J. Syst. Softw. **110**, 253–263 (2015)
45. Borràs, J., Moreno, A., Valls, A.: Intelligent tourism recommender systems: a sur vey. Expert Syst. Appl. **41**(16), 7370–7389 (2014)
46. Clohessy, T., Acton, T., Morgan, L.: Smart city as a service (SCaaS): a future roadmap for e-government smart city cloud computing initiatives. In: Proceedings of the 7th IEEE/ACM International Conference on Utility and Cloud Computing, UCC 2014, pp. 836–841, London, UK, December 2014
47. Ali, S.M., Mehmood, C.A., Khawja, A., et al.: Micro-controller based smart electronic voting machine system. In: Proceedings of the 2014 IEEE International Conference on Electro/Information Technology, EIT 2014, pp. 438–442, Milwaukee, Wis, USA, June 2014
48. Janssen, M., van den Hoven, J.: Big and Open Linked Data (BOLD) in government: a challenge to transparency and privacy? Gov. Inf. Q. **32**(4), 363–368 (2015)
49. Perng, S.Y., Kitchin, R., Donnelly, D.M.: Hackathons, entrepreneurship and the passionate making of smart cities (2017)

Applications of AI systems in Smart Cities (APAISC)

Intelligent System for Switching Modes Detection and Classification of a Half-Bridge Buck Converter

Luis-Alfonso Fernandez-Serantes[1](✉), José-Luis Casteleiro-Roca[1], Paulo Novais[2], and José Luis Calvo-Rolle[1]

[1] CTC, CITIC, Department of Industrial Engineering, University of A Coruña, Ferrol, A Coruña, Spain
luis.alfonso.fernandez.serantes@udc.es

[2] Algoritmi Center, Department of Informatics, University of Minho, Braga, Portugal

Abstract. The present research shows the implementation of a classification algorithm applied to power electronics with the aim of detection different operation modes. The analysis of a half-bridge buck converter is done, showing two different working state: hard-switching and soft-switching. A model based on classification methods through intelligence techniques is implemented. This intelligent model is able to differentiate between the two operation modes. Very good results were obtained and high accuracy is achieved with the proposed model.

Keywords: Hard-switching · Soft-switching · Half-bridge buck · Power electronics · Classification

1 Introduction

Giants steps have been done in the last years to increase efficiency and size reduction in power electronics. The last big move is towards the used of new materials for transistors and diodes: the introduction of wide band-gap materials in power electronics. Silicon carbide (SiC) and Gallium nitride (GaN) are already replacing silicon power devices in this field [2,6], even the production prices of this new technology have been decreasing in the last years, making them more competitive in every sense. Furthermore, with the introduction of these new materials, the use of soft-switching techniques are more attractive due to the better characteristics of them [10,21].

Moreover, and starting slower to take part in this field, is the use of Artificial intelligence (AI) in the design of components [3,24], control strategies [16,23]. AI has a big potential in this field, it helps the designers to select the appropriate components and to, from a plenty of solutions, to choose the components to achieve the best performance.

J. M. Corchado and S. Trabelsi (Eds.): SSCTIC 2021, LNNS 253, pp. 229–239, 2022.
https://doi.org/10.1007/978-3-030-78901-5_20

With the aim to continue the research in the introduction of the AI in power electronics, this paper is focused in the detection of hard- versus soft-switching mode detection. The analysis of a half-bridge configuration in a synchronous buck converter is explained in the case study Sect. 2. The model approach describes how the AI can be used to identify the operation mode. Also, in this Sect. 3 the used data-set and models are described. Then, in the results Sect. 4, the performance of the model is presented and finally, conclusions are drawn, Sect. 5.

2 Case Study

The case study is a half-bridge buck converter, as shown in the Fig. 1, which is being analyzed. This topology is chosen as it is the base component of most of the power converters, such as buck, boost, full-bridge, etc. The half-bridge topology is integrated by two transistors that operate complementary, when one is conduction, the other one is not. Additionally, to filter the pulsed voltage, the switching node (Vsw in Fig. 1) of the half-bridge is connected to a inductor-capacitor filter, which is filtering the high frequency to get a constant output voltage at the load.

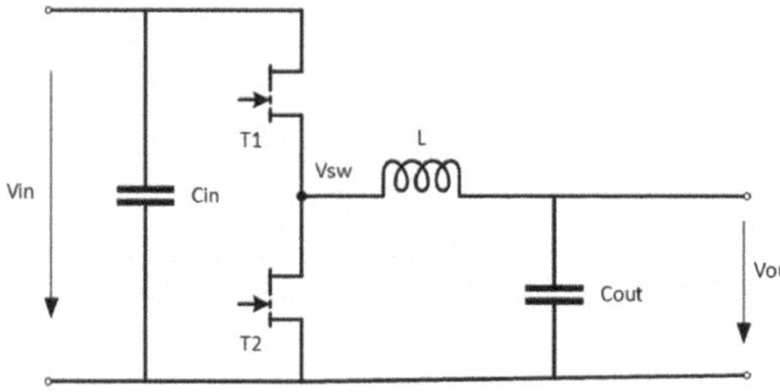

Fig. 1. Synchronous buck converter (half-bridge).

The Fig. 2a shows the voltage and current flowing through the top switch. As it can be seen in the figure, in a first state, the switch is blocking the voltage. When a gate signal is applied, the turn-on commutation starts taking place: the channel of the transistor starts conducting and, therefore, the voltage across the device starts decreasing while the current starts rising. Afterwards, the device is in saturation mode and it is conducting the current while the voltage keeps low, just there is the voltage drop caused by the on-state resistance of the device.

When the voltage in the gate decreases, so the devices is turned-off, the channel of the transistor starts to close, to become high impedance, causing the current to decrease while the voltage rises. Thus, the device blocks the applied voltage. This mode is called Hard-Switching (HS) mode.

On the other hand, the proposed converter can also operated in other different mode, called Soft-Switching (SS) mode. In this case, as shown in the Fig. 2b, the turn-on and/or turn-off transitions are done under a certain condition: Zero

Current Switching (ZCS) or Zero Voltage Switching (ZVS). As in the example, when the gate signal is applied, the voltage starts decreasing while the current keeps at zero. Once the voltage has dropped, the current starts rising until its maximum value. Then, during turn-off, a similar situation is happening: the current falls to zero before the voltage starts rising. These condition of ZCS or ZVS is mainly achieved by the resonance of components in the circuit that makes the voltage/current to drop or additional components are added to the circuit to achieve this conditions, such as snubber capacitors, resonance LC tanks, etc.

The aim of forcing a soft-switching condition is to decrease the power losses during the switching transitions. As it can be seen in the Fig. 2a and 2b that shows the power in hard- and soft-switching modes, thanks to the shift of the current or voltage during the commutation the power losses, which are equal to $P(t) = v(t) \cdot i(t)$, can be significantly reduce.

More in specifically to this research, the resonance components that take place in the proposed circuit are the filter inductor and the parasitic output capacitance of the transistor (Coss), which can be used as a non-disipative snubber, discharging the transistor's channel before the commutation happens and making the circuit to operate in soft-switching mode [20].

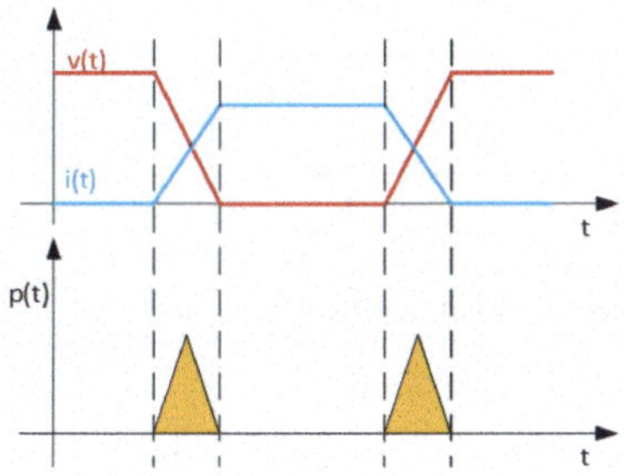

(a) Hard-switching transitions.

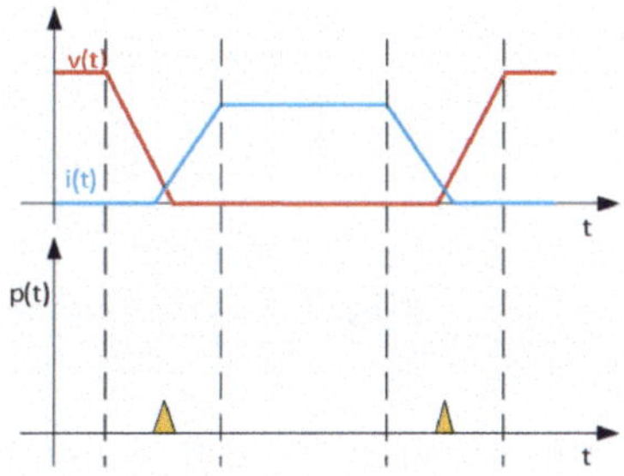

(b) Soft-switching transitions.

Fig. 2. Hard- vs Soft-switching transitions.

As mentioned before, the selection of the filter inductor plays an important role to operate the converter in the different modes. Its dimensioning determines the current ripple value as well as the current flowing through the transistors, and in this way allowing to make ZCS.

Traditionally, in the buck converter as in other topologies, the ripple current in the inductor is tried to be kept low, varying around 5% to 20% of the average current. The design of the inductance is done accordingly to the Eq. 1, where the output is a function of the inductance and switching frequency.

$$I_{ripple} = \frac{(V_{in} - V_{out}) \cdot D}{f \cdot L} \quad (1)$$

where I_{ripple} is the current ripple in the inductor, V_{in} is the input voltage to the circuit, V_{out} is the output voltage from the converter, D is the duty cycle, f is the switching frequency and L is the inductance value of the inductor.

The low ripple current in the inductor makes the output filtering of the voltage easier to the capacitor, as the ripple current absorbed by it is lower and, therefore, the output current is more stable. Nowadays, thanks to the emergence of new materials for the transistors, the traditional way of designing a filter inductor needs to be reconsidered. As explained in [9,14,21], the Triangular Current Mode (TCM) can be beneficial to the converter allowing the soft-switching of the transistors. As shown in Fig. 3, the selection of the filter inductance affects the current ripple and, therefore, the current flowing through the transistors. When the inductance value is high, the current ripple can be kept low, as it have been done until nowadays. On the other hand, if the inductance value is kept low, the current ripple is very high, even crossing zero current, thus allowing to turn-on or turn-off the switching devices in this exact moment. In this way, the switching losses can be highly reduced.

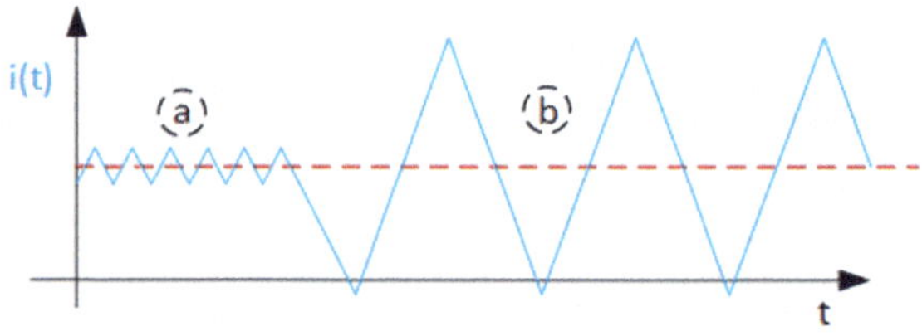

Fig. 3. Current ripple with different filter inductors.

In this research, the soft-switching method is based on a zero-crossing current ripple, operating the converter with a high current ripple at the inductor. The current ripple is always allowed to cross the zero current and slightly flow in the other direction. As explained previously, this operating mode allows a reduction of the switching losses, but as drawback the Root Means Square (RMS) current in the inductor is increased and so in the transistors. This increase of the RMS current causes higher on-state losses. To avoid a extremely high increase of the conduction losses and thanks that the switching losses are highly reduce, the converter would need to operate at higher frequencies. Thus, the increase of the switching frequency allows a reduction of the filter components, main volume parts in the converters, and increase the power densities.

3 Model Approach

The main objective of the research is the detection and classification of the different switching modes, hard- and soft-switching mode, of the half-bridge buck converter. With the aim of classifying the modes, four different techniques have been applied to the obtained data-set.

As shown in Fig. 4, the simulation data is first pre-processed to transform the raw data into more representative parameters that can provide information about the behavior of the converter.

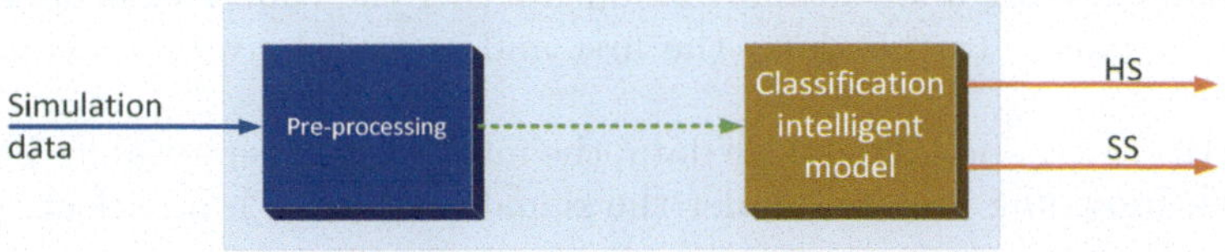

Fig. 4. Model approach.

3.1 Dataset

As mentioned above, the LTSpice simulation tool is used to obtain the data. The circuit has been simulated at different operation points, where the load is varying at a specific output voltage. In order to have a consistent data, the circuit keep unchanged during the whole recompilation of data. In total, 80 simulation results have been obtained, a combination of both hard- and soft-switching data, 50% each type. In this circuit, different variables are measured to obtain the dataset:

- Input voltage: the applied input voltage is kept constant at 400 V.
- Output voltage: the output voltage is controlled to be 200 V and has a ripple of maximum 5%, so varying from 190 V to 210 V.
- Switching node voltage (Vsw node Fig. 4): the voltage at the switching node is a square signal that varies from 0 up to 400 V. The frequency of this signal is also variable, depending on the simulation case, varying from 80 kHz to 2 MHz.
- Inductor current: the current at the inductor has a triangular shape. The current varies accordingly to the load and the switching frequency. In hard-switching, the ripple is kept at 20% of the output current while in soft-switching, the ripple is around 2 times the output ripple, to ensure that the current drops to 0 A.
- Output current: the output current depends on the output load, which is a parameter that changes for each of the simulations. Its value is constant with a ripple of 5%, as the output voltage.

After the data is obtained from the simulation, it is analyzed. From the measured signals, all have been taking into account but taking a special look in the analysis is the switching node voltage, as it reflects how the transition occurs.

From this signal, to the raw data of the switching voltage (Vsw) is used as a base. The first and second derivative are done to the signal. In this way, the on- and off-states of the signal are removed, as their derivative is 0, while the information of the transitions remains.

Moreover, the rising and falling edge of Vsw are separated. These transitions are carefully analyzed, as they provide information of how the commutation is happening. As shown in Fig. 5, the transitions provide information of the rising and falling times (*tr* and *tf*, respectively), which are different when hard- and soft-switching, giving a lot of information whether the converter is operation in one or other mode. To this data, the first and second derivative has also been calculated.

From the rising and falling edge data, the integral of these variables have been calculated, providing the area under the signal (ar and af), a perfect indicator of the transitions.

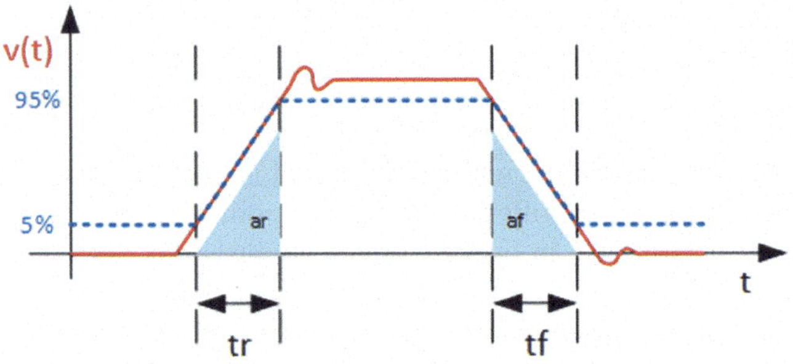

Fig. 5. Rising and falling edge of the switching node voltage, in dashed blue, and the original signal in continuous red.

In total, 8 signals derived from the Vsw are used for each of the simulations: the raw data (red signal in Fig. 5), the first and the second derivatives of the raw data, the rising/falling edge data (dotted blue signal in Fig. 5), the first and second derivatives of rising/falling edge data, the rising edge integral (area at the rising edge, *ar*, in Fig. 5) and the falling edge integral (area at the falling edge, *af*, in Fig. 5).

In order to make the data more significant and easier to analyzed, the following statistics have been calculated for each of the 8 variables: average, standard deviation, variance, co-variance, Root Mean Square (RMS) and Total Harmonic Distortion (THD). Resulting in a matrix of 8×6 for each of the 80 simulations.

3.2 Methods

The classification algorithms used in this research are the Multilayer Perceptron (MLP), the Support Vector Machine (SVM), the Linear Discrimination Analysis (LDA) and the ensemble classifier. These methods are described bellow.

Multilayer Perceptron. A perceptron is an artificial neural network that has only one hidden layer of neurons. When the same structure is used, but with multiple hidden layer, we refer as a multilayer perceptron. The structure is the following: one input layer, which consists of input features to the algorithm, then the multiple hidden layer which have neurons with an activation function, and one output layer, which number of neurons depends on the desired outputs. All these layer are connected in between by weighted connections. These weights are tuned with the aim of decreasing the error of the output [11,23].

The MLP was used for classification, in this case the operation mode. Thus, the output layer provides a discrete output, dividing the data into the two groups. One hidden layer have been used with different number of neurons, that varies from 1 to 10. The Levenberg-Marquardt was used as training function.

Linear Discrimination Analysis. Another method used for classification is the Linear Discrimination Analysis. This method is based on a dimension reduction, projecting the data from a high dimensional space into a low-dimensional space, where the separation of the classes is done. This method uses a weight vector W, which projects the given set of data vector E in such a way that maximizes the class separation of the data but minimizes the intra-class data [12,25]. The projection is done accordingly to the Eq. 2. The separation is good when the projections of the class involves exposing long distance along the direction of vector W.

$$P_i = W^T E_i \tag{2}$$

The LDA provides each sample with its projection and the class label. Two outputs are provided by the analysis, first a gradual decision which is then converted into a binary decision. This method maximizes ratio between the inter-class variance to the intra-class variance, finding the best separation of the classes. The performance of the method increases with the distance between the samples [13,19].

Support Vector Machine. A common used method in classification is the support vector machine, which is a supervised algorithm of machine learning [4,18]. The algorithm tries to find two parallel hyper-planes that maximize the minimum distance between two class of samples [18]. Therefore, the vectors are defined as training instances presented near the hyperplane and the projection of the dataset is done in a high dimensional feature space using a kernel operator.

Ensemble. The term ensemble is used to define multiple classification methods which are used in combination with the aim of improving the performance over single classifiers [1,22]. They are commonly used for classification tasks. The ensemble does a regularization, process of choosing fewer weak learners in order to increase predictive performance [26,27].

3.3 Measurement of the Classification Performance

Once a model was created for each of the previous methods, the models are validated. The predicted outputs from the models are compared with data that has been correctly classified and this comparison is summarize in a confusion matrix. The confusion matrix is a commonly used method to asses the quality of a classifier. In this matrix, the classes serve as column and row labels of a square matrix. The entry to this matrix are the true values and the predicted values, where the true values compound the columns while the predicted ones the rows. As the minimum classification that can be done is a binary classification, the simplest confusion matrix is a 2×2 matrix [7,8].

Usually, the entries to the matrix are two decision classes, positives (P) and negatives (N), and the table entries are called true positives (TP), false positives (FP), true negatives (TN) or false negatives (FN) [5,17].

Once the confusion matrix have been created, there are 5 indicators that are used to analyzed the performance of the models and compare the results between them. These statistics are the followings: SEnsitivity (SE), SPeCificity (SPC), Positive Prediction Value (PPV), Negative Prediction Value (NPV), and Accuracy (ACC) [5,15].

3.4 Experiments Description

In this section, the experiments that had been carried out are explained step by step. From the initial dataset, the data is divided into two sub-sets, one used to train the different models and other set used to validate the model. In this case, the data is divided in 75–25%, where 75% of the data is used to train the model and the rest is used to validate it. The division of the dataset is done randomly.

Once the data is grouped in two sets, the different models are trained:

- MLP: the chosen algorithm for the MLP is the Levenberg-Marquardt backpropagation. This algorithm has been trained from 1 to 10 neurons on the hidden layers.
- LDA: the discriminant type is the regularized LDA, in which all classes have the same co-variance matrix
- SVM: The SVM has been trained using the linear kernel function, commonly used for two-class learning. Also, it trains the classifier using the standardized predictors and centers and scales each predictor variable by the corresponding weighted column mean and standard deviation.
- Ensemble: the used ensemble method is an adaptive logistic regression which is used for binary classification. The number of cycles of the ensemble vary from 10 to 100 in steps of 10. The weak-learners used function is the decision tree.

Then, the validation data is used to check if the models have been correctly trained. The predictions obtained from the models are compared with the verified data, using a confusion matrix and the different statistics are calculated.

4 Results

The obtained results from the models are shown in this section.

As a summary from all the models, the Table 1 shows the statistics obtained from every model, allowing the comparison between them.

As it can be seen in Table 1, the best result is achieved by the MLP7 with an accuracy of 0.97059 and a sensitivity of 0.94444. In general terms, the MLP method provides better performance in many configurations. In contrast, the worst result is obtained by the ensemble, with an accuracy of 0.51471. The other two methods, LDA and SVM provides medium performance, with accuracy of 0.60294 and 0.77941, respectively. In conclusion, the model that achieves a higher accuracy is the MLP7.

Table 1. Summary of the results.

	Sensitivity	Specificity	Negative prediction value	Positive prediction value	Accuracy
MLP1	0.72727	1	0.63636	1	0.81538
MLP2	0.65385	1	0.47059	1	0.73529
MLP3	0.63462	0	0	0.97059	0.62264
MLP4	0.53125	1	0.11765	1	0.55882
MLP5	0.70833	1	0.58824	1	0.79412
MLP6	0.55738	1	0.20588	1	0.60294
MLP7	0.94444	1	0.94118	1	0.97059
MLP8	0.65385	1	0.47059	1	0.73529
MLP9	0.7907	1	0.73529	1	0.86765
MLP10	0.51515	1	0.058824	1	0.52941
SVM	0.69388	1	0.55882	1	0.77941
LDA	0.55738	1	0.20588	1	0.60294
Ensemble10–100	0.50746	1	0.029412	1	0.51471

5 Conclusions and Future Works

This paper proposed a novel method to detect whether a half-bridge buck converter is operating in hard- or soft-switching mode. The method is based on classification intelligent models to predict the operation mode. In this research, a simulated circuit of the half-bridge buck converter is studied to obtain the significant variables for the detection. Then, a model is created to distinguish between the operation modes. The model inputs is simulation data, which are divided into 5 main variables: input voltage, output voltage, switching node voltage, output current and inductor current.

Overall, most of the models are able to predict the operation modes accurately. The best performance is 100% of correct classification, achieve by most of the MLP, SVM and LDA, and the worse with 90%, from the Ensemble models.

The use of the proposed model in this work can be used as a very useful tool for the detection of the operation modes in power converters and, therefore, helping the design and the increase of efficiency in them.

Future work in this area will be oriented to the development a hybrid intelligent model with the aim of improving the classification method up to a 100% accuracy. Then, in a second stage, the design of the circuit will be done with the idea of applying this method with real measured data from a circuit.

References

1. Agrawal, U., et al.: Combining clustering and classification ensembles: a novel pipeline to identify breast cancer profiles. Artif. Intell. Med. **97**, 27–37 (2019)
2. Al-bayati, A.M.S., Alharbi, S.S., Alharbi, S.S., Matin, M.: A comparative design and performance study of a non-isolated DC-DC buck converter based on Si-MOSFET/Si-Diode, SiC-JFET/SiC-Schottky diode, and GaN-transistor/SiC-Schottky diode power devices. In: 2017 North American Power Symposium (NAPS), pp. 1–6 (2017)
3. Aláiz-Moretón, H., Castejón-Limas, M., Casteleiro-Roca, J.L., Jove, E., Fernández Robles, L., Calvo-Rolle, J.L.: A fault detection system for a geothermal heat exchanger sensor based on intelligent techniques. Sensors **19**(12), 2740 (2019)
4. Basurto, N., Arroyo, Á., Vega, R., Quintián, H., Calvo-Rolle, J.L., Herrero, Á.: A hybrid intelligent system to forecast solar energy production. Comput. Electr. Eng. **78**, 373–387 (2019)
5. Calvo-Rolle, J.L., Corchado, E.: A bio-inspired knowledge system for improving combined cycle plant control tuning. Neurocomputing **126**, 95–105 (2014)
6. Casteleiro-Roca, J.L., Javier Barragan, A., Segura, F., Luis Calvo-Rolle, J., Manuel Andujar, J.: Intelligent hybrid system for the prediction of the voltage-current characteristic curve of a hydrogen-based fuel cell. Revista Iberoamericana de Automática e Informática industrial **16**(4), 492–501 (2019)
7. Casteleiro-Roca, J.L., Quintián, H., Calvo-Rolle, J.L., Méndez-Pérez, J.A., Perez-Castelo, F.J., Corchado, E.: Lithium iron phosphate power cell fault detection system based on hybrid intelligent system. Log. J. IGPL **28**(1), 71–82 (2020)
8. Düntsch, I., Gediga, G.: Indices for rough set approximation and the application to confusion matrices. Int. J. Approx. Reason. **118**, 155–172 (2020)
9. Fernandez-Serantes, L.A., Berger, H., Stocksreiter, W., Weis, G.: Ultra-high frequent switching with gan-hemts using the coss-capacitances as non-dissipative snubbers. In: PCIM Europe 2016, International Exhibition and Conference for Power Electronics, Intelligent Motion, Renewable Energy and Energy Management, pp. 1–8. VDE (2016)
10. Fernández-Serantes, L.A., Estrada Vázquez, R., Casteleiro-Roca, J.L., Calvo-Rolle, J.L., Corchado, E.: Hybrid intelligent model to predict the SOC of a LFP power cell type. In: Polycarpou, M., de Carvalho, A.C.P.L.F., Pan, J.-S., Woźniak, M., Quintian, H., Corchado, E. (eds.) HAIS 2014. LNCS (LNAI), vol. 8480, pp. 561–572. Springer, Cham (2014). https://doi.org/10.1007/978-3-319-07617-1_49
11. Jove, E., Casteleiro-Roca, J., Quintián, H., Méndez-Pérez, J., Calvo-Rolle, J.: Anomaly detection based on intelligent techniques over a bicomponent production plant used on wind generator blades manufacturing. Revista Iberoamericana de Automática e Informática industrial **17**(1), 84–93 (2020)
12. Jove, E., Aláiz-Moretón, H., Casteleiro-Roca, J.L., Corchado, E., Calvo-Rolle, J.L.: Modeling of bicomponent mixing system used in the manufacture of wind generator blades. In: Corchado, E., Lozano, J.A., Quintián, H., Yin, H. (eds.) IDEAL 2014. LNCS, vol. 8669, pp. 275–285. Springer, Cham (2014). https://doi.org/10.1007/978-3-319-10840-7_34

13. Jove, E., et al.: Missing data imputation over academic records of electrical engineering students. Log. J. IGPL **28**(4), 487–501 (2020)
14. Jove, E., Casteleiro-Roca, J.L., Quintián, H., Méndez-Pérez, J.A., Calvo-Rolle, J.L.: A fault detection system based on unsupervised techniques for industrial control loops. Expert Syst. **36**(4), e12395 (2019)
15. Jove, E., Casteleiro-Roca, J.L., Quintián, H., Méndez-Pérez, J.A., Calvo-Rolle, J.L.: Virtual sensor for fault detection, isolation and data recovery for bicomponent mixing machine monitoring. Informatica **30**(4), 671–687 (2019)
16. Jove, E., Casteleiro-Roca, J.L., Quintián, H., Simić, D., Méndez-Pérez, J.A., Luis Calvo-Rolle, J.: Anomaly detection based on one-class intelligent techniques over a control level plant. Log. J. IGPL **28**(4), 502–518 (2020)
17. Jove, E., et al.: Modelling the hypnotic patient response in general anaesthesia using intelligent models. Log. J. IGPL **27**(2), 189–201 (2019)
18. Liu, M.Z., Shao, Y.H., Li, C.N., Chen, W.J.: Smooth pinball loss nonparallel support vector machine for robust classification. Appl. Soft Comput. **98**, 106840 (2020)
19. Marchesan, G., Muraro, M., Cardoso, G., Mariotto, L., da Silva, C.: Method for distributed generation anti-islanding protection based on singular value decomposition and linear discrimination analysis. Electr. Power Syst. Res. **130**, 124–131 (2016)
20. Mohan, N., Undeland, T.M., Robbins, W.P.: Power Electronics: Converters, Applications, and Design. Wiley, New York (2003)
21. Neumayr, D., Bortis, D., Kolar, J.W.: The essence of the little box challenge-part a: key design challenges solutions. CPSS Trans. Power Electr. Appl. **5**(2), 158–179 (2020)
22. Quintián, H., Casteleiro-Roca, J.-L., Perez-Castelo, F.J., Calvo-Rolle, J.L., Corchado, E.: Hybrid intelligent model for fault detection of a lithium iron phosphate power cell used in electric vehicles. In: Martínez-Álvarez, F., Troncoso, A., Quintián, H., Corchado, E. (eds.) HAIS 2016. LNCS (LNAI), vol. 9648, pp. 751–762. Springer, Cham (2016). https://doi.org/10.1007/978-3-319-32034-2_63
23. Tahiliani, S., Sreeni, S., Moorthy, C.B.: A multilayer perceptron approach to track maximum power in wind power generation systems. In: TENCON 2019–2019 IEEE Region 10 Conference (TENCON), pp. 587–591 (2019)
24. Liu, T., Zhang, W., Yu, Z.: Modeling of spiral inductors using artificial neural network. In: Proceedings 2005 IEEE International Joint Conference on Neural Networks, vol. 4, pp. 2353–2358 (2005)
25. Thapngam, T., Yu, S., Zhou, W.: DDoS discrimination by linear discriminant analysis (LDA). In: 2012 International Conference on Computing, Networking and Communications (ICNC), pp. 532–536. IEEE (2012)
26. Tomás-Rodríguez, M., Santos, M.: Modelling and control of floating offshore wind turbines. Revista Iberoamericana de Automática e Informática Industrial **16**(4) (2019)
27. Uysal, I., Gövenir, H.A.: An overview of regression techniques for knowledge discovery. Knowl. Eng. Rev. **14**, 319–340 (1999)

A Virtual Sensor for a Cell Voltage Prediction of a Proton-Exchange Membranes Based on Intelligent Techniques

Esteban Jove[1(✉)], Antonio Lozano[2], Ángel Pérez Manso[3], Félix Barreras[2], Ramon Costa-Castelló[4], and José Luis Calvo-Rolle[1]

[1] Department of Industrial Engineering, University of A Coruña, CTC, CITIC, Ferrol, A Coruña, Spain
{esteban.jove,jlcalvo}@udc.es

[2] LIFTEC, CSIC-University of Zaragoza, C/ María de Luna, 10. 50018 Zaragoza, Spain
{alozano,felix}@litec.csic.es

[3] Escuela de Ingeniería de Guipúzcoa, University of the Basque Country, UPV-EHU, Plaza de Europa, 1. 200018 San Sebastián, Spain
angel.perez@ehu.eus

[4] Institut de Robótica i Informática Industrial, CSIC-UPC, C/ Llorens i Artigas 4-6, 08028 Barcelona, Spain
ramon.costa@upc.edu

Abstract. The use of Proton-Exchange Membranes Fuel Cells is presented as a key alternative to face the increasing and concerning problems related to global warming. The international expansion of green policies, has resulted in the need of ensuring their quality and reliability performance. Although fuel cells can get to play a significant role, this technology is still under development, paying special attention to the problems related to gas starvation and degradation. In this context, the present work deals with the virtual sensor implementation of one of the voltage cells present in a stack, whose operation is subjected to several degradation cycles. The proposal predicts indirectly the voltage of one cell from the current state of the rest of the cells by means of an intelligent model.

Keywords: Fuel cell · Virtual sensor · Intelligent modelling · MLP

1 Introduction

During last century, the global society has faced a significant development in terms of technology, industry and life standard, among others. However, this has resulted in an increase of greenhouse gasses emission derived by fossil fuels use. Then, modern societies started to focus their efforts on palliating this critical

J. M. Corchado and S. Trabelsi (Eds.): SSCTIC 2021, LNNS 253, pp. 240–248, 2022.
https://doi.org/10.1007/978-3-030-78901-5_21

problem, and, consequently, most governments promoted green policies [6,18] and, nowadays, there is a legislation whose restriction level tend to be higher [4,17,28].

Given this situation, different alternatives are presented to contribute for preserving the environment by slowing down the climate change [7,8,10]. A common solution consists of the use of renewable energy sources such as solar, wind, hydraulic or even ocean energy [7,11,20]. On the other hand, an interesting research line is focused on Proton-Exchange Membranes Fuel Cells (PEMFC) [19,21]. This technology can be efficiently used in micro-combined heat and power units (CHP) and electric vehicles [3,27].

Although the PEMFC can get to play a significant role, its technology is still under development [2,3]. One of the main research lines is related to the development of devices which works over 100 °C, also denoted as High-Temperature PEM fuel cells (HT-PEMFC). The High-Temperature Proton-Exchange Membrane Power Cell commonly works with a phosphoric acid doped polybenzimidazole (PBI) membrane [30], that raises the allowed temperature to a range between 120 °C and 180 °C. This features give significant advantages compared to the low-temperature PEMFCs [5,30]. First, it is important to emphasise the CO tolerance and the simplification of water management systems [26]. Furthermore, it is demonstrated that the electrochemical kinetics of cathode and anode reactions are enhanced [26]. Finally, these high-temperature devices are simpler and more reliable since sophisticated humidification subsystems can be dispensable. Hence, cooling systems are highly simplified due to the increase in temperature gradient between fuel cell stack and coolant [26].

One of the common problems of power cells is the degradation induced by gas starvation, considerably decreasing the durability of the electrodes [9]. In this context, having an accurate model of a power cell could present a useful tool to determine normal operation of the device. This work deals with the implementation of a virtual sensor to determine the voltage cell, which offers the possibility of estimating the current state of the system. To achieve this goal, an empirical dataset registered from different operating conditions and degradation levels, ensuring a good generalisation [1].

The present document is structured as follows. After this introduction, a brief description of the case of study is presented in next section. Then, the model approach is detailed, followed by the Experiments and Results section. Finally, the reached conclusions are exposed in the last section.

2 Case Study

In spite of the significant breakthroughs made in PEMFC technology, some key aspects regarding performance durability and degradation are still under development [22]. In this sense, several research works are focused on improving lifetime of HT-PEMFCs under variable load conditions [22]. This work deals with a laboratory equipment to test the battery degradation due to gas starvation, whose main features are described in this section.

2.1 Physical System

The starvation experiments have been developed in a stack consisting of five different cells whose gas supply is independent. It has six JP-945 graphite bipolar plates of 280 mm x 195 mm x 5 mm whose temperature can reach up to 200 °C. Furthermore, there are two more plates, made of stainless steal, in charge of connecting the reactant gasses lines (H_2 and O_2). The flowfield geometry of both cathode and anode sides consisted of straight parallel channels with a land-to-channel ratio of 1, as recommended by the manufacturer. The cathode side flowfield geometry consisted of 87 channels with a width of 1 mm and a depth of 2 mm, and a total length of 120 mm. The anode side was formed by 47 channels with a width of 1 mm, a depth of 1.5 mm, and a total length of 210 mm. Figure 1 shows the 3D design of the fuel cell and the physical system.

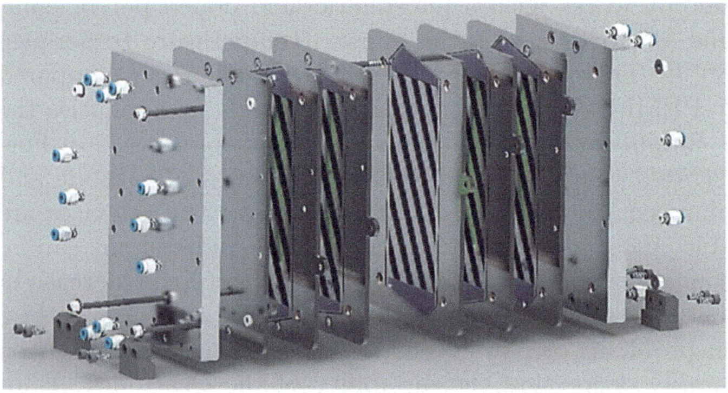

a)

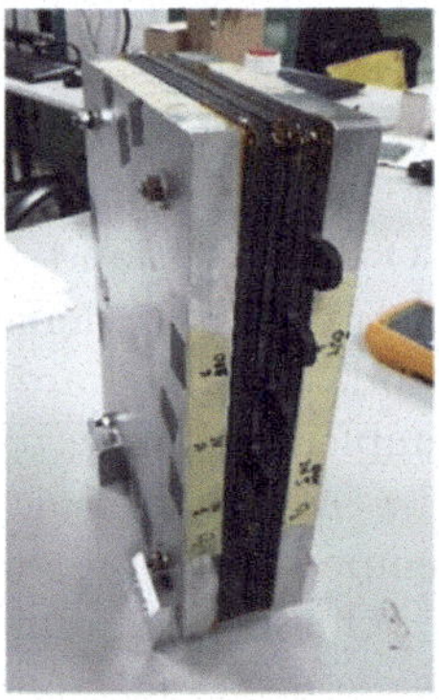

b)

Fig. 1. Fuel cell 3D design (a) and physical system (b)

The stack was assembled inside a greenhouse in the laboratory to guarantee adequate humidity conditions. To decrease the relative humidity inside the greenhouse, dry compressed air was injected using two pipes connected to the main air pressure line. A complete description and illustrative photos of this facility can be consulted in [3].

2.2 Dataset Description

To check the HT-PEMFC degradation performance and durability, the stack was subject to a series of starvation cycles. These tests followed the next structure:

- **Day 1**
    1. The stack is warmed up to 160 °C.
    2. A constant current of 32.7 A is demanded for 3 h.

   3. The flow rate of one cell is reduced a 20% of that corresponding to stoichiometric flow conditions for 30 min.
- **Day 2**
   1. The stack is warmed up to 160 °C.
   2. A constant current of 32.7 A is demanded for 3 h.
   3. The flow rate of one cell is reduced a 50% of that corresponding to stoichiometric flow conditions for 30 min.
- **Day 3**
   1. The stack is warmed up to 160 °C.
   2. A constant current of 32.7 A is demanded for 3 h.
   3. The flow rate of one cell is reduced a 50% of that corresponding to stoichiometric flow conditions for 120 min.
- **Day 4**
   1. The stack is warmed up to 160 °C.
   2. A constant current of 32.7 A is demanded for 1 h.

During the operation, the voltage measured at each of the 5 fuel cells is registered with a sample rate of five minutes. The data available are the following:

- Operation at constant current from days 1, 2, 3 and 4: 130 samples.
- Degradation stage of day 1: 60 samples.
- Degradation stage of day 2: 40 samples.
- Degradation stage of day 3: 55 samples.

3 Model Approach

As the main goal of this research is the implementation of a virtual sensor capable of estimating the voltage value of one cell from the values of the other cells, the topology of the proposal and the technique applied are described in this section.

3.1 Model Topology

The general topology of the proposed model is shown in Fig. 2. This consists in the prediction of voltage in cell 1 from the voltage of cells 2, 4 and 5. Furthermore, to incorporate the system dynamic to the model, the possibility of adding the previous states of the measurements is considered.

In this case, the voltage of cell 3 is not considered because it is the one subjected to a gas starvation during the test. The idea is to check if it is possible to model the behaviour of one cell from data measured on other cells.

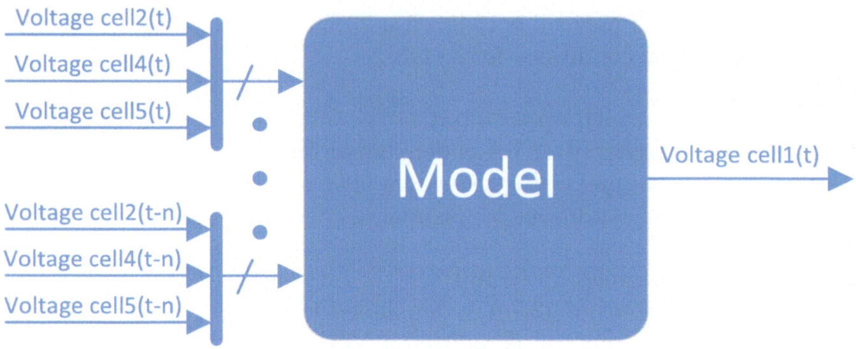

Fig. 2. Model topology

3.2 Multilayer Perceptron

The modelling process is carried out using the Multilayer Perceptron (MLP) technique, which is one of the most used supervised learning ANN due to its simple structure and its robustness. An artificial neural network (ANN) is a system designed to emulate the brain operation in a specific functions of interest or tasks. In this century, ANN have been applied successfully to solve real and challenging problems [16,29]. The MLP presents the structure shown in Fig. 3, which presents one input layer, one output layer and one or more hidden layers. These layers are made of neurons and weighted connections links different layer neurons [12,13,15,23]. The values of the weights are adjusted following an error reduction criteria, being the error the difference between real and estimated output. In the most common configuration, the same activation function is assigned to all neurons from a layer. The activation function can be linear, tan-sigmoid or log-sigmoid.

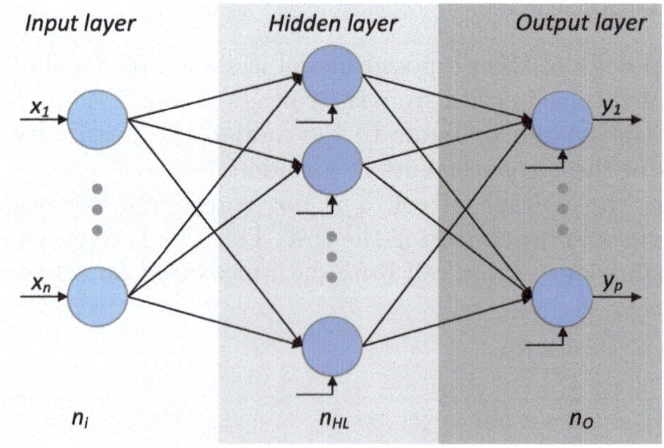

Fig. 3. MLP structure

The employed learning algorithm was Gradient descent, and the algorithm for model training was Levenberg-Marquardt. Also, to measure the network performance, the MAE (Mean Absolute Error) method was applied.

4 Experiments and Results

To achieve the best model performance, a wide range of configurations is checked. The MLP parameters were swept according the following configurations:

- Three MLP topologies are tested: the current state of the inputs, the current and one previous state of the inputs and the current and two previous states of the inputs.
- The activation function of the hidden layer was set to linear, tan-sigmoid or log-sigmoid.
- The number of neurons in the hidden layer was tested from 1 to 50.

From the combination of all these configurations, a total amount of 450 models have been implemented and validated. To validate the proposal, a k-fold cross validation with k = 5 is implemented. The process followed by this method is depicted in Fig. 4.

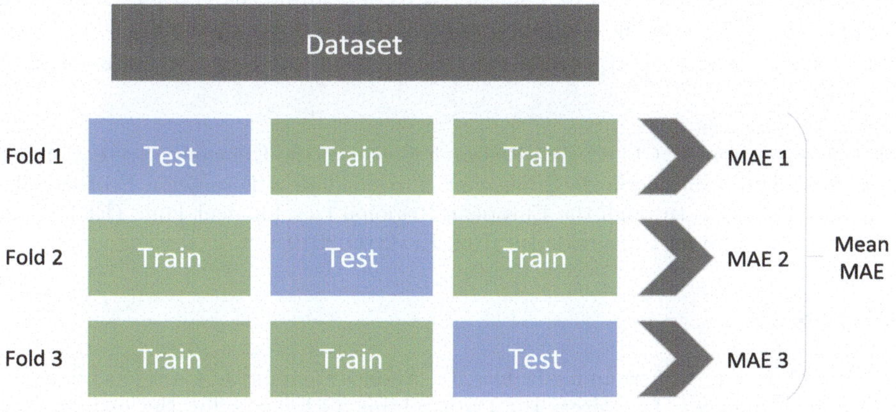

Fig. 4. Implementation of k-fold with k = 3

The results obtained for each MLP topology and their corresponding configuration are summarised in Table 1.

It is important to remark that the best configuration achieves an error of 1.1029 mV in the prediction, which is a significantly low value. This result is reached with a log-sigmoid activation function in the hidden layer and 22 neurons and taking into consideration only the current state of the rest of cells.

Table 1. Best results for each tolopogy

Input states	Activation function	Hidden layer size	MAE (mV)
Current	log-sigmoid	22	1.1029
Current and previous	tan-sigmoid	16	1.3690
Current and two previous	log-sigmoid	24	1.2258

5 Conclusions and Future Works

The present work dealt with the voltage prediction of a cell located in a HT-PEMFC that has been subjected to different gas degradation process. The model proposed gives successful results, especially when the model takes into consideration the current state and previous state of the inputs.

The proposal is presented is a useful tool to estimate the real state of the fuel cell as a previous step to detect anomalous situations. The difference between the real and predicted values is presented as a good indicator about the correct performance of the fuel cell. Furthermore, the model is trained taking into consideration different degradation cycles, so it represents a wide range of cell operation.

As future works, in spite of using data from steady state, the use of data from the degradation cycles could be considered to determine the voltage value at the degraded cell. This model, combined with expert system knowledge [24,25] and imputation techniques [14], could help to determine the degradation level of the fuel cell.

Acknowledgements. CITIC, as a Research Center of the University System of Galicia, is funded by Consellería de Educación, Universidade e Formación Profesional of the Xunta de Galicia through the European Regional Development Fund (ERDF) and the Secretaría Xeral de Universidades (Ref. ED431G 2019/01).

References

1. Alaiz-Moreton, H., Fernández-Robles, L., Alfonso-Cendón, J., Castejón-Limas, M., Sánchez-González, L., Pérez, H.: Data mining techniques for the estimation of variables in health-related noisy data. In: Pérez García, H., Alfonso-Cendón, J., Sánchez González, L., Quintián, H., Corchado, E. (eds.) SOCO/CISIS/ICEUTE -2017. AISC, vol. 649, pp. 482–491. Springer, Cham (2018). https://doi.org/10.1007/978-3-319-67180-2_47
2. Alaiz-Moretón, H., et al.: Bioinspired hybrid model to predict the hydrogen inlet fuel cell flow change of an energy storage system. Processes **7**(11), 825 (2019)
3. Barreras, F., Lozano, A., Roda, V., Barroso, J., Martín, J.: Optimal design and operational tests of a high-temperature PEM fuel cell for a combined heat and power unit. Int. J. Hydrogen Energy **39**(10), 5388–5398 (2014)
4. Baruque, B., Porras, S., Jove, E., Calvo-Rolle, J.L.: Geothermal heat exchanger energy prediction based on time series and monitoring sensors optimization. Energy **171**, 49–60 (2019)

5. Casteleiro-Roca, J.L., Barragán, A.J., Segura, F., Calvo-Rolle, J.L., Andújar, J.M.: Fuel cell output current prediction with a hybrid intelligent system. Complexity **2019**, Article ID 6317270 (2019)
6. Casteleiro-Roca, J.L., Gómez-González, J.F., Calvo-Rolle, J.L., Jove, E., Quintián, H., Gonzalez Diaz, B., Mendez Perez, J.A.: Short-term energy demand forecast in hotels using hybrid intelligent modeling. Sensors **19**(11), 2485 (2019)
7. Casteleiro-Roca, J.L., Javier Barragan, A., Segura, F., Luis Calvo-Rolle, J., Manuel Andujar, J.: Intelligent hybrid system for the prediction of the voltage-current characteristic curve of a hydrogen-based fuel cell. Revista Iberoamericana de Automática e Informática Industrial **16**(4), 492–501 (2019)
8. Casteleiro-Roca, J.L., Jove, E., Sánchez-Lasheras, F., Méndez-Pérez, J.A., Calvo-Rolle, J.L., de Cos Juez, F.J.: Power cell SOC modelling for intelligent virtual sensor implementation. J. Sensors **2017**, Article ID 9640546 (2017)
9. Chen, Y.S., Yang, C.W., Lee, J.Y.: Implementation and evaluation for anode purging of a fuel cell based on nitrogen concentration. Appl. Energy **113**, 1519–1524 (2014)
10. Chukwuka, C., Folly, K.: Batteries and super-capacitors. In: Power Engineering Society Conference and Exposition in Africa (PowerAfrica). IEEE, pp. 1–6, July 2012. https://doi.org/10.1109/PowerAfrica.2012.6498634
11. Fernández-Serantes, L.A., Estrada Vázquez, R., Casteleiro-Roca, J.L., Calvo-Rolle, J.L., Corchado, E.: Hybrid intelligent model to predict the SOC of a LFP power cell type. In: Polycarpou, M., de Carvalho, A.C.P.L.F., Pan, J.-S., Woźniak, M., Quintian, H., Corchado, E. (eds.) HAIS 2014. LNCS (LNAI), vol. 8480, pp. 561–572. Springer, Cham (2014). https://doi.org/10.1007/978-3-319-07617-1_49
12. González, S., Sedano, J., Villar, J.R., Corchado, E., Herrero, Á., Baruque, B.: Features and models for human activity recognition. Neurocomputing **167**, 52–60 (2015). https://doi.org/10.1016/j.neucom.2015.01.082
13. Jove, E., Aláiz-Moretón, H., Casteleiro-Roca, J.L., Corchado, E., Calvo-Rolle, J.L.: Modeling of bicomponent mixing system used in the manufacture of wind generator blades. In: Corchado, E., Lozano, J.A., Quintián, H., Yin, H. (eds.) IDEAL 2014. LNCS, vol. 8669, pp. 275–285. Springer, Cham (2014). https://doi.org/10.1007/978-3-319-10840-7_34
14. Jove, E., et al.: Missing data imputation over academic records of electrical engineering students. Logic J. IGPL **28**(4), 487–501 (2020)
15. Jove, E., Casteleiro-Roca, J.L., Quintián, H., Méndez-Pérez, J.A., Calvo-Rolle, J.L.: A fault detection system based on unsupervised techniques for industrial control loops. Expert Syst. **36**(4), e12395 (2019)
16. Jove, E., Casteleiro-Roca, J.L., Quintián, H., Méndez-Pérez, J.A., Calvo-Rolle, J.L.: Virtual sensor for fault detection, isolation and data recovery for bicomponent mixing machine monitoring. Informatica **30**(4), 671–687 (2019)
17. Jove, E., Casteleiro-Roca, J.L., Quintián, H., Méndez-Pérez, J.A., Calvo-Rolle, J.L.: A new method for anomaly detection based on non-convex boundaries with random two-dimensional projections. Inf. Fusion **65**, 50–57 (2021)
18. Jove, E., Casteleiro-Roca, J.L., Quintián, H., Simić, D., Méndez-Pérez, J.A., Luis Calvo-Rolle, J.: Anomaly detection based on one-class intelligent techniques over a control level plant. Logic J. IGPL **28**(4), 502–518 (2020)
19. Jove, E., Casteleiro-Roca, J.L., Quintián, H., Zayas-Gato, F., Vercelli, G., Calvo-Rolle, J.L.: A one-class classifier based on a hybrid topology to detect faults in power cells. Logic Journal of the IGPL (2021)
20. Kaltschmitt, M., Streicher, W., Wiese, A.: Renewable Energy. Springer, Heidelberg (2007)

21. Niakolas, D.K., Daletou, M., Neophytides, S.G., Vayenas, C.G.: Fuel cells are a commercially viable alternative for the production of "clean" energy. Ambio **45**(1), 32–37 (2016)
22. Park, J., Oh, H., Ha, T., Lee, Y.I., Min, K.: A review of the gas diffusion layer in proton exchange membrane fuel cells: durability and degradation. Appl. Energy **155**, 866–880 (2015)
23. Pinzón, C., de Paz, J.F., Bajo, J., Herrero, Á., Corchado, E.: AIIDA-SQL: an adaptive intelligent intrusion detector agent for detecting SQL injection attacks. In: 10th International Conference on Hybrid Intelligent Systems (HIS 2010), Atlanta, 23–25 August 2010, pp. 73–78 (2010). https://doi.org/10.1109/HIS.2010.5600026
24. Quintián, H., Corchado, E.: Beta Hebbian learning as a new method for exploratory projection pursuit. Int. J. Neural Syst. **27**(06), 1750024 (2017)
25. Quintián, H., Corchado, E.: Beta scale invariant map. Eng. Appl. Artif. Intell. **59**, 218–235 (2017)
26. Rosli, R., Sulong, A., Daud, W., Zulkifley, M., Husaini, T., Rosli, M., Majlan, E., Haque, M.: A review of high-temperature proton exchange membrane fuel cell (HT-PEMFC) system. Int. J. Hydrogen Energy **42**(14), 9293–9314 (2017)
27. Sebastián, D., Baglio, V.: Advanced materials in polymer electrolyte fuel cells (2017)
28. Contribution of Working Group Team: Contribution of working groups i, ii and iii to the fourth assessment report of the intergovernmental panel on climate change. In: IPCC 2007: Climate Change 2007: Synthesis Report 104 (2007)
29. Wasserman, P.: Advanced Methods in Neural Computing, 1st edn. Wiley, New York (1993)
30. Zhang, L., Chae, S.R., Hendren, Z., Park, J.S., Wiesner, M.R.: Recent advances in proton exchange membranes for fuel cell applications. Chem. Eng. J. **204**, 87–97 (2012)

Intrusion Detection System for MQTT Protocol Based on Intelligent One-Class Classifiers

Esteban Jove[1(✉)], Jose Aveleira-Mata[3], Héctor Alaiz-Moretón[3], José-Luis Casteleiro-Roca[1], David Yeregui Marcos del Blanco[2], Francisco Zayas-Gato[1], Héctor Quintián[1], and José Luis Calvo-Rolle[1]

[1] CTC, CITIC, Department of Industrial Engineering, University of A Coruña, Avda. 19 de Febrero s/n, 15405 Ferrol, A Coruña, Spain
{esteban.jove,jose.luis.casteleiro,f.zayas.gato,hector.quintian, jlcalvo}@udc.es

[2] University of León, Avd. Facultad, 24071 León, Spain
dmarcb01@estudiantes.unileon.es

[3] Research Institute of Applied Sciences in Cybersecurity (RIASC) MIC, Universidad de León, 24071 León, Spain
{jose.aveleira,hector.moreton}@unileon.es

Abstract. The significant advance in smart devices connected to Internet has promoted the "Internet of Things" technology. However, the success of this term comes with the need of implementing Intrusion Detection Systems to face possible attacks. The present research deals with the intrusion detection in a network with Message Queuing Telemetry Transport protocol. To achieve this goal, different one-class classifiers have been implemented from a real dataset, achieving good performance in the detection of intrusion attacks.

Keywords: MQTT · One-class · PCA · APE · NCBoP · K-Means · SVM

1 Introduction

The "Internet of Things" (IoT) makes reference to a technology based on smart devices connected to Internet, interacting with environments changing information between these and external systems. Actually 5 billion IoT devices are connected to Internet [15]. This connectivity applied to industry is known as industry 4.0 [42], providing connection to automated control systems, together with remote monitoring, making it possible to manage equipment in real time from any client device whether it is a PC, Tablet, or Smartphone that has a connection to the Internet. In addition, a higher level of interconnection and automation can be implemented using the collected information and cloud-based processing [24,34].

The original version of this chapter was revised: The name of the author has been corrected from Francico Zayas-Gato to Francisco Zayas-Gato. The correction to this chapter is available at https://doi.org/10.1007/978-3-030-78901-5_30

J. M. Corchado and S. Trabelsi (Eds.): SSCTIC 2021, LNNS 253, pp. 249–260, 2022.
https://doi.org/10.1007/978-3-030-78901-5_22

Due to low price, this kind of devices have limited the computing capacity, this makes that security primitives cannot be implemented in the IoT devices. Its true that they are able to support some security policies however there are not enough for the last well know attacks. Protocols associated to IoT environments make that, these system can work efficiently but compromising the security of the networks where IoT devices are deployed. The main feature of IoT communication protocols is their lightweight nature, in order to consume not too many resources.

Most popular IoT protocols in the link and physical layer are LORA [53], zigbee [5] or Bluethooth [3]. On the other hand the most famous application layer IoT protocols can be MQTT (Message Queuing Telemetry Transport) and CoAP (Constrained Application Protocol), but one more time the heterogeneous nature of this set of protocols make that the policies in cybersecurity being critical [14,31,43].

A way to protect an IoT network can be the utilization of the IDS (Intrusion Detection Systems) approach [17,39,52]. This solution is based on matching patterns of well known attacks. Thus, the stranger behaviours of an IoT environment can be detected [7].

There are two approaches for implanting an IDS, the first one is oriented to find threats thank to real time monitoring of the environment while this is working, making an comparison between the real time status and the normal status previously known [40]. The second approach is based on signature detection, finding patterns for well know attacks monitoring the network traffic [45].

For the second approach is necessary to develop a model for being deployed in the IDS. The deep learning and machine learning techniques can be utilized for creating and improving traditional detection methods.

Good results have been achieved with machine learning algorithms such as Support Vector Machine and Random Forest [18] [35] as well as with not unsupervised methods based on clustering [11].

From the Deep learning point of view, there are some solutions that can be included in the IDS. Thus, the use of Auto-encoders and Deep Belief Networks (DBN) is a very interesting approach for reducing dimension purposes, due to this is the first steep when a optimal classification model is going to be made [33,36,49]. More complex architectures as Long Short Term Memory (LSTM) have got good results for implementing attack detection models [4,29].

To achieve a useful model is critical to have high quality datasets [8]. Around the world we can find data-sets that contains traffic data about TCP/IP frames, being the most famous the KDD99 data-set [48], NSL-KDD Dataset [50] and AWID [30]. Also, a different IoT type of data-set is available, these contains information about values and status of the sensors included in the IoT network (temperature, humidity, state of a actuator).

The main objective of this work is the implementation of one-class classification method oriented to achieve an useful model for detecting if the protocol MQTT is threatened by an intrusion attack.

This work is structured as follows: the case study is described in the next section. The one-class classifier approach used is introduced in Sect. 3. Experiments and their results are described in Sect. 4, and the conclusions and some future works are showed in Sect. 5.

2 Case Study

MQTT protocol is widely used in IoT [46] and industry [42]. It is a publication/subscription messaging protocol designed for lightweight machine-to-machine (M2M) communications and ideal for connecting small devices to networks with minimal bandwidth.

The architecture of a MQTT systems follows a star topology [16] with a central node acting as a server called a broker. The broker is responsible for managing the network and transmits the different messages in real time. Communication is based on topics created by the client that publishes the message and the nodes that wish to receive it must subscribe to the topic, so that communication can be one-to-one or one-to-many.

The use of MQTT in devices with low computational capacity and communication is a priority, which makes it difficult to use encryption systems. This makes it vulnerable to intrusion attacks by agents outside the system. An attacker can see what is happening in all the publications and even interact with the system sending his own code in publications or creating new topics.

An attacker can access the system from outside the network if no password is available by using Shodan network scanning on port 1883 [6] to connect to the system (this is the default port used by the protocol). From inside the network the attacker can see the password that is in plain text [12] by sniffing the packets, in this way be able to connect as another client of the system. Once connected to the system, the attacker can watch all messages from the broker with the special character '#' used for protocol management purposes, and interact with the topics.

Due to the importance of datasets to create models using IDS, it is developed an IoT environment using the MQTT protocol in the following way:

- **The server** that hosts the web application to interact with the system and is also the MQTT broker has implemented in node.js. Using the library "Aedes" [2] to work as MQTT broker.
- **Web application** to interact with different sensors and actuators. A web application has been developed in Angular.js. This application connects to the MQTT broker like another client.
- **Actuators and sensors** This uses two integrated boards NodeMCU, which consist of a low-power microcontroller with an interface for wifi connection to a ESP8266 chip [37]. The boards have GPIO connections to connect a distance sensor HC-SR04, and an actuator, a relay that turns on and off a light bulb.
- **System clients** the system includes a PC and a smartphone that interact with the sensors and actuators, using the web application, connected by WiFi, also generate network traffic to the internet.

The intrusion attack is performed on this environment from a client outside the IoT system with the mosquito software [1]. Using the special character, it discovers the topics and the traffic of the system. After this, the sensor is modified to modify the temperatures and send actions on/off to the actuator. To generate more network frames, a Power Shell script is generated that performs these intrusions randomly, All traffic generated by the IoT under attack and normal traffic and Internet browsing traffic is collected in a PCAP file from a router configured with OpenWRT [38], a linux-based operating system that, among other tools, provides to allow the capture command "tcpdump".

To simplify the PCAP files the traffic is dissected considering all the fields of the MQTT protocol and common fields to all the frames as the timestamp, ports and ip addresses according to the work of AWID dataset, also a tagging field is added to the frames indicating if they are under attack or not, obtaining a dataset in CSV format, the dataset has 80.893 frames of them 1.898 under attack frames and 78.995 normal frames. The described process can be see in Fig. 1.

3 Intrusion Detection Classifier

To tackle the intrusion problem in MQTT protocol environment, a one-class topology is proposed. The followed approach and the different techniques used to achieve the objective are detailed in this section.

3.1 Classifier Approach

The use of one-class classifier is based on the prior knowledge of data corresponding to situations without intrusion attacks. Then, once the classifier is trained, and the normal situations are learned, it is tested using the both normal and attacks instances. Figures 2 and 3 represent the training and testing phases, where the classifier inputs are the 42 variables registered by the MQTT environment.

3.2 Methods

This subsection describes the five different methods used to determine the anomaly detection.

Approximate Polytope Ensemble. The Approximate Polytope Ensemble (APE) is a boundary one-class classification technique, whose good performance has been proved over many different applications [10]. The main idea of APE is to achieve an approximation of the limits of a dataset $D \in \mathbb{R}^n$ by means of the calculation of its convex limits. However, since the convex hulls of D with N samples and d variables has a computational cost of $O(N^{(d/2)+1})$ [10]. Then, the convex hull is approximated using p random projections on $2D$ planes and determine their convex limits on that plane with the significant computational effort reduction [26].

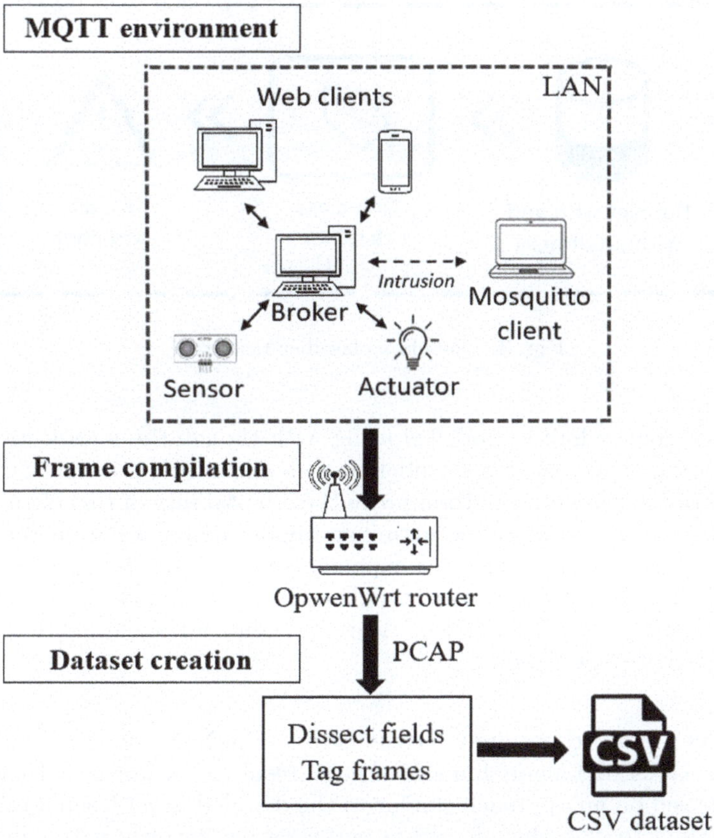

Fig. 1. Steps for the creation intrusion dataset

Fig. 2. One-class classifier implementation

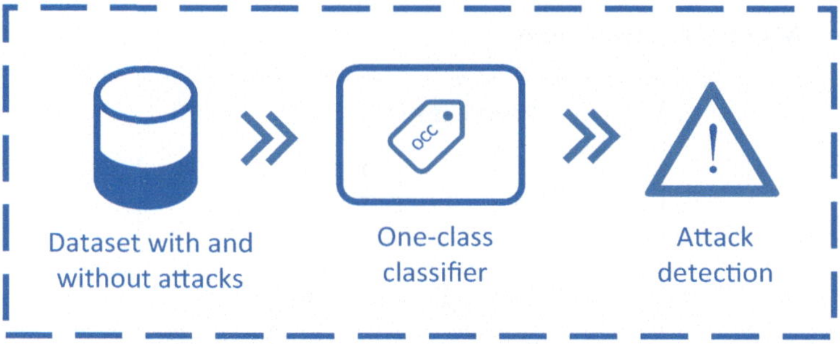

Fig. 3. One-class classifier test stage

Once the convex hull is modelled using APE though the p projections, when a new test sample arrives, it is considered as an anomaly if it is out of the convex hull in any of the projections. To improve the adaptability of this algorithm over different datasets, avoiding underfitting and underfitting, a parameter α can be defined as a factor that reduces or expands convex limits from the centroid of each projection. When this factor is higher than 1, the limits are widened, and the are narrowed when it is below 1. A proper value of the parameter α depends on the nature of the dataset.

Non Convex Boundary over Projections. The Non Convex Boundary over Projections (NCBoP) algorithm shown good results over different UCI datasets [25]. It is based on an approach similar to the detailed in APE subsection. However, according to [25], this algorithm performs significantly better in non convex datasets. This technique proposes the computation of the dataset boundaries through non-convex limits, avoiding false positive classification when the anomalies are inside the convex hull.

Although this technique has been proved to perform better than APE, its computational effort is significantly higher, so its adequacy depends on the shape and complexity of the dataset to be modeled. Furthermore, the possibility of avoiding overfitting and underfitting is achieved through a parameter α [25].

K-Means. The K-Means algorithm has been widely used for data clustering. This unsupervised technique [27,28] groups the initial dataset in the number clusters defined by the user. To do so, the clustering error, which is the sum of distances from each point to its cluster centroid, is minimized.

Once the centroids are calculated from the training set, this technique can be used for one-class purposes taking into consideration the distance of a test data to its closest centroid. If this distance is higher than the distance of every cluster data to the centroid, the anomaly is detected.

Principal Component Analysis. The well known Principal Component Analysis (PCA) algorithm has been commonly applied especially for dimension reduction problems [23,51]. However, it offered effectiveness in one-class classification problems and anomaly detection tasks [19,22].

The PCA technique calculates the directions with higher data variability by means of the eigenvectors of the covariance matrix. Once these directions, known as principal components, are calculated, they are used to perform linear projections of the original data. Consequently, the data projected has lower dimensions. The one-class criteria is based on the distance between the original data and the data projected, also known as reconstruction error. If a test data presents a reconstruction error greater than the registered in the training process, the anomaly is detected.

One-Class Support Vector Machine. The use of Support Vector Machine (SVM) for One-Class (OCSVM) tasks is one of the most common used technique, whose performance has offered successful results in many different applications [32,41].

The aim of this supervised technique is to map the dataset into a high dimensional space using a kernel function. Then, a hyperplane that maximizes the distance between the mapped points and the origin is constructed [44]. The support vectors are defined as the instances placed close to the hyperplane.

After the training process, when a new data arrives to the implemented classifier, it gives the distance from this data to the high dimensional plane. If this distance is positive, the data belongs to the target class and it is considered an anomaly otherwise.

4 Experiments and Results

4.1 Experiments Setup

The different experiments carried out are detailed in this subsection.

Techniques Configuration. To evaluate the performance of each technique, the following features were tested.

APE and NCBoP. The number of projections checked was 10, 50, 100 and 500. The parameter α was set to 0.8, 0.9, 1, 1.1 and 1.2 to evaluate different restriction boundaries.

K-Means. The number of cluster in which the dataset is divided varied from 1 to 15. The possibility of adding an outlier fraction in the training data is taken into consideration. This was set to 0, 0.05, 0.1, 0.15, 0.20 and 0.25.

PCA. The number of principal components was varied from 1 to $n-1$, being n the number of variables of the training set. An outlier fraction is considered as well as the K-Means technique.

SVM. An outlier percentage is considered in the same way as K-Means and PCA.

Data Preprocessing. The categorical variables were replaced by numerical ones. With the aim of obtaining the best results, the dataset was normalized using a 0 to 1 interval and also using the z-score method [47]. Furthermore, the dataset without any kind of normalization was tested.

Performance Measurement. The evaluation of the classifier performance has been done though the Area Under the Receiver Operating Characteristic Curve (AUC) [9]. This parameter, in percentage, has offered a really interesting performance, with the particular feature of being insensitive to changes in class distribution, which is a key factor in one-class classification tasks [13]. Furthermore, the time needed to train each classifier is also presented as a measure of the computational cost. The tests were validated following a $k-fold$ cross-validation method, using k = 10.

4.2 Results

The experiments detailed in previous resulted in the different performance shown in Table 1. This Table shown the greatest AUC values obtained for each technique, as well as the corresponding configuration.

Table 1. Best AUC results for each classifier

Technique	Features	Normalization	AUC (%)	Time (s)
APE	500 projections $\alpha = 1$	Zscore	52.063	3.11
NCBoP	500 projections $\alpha = 1$	No	60.317	1211.73
K-Means	Clusters = 10 Outlier fraction = 15%	No	76.496	0.84
PCA	Components = 15 Outlier fraction = 20%	0–1	89.296	0.21
SVM	Outlier fraction = 0%	No	69.235	285.56

5 Conclusions and Futures Works

The results achieved by this work reveal that PCA is the best technique, with 89.27% of AUC and also the lowest training time. This dimensional reduction technique has offer the highest AUC value and the lowest computational time.

Then, the proposed approach would be presented as a really interesting tool to determine the appearance of intrusion attacks in MQTT protocols. K-Means and SVM have shown also interesting performance, especially the first one, but far from PCA's behaviour. In case of APE and NCBoP, the performance is significantly better with the second technique, which might be caused by the non convex shape of the set. However, NCBoP algorithm presents more computational cost than APE.

As PCA is the one with best performance, the use of dimensional reduction techniques could be considered prior to the classifier implementation in future works. In this case, reducing the data dimension could lead to better results. Furthermore, additional one-class techniques or data imputation techniques [20, 21] may be considered in future works.

Acknowledgements. Spanish National Cybersecurity Institute (INCIBE) and developed Research Institute of Applied Sciences in Cybersecurity (RIASC).

CITIC, as a Research Center of the University System of Galicia, is funded by Consellería de Educación, Universidade e Formación Profesional of the Xunta de Galicia through the European Regional Development Fund (ERDF) and the Secretaría Xeral de Universidades (Ref. ED431G 2019/01).

References

1. Light, R.A.: Mosquitto: server and client implementation of the MQTT protocol. J. Open Source Softw. **2**(13), 265 (2017). https://doi.org/10.21105/joss.00265
2. aedes: GitHub - moscajs/aedes: Barebone MQTT broker that can run on any stream server, the node way. https://github.com/moscajs/aedes
3. Al-sarawi, S., Anbar, M., Alieyan, K., Alzubaidi, M.: Internet of Things (IoT) communication protocols: Review. In: 2017 8th International Conference on Information Technology (ICIT), pp. 685–690 (2017)
4. Alaiz-Moreton, H., Aveleira-Mata, J., Ondicol-Garcia, J., Muñoz-Castañeda, A.L., García, I., Benavides, C.: Multiclass classification procedure for detecting attacks on MQTT-IoT protocol. Complexity **2019**, 1–11 (2019). https://doi.org/10.1155/2019/6516253
5. Alobaidy, H.A.H., Mandeep, J.S., Nordin, R., Abdullah, N.F.: A review on ZigBee based WSNs: concepts, infrastructure, applications, and challenges. Int. J. Electr. Electron. Eng. Telecommun. **9**(3), 189–198 (2020). https://doi.org/10.18178/ijeetc.9.3.189-198
6. Andy, S., Rahardjo, B., Hanindhito, B.: Attack scenarios and security analysis of MQTT communication protocol in IoT system. In: 2017 4th International Conference on Electrical Engineering, Computer Science and Informatics (EECSI), pp. 1–6 (2017). https://doi.org/10.1109/EECSI.2017.8239179
7. Ben-Asher, N., Gonzalez, C.: Effects of cyber security knowledge on attack detection. Comput. Hum. Behav. **48**, 51–61 (2015). https://doi.org/10.1016/j.chb.2015.01.039
8. Bhuyan, M.H., Bhattacharyya, D.K., Kalita, J.K.: Towards generating real-life datasets for network intrusion detection. Int. J. Netw. Secur. **17**(6), 683–701 (2015)
9. Bradley, A.P.: The use of the area under the ROC curve in the evaluation of machine learning algorithms. Pattern Recogn. **30**(7), 1145–1159 (1997). https://doi.org/10.1016/S0031-3203(96)00142-2

10. Casale, P., Pujol, O., Radeva, P.: Approximate convex hulls family for one-class classification. In: Sansone, C., Kittler, J., Roli, F. (eds.) MCS 2011. LNCS, vol. 6713, pp. 106–115. Springer, Heidelberg (2011). https://doi.org/10.1007/978-3-642-21557-5_13
11. Chakrabarty, B., Chanda, O., Saiful, M.: Anomaly based intrusion detection system using genetic algorithm and K-centroid clustering. Int. J. Comput. Appl. **163**(11), 13–17 (2017). https://doi.org/10.5120/ijca2017913762
12. Dinculeană, D., Cheng, X.: Vulnerabilities and limitations of MQTT protocol used between IoT devices. Appl. Sci. **9**(5), 848 (2019). https://doi.org/10.3390/app9050848, http://www.mdpi.com/2076-3417/9/5/848
13. Fawcett, T.: An introduction to ROC analysis. Pattern Recogn. Lett. **27**(8), 861–874 (2006)
14. Ordóñez Galán, C., Sánchez Lasheras, F., de Cos Juez, F.J., Bernardo Sánchez, A.: Missing data imputation of questionnaires by means of genetic algorithms with different fitness functions. J. Comput. Appl. Math. **311**(C), 704–717 (2017). https://doi.org/10.1016/j.cam.2016.08.012
15. Green, J.: The internet of things reference model. In: Internet of Things World Forum, pp. 1–12 (2014)
16. Gupta, A.B., R.: MQTT version 3.1.1. OASIS Standard (2014). http://docs.oasis-open.org/mqtt/mqtt/v3.1.1/os/mqtt-v3.1.1-os.html
17. Hamed, T., Ernst, J.B., Kremer, S.C.: A survey and taxonomy of classifiers of intrusion detection systems. In: Daimi, K. (ed.) Computer and Network Security Essentials, pp. 21–39. Springer, Cham (2018). https://doi.org/10.1007/978-3-319-58424-9_2
18. Hasan, M., Nasser, M., Pal, B., Ahmad, S.: Support vector machine and random forest modeling for intrusion detection system (IDS). J. Intell. Learn. Syst. Appl. **2014**(February), 45–52 (2014). https://doi.org/10.4236/jilsa.2014.61005
19. Jove, E., Aláiz-Moretón, H., Casteleiro-Roca, J.L., Corchado, E., Calvo-Rolle, J.L.: Modeling of bicomponent mixing system used in the manufacture of wind generator blades. In: Corchado, E., Lozano, J.A., Quintián, H., Yin, H. (eds.) IDEAL 2014. LNCS, vol. 8669, pp. 275–285. Springer, Cham (2014). https://doi.org/10.1007/978-3-319-10840-7_34
20. Jove, E., et al.: Attempts prediction by missing data imputation in engineering degree. In: Pérez García, H., Alfonso-Cendón, J., Sánchez González, L., Quintián, H., Corchado, E. (eds.) SOCO/CISIS/ICEUTE -2017. AISC, vol. 649, pp. 167–176. Springer, Cham (2018). https://doi.org/10.1007/978-3-319-67180-2_16
21. Jove, E., et al.: Missing data imputation over academic records of electrical engineering students. Logic J. IGPL **28**(4), 487–501 (2020)
22. Jove, E., Casteleiro-Roca, J.L., Quintián, H., Méndez-Pérez, J.A., Calvo-Rolle, J.L.: Anomaly detection based on intelligent techniques over a bicomponent production plant used on wind generator blades manufacturing. Revista Iberoamericana de Automática e Informática industrial (2019)
23. Jove, E., Casteleiro-Roca, J.L., Quintián, H., Méndez-Pérez, J.A., Calvo-Rolle, J.L.: A fault detection system based on unsupervised techniques for industrial control loops. Expert Syst. **36**(4), e12395 (2019)
24. Jove, E., Casteleiro-Roca, J.L., Quintián, H., Méndez-Pérez, J.A., Calvo-Rolle, J.L.: Virtual sensor for fault detection, isolation and data recovery for bicomponent mixing machine monitoring. Informatica **30**(4), 671–687 (2019)
25. Jove, E., Casteleiro-Roca, J.L., Quintián, H., Méndez-Pérez, J.A., Calvo-Rolle, J.L.: A new method for anomaly detection based on non-convex boundaries with random two-dimensional projections. Inf. Fusion **65**, 50–57 (2021)

26. Jove, E., Casteleiro-Roca, J.L., Quintián, H., Simić, D., Méndez-Pérez, J.A., Luis Calvo-Rolle, J.: Anomaly detection based on one-class intelligent techniques over a control level plant. Logic J. IGPL **28**(4), 502–518 (2020)
27. Jove, E., Casteleiro-Roca, J.L., Quintián, H., Zayas-Gato, F., Vercelli, G., Calvo-Rolle, J.L.: A one-class classifier based on a hybrid topology to detect faults in power cells. Logic J. IGPL (2021, online)
28. Jove, E., López, J.A.V., Fernández-Ibáñez, I., Casteleiro-Roca, J.L., Calvo-Rolle, J.L.: Hybrid intelligent system to predict the individual academic performance of engineering students. Int. J. Eng. Educ. **34**(3), 895–904 (2018)
29. Kim, J., Kim, J., Thu, H.L.T., Kim, H.: Long short term memory recurrent neural network classifier for intrusion detection. In: 2016 International Conference on Platform Technology and Service (PlatCon), pp. 1–5 (2016). https://doi.org/10.1109/PlatCon.2016.7456805, http://ieeexplore.ieee.org/document/7456805/
30. Kolias, C., Kambourakis, G., Stavrou, A., Gritzalis, S.: Intrusion detection in 802.11 networks: empirical evaluation of threats and a public dataset. IEEE Commun. Surv. Tutorials **18**(1), 184–208 (2016). https://doi.org/10.1109/COMST.2015.2402161
31. Kolias, C., Kambourakis, G., Stavrou, A., Voas, J.: DDoS in the IoT: Mirai and other botnets. Computer **50**(7), 80–84 (2017). https://doi.org/10.1109/MC.2017.201
32. Li, K.L., Huang, H.K., Tian, S.F., Xu, W.: Improving one-class SVM for anomaly detection. In: 2003 International Conference on Machine Learning and Cybernetics, vol. 5, pp. 3077–3081. IEEE (2003)
33. Li, Y., Ma, R., Jiao, R.: A hybrid malicious code detection method based on deep learning. Int. J. Secur. Appl. **9**(5), 205–216 (2015). https://doi.org/10.14257/ijsia.2015.9.5.21
34. Wollschlaeger, M., Sauter, T., Jasperneite, J.: The future of industrial communication. IEEE Ind. Electron. Mag. **11**(4), 17–27 (2017). https://doi.org/10.1109/MIE.2017.2649104
35. Nieto, P.G., Fernández, J.A., Lasheras, F.S., de Cos Juez, F., Muñiz, C.D.: A new improved study of cyanotoxins presence from experimental cyanobacteria concentrations in the Trasona reservoir (Northern Spain) using the MARS technique. Sci. Total Environ. **430**, 88–92 (2012)
36. Nieto, P.G., Torres, J.M., de Cos Juez, F.J., Lasheras, F.S.: Using multivariate adaptive regression splines and multilayer perceptron networks to evaluate paper manufactured using eucalyptus globulus. Appl. Math. Comput. **219**(2), 755–763 (2012)
37. NodeMCU: NodeMCU - an open-source firmware based on ESP8266 WIFI-SOC (2014). http://www.nodemcu.com/index_en.html
38. openwrt: openwrt.org, https://openwrt.org/
39. Perdisci, R., Ariu, D., Fogla, P., Giacinto, G., Lee, W.: McPAD: a multiple classifier system for accurate payload-based anomaly detection. Comput. Netw. **53**(6), 864–881 (2009). https://doi.org/10.1016/j.comnet.2008.11.011
40. Prabha, K., Sudha, S.: A survey on IPS methods and techniques. Int. J. Comput. Sci. Issues **13**(2), 38–43 (2016). https://doi.org/10.20943/01201602.3843
41. Quintián, H., Corchado, E.: Beta scale invariant map. Eng. Appl. Artif. Intell. **59**, 218–235 (2017)
42. Ramamoorthy, K., Karthikeyan, S., Chelladurai, T.: An investigation on industrial internet of things for mission critical things in industry 4. 0 2. literature review. Seybold Rep. **15**(9), 3294–3300 (2020)

43. Razzaq, M.A., Gill, S.H., Qureshi, M.A., Ullah, S.: Security issues in the internet of things (IoT): a comprehensive study. Int. J. Adv. Comput. Sci. Appl. **8**(6), 383 (2017). https://doi.org/10.14569/IJACSA.2017.080650
44. Rebentrost, P., Mohseni, M., Lloyd, S.: Quantum support vector machine for big data classification. Phys. Rev. Lett. **113**, 130503 (2014). https://doi.org/10.1103/PhysRevLett.113.130503
45. Samrin, R., Vasumathi, D.: Review on anomaly based network intrusion detection system. In: International Conference on Electrical, Electronics, Communication Computer Technologies and Optimization Techniques, ICEECCOT 2017, pp. 141–147 (2018). https://doi.org/10.1109/ICEECCOT.2017.8284655
46. Sethi, P., Sarangi, S.R.: Internet of things: architectures, protocols, and applications. J. Electr. Comput. Eng. **2017** (2017)
47. Shalabi, L.A., Shaaban, Z.: Normalization as a preprocessing engine for data mining and the approach of preference matrix. In: 2006 International Conference on Dependability of Computer Systems, pp. 207–214, May 2006. https://doi.org/10.1109/DEPCOS-RELCOMEX.2006.38
48. Stolfo, S.J.: KDD cup 1999 dataset (1999). http://kdd.ics.uci.edu/
49. Tao, X., Kong, D., Wei, Y., Wang, Y.: A big network traffic data fusion approach based on fisher and deep auto-encoder. Information **7**(2), 20 (2016). https://doi.org/10.3390/info7020020
50. Tavallaee, M., Bagheri, E., Lu, W., Ghorbani, A.A.: A detailed analysis of the KDD cup 99 data set. In: 2009 IEEE Symposium on Computational Intelligence for Security and Defense Applications, pp. 1–6 (2009). https://doi.org/10.1109/CISDA.2009.5356528
51. Wu, J., Zhang, X.: A PCA classifier and its application in vehicle detection. In: International Joint Conference on Neural Networks Proceedings, IJCNN 2001, (Cat. No. 01CH37222), vol. 1, pp. 600–604. IEEE (2001)
52. Zhou, Q., Pezaros, D.: Evaluation of machine learning classifiers for Zero-Day intrusion detection - an analysis on CIC-AWS-2018 dataset, May 2019. http://arxiv.org/abs/1905.03685
53. Zorbas, D., Abdelfadeel, K., Kotzanikolaou, P., Pesch, D.: TS-LoRa: time-slotted LoRaWAN for the industrial internet of things. Comput. Commun. **153**, 1–10 (2020). https://doi.org/10.1016/j.comcom.2020.01.056

Smart Mobility for Smart Cities (SMSC)

Infrastructure for the Enhancement of Urban Fleet Simulation

Pasqual Martí[1(✉)], Jaume Jordán[1], Fernando De la Prieta[2], Holger Billhardt[3], and Vicente Julian[1]

[1] Valencian Research Institute for Artificial Intelligence (VRAIN), Universitat Politècnica de València, Camino de Vera S/n, 46022 Valencia, Spain
pasmargi@vrain.upv.es, {jjordan,vinglada}@dsic.upv.es

[2] BISITE Research Group, University of Salamanca, Calle Espejo s/n. Edificio Multiusos I+D+i, 37007 Salamanca, Spain
fer@usal.es

[3] Centre for Intelligent Information Technologies (CETINIA), University Rey Juan Carlos, 28933 Móstoles, Madrid, Spain
holger.billhardt@urjc.es

Abstract. When it comes to urban fleet simulation, there are many factors which determine the quality of the outcome. Without real-world data on which to ground the setup, the results are not guaranteed to be useful. In addition, the coordination mechanisms for agents must be flexible and give the chance to agents to act following their own interests, as most of the urban traffic system users do. In this work we present an infrastructure for the simulation of urban fleets which deals with two challenges: realistic data generation, and self-interested agent coordination. Our infrastructure aims to ease the setup and execution of more realistic simulations in the urban traffic domain.

Keywords: Simulation · Transportation · Electric vehicle · Planning · Smart city · Urban fleets

1 Introduction

With more than half of the world's population living in cities, the list of challenges for keeping them sustainable has grown. "A smart sustainable city is an innovative city that uses ICTs (Information and Communication Technologies) to improve quality of life, the efficiency of urban operations and services and competitiveness while ensuring that it meets the needs of present and future generations concerning economic, social, environmental and cultural aspects"[1]. In this line, new concerns have awakened among citizens and city councils. On the one hand, they want to reduce air pollution by promoting the use of bicycles,

[1] This definition was provided by the International Telecommunication Union (ITU) and United Nations Economic Commission for Europe (UNECE) in 2015.

J. M. Corchado and S. Trabelsi (Eds.): SSCTIC 2021, LNNS 253, pp. 263–273, 2022.
https://doi.org/10.1007/978-3-030-78901-5_23

public transport and even electric vehicles instead of the conventional gasoline-powered car. On the other hand, the existence of green areas throughout the city is valued; areas that beautify the appearance of the city and are related to a better quality of life. All of this seems to have influenced urban traffic, making it evolve to focus on the people rather than the vehicles. For instance, many municipalities are restricting the traffic inside their town's center, increasing the space available for pedestrians to walk, as well as air quality.

The urban traffic system, which was already complex, is therefore becoming more entangled. The traffic interactions that arise are not trivially sorted out which causes experts to be constantly researching for new ways of optimizing the traffic flow in urban settlements. Meanwhile, cities are evolving into smart cities that control parameters such as traffic status, the influx of people in different areas or on public transport and even the quality of their air, in real time. Then, as more city services become intelligent, we have more and more data that we can use both to better characterize the urban traffic problems and to advance in their resolution. Through the use of artificial intelligence, we can give more potential to such data by using it in a more problem-oriented way. While the handling of such data can lead to solutions, the changes these imply cannot be applied without considering the impact they may have on the city's inhabitants. In addition, the variety of scenarios that urban traffic offers is massive, since the urban domain counts with many different users. From fleets of taxis to privately owned vehicles, all elements must be taken into account when researching urban traffic optimization. Through the use of simulators we manage to reproduce a small part of the real world in a virtual way, which allows us to modify it as we wish. All kinds of changes or improvements can be tested without the need to implement them or the risk of causing negative effects on people's lives. This offers a perfect working area for exploring solutions to the problem of urban traffic, as these are often expensive and costly to implement.

For an accurate simulation, however, the interactions between the elements of the system to replicate must be accurately reproduced. For this, we make use of multi-agent systems. Agents are pieces of software that are inspired by human reasoning: capturing signals from their environment and reacting appropriately, communicating with other agents, making their own decisions, etc. This makes the agents suitable for modeling the different users of a city. All users can be represented by an intelligent agent that adapts its actions and its way of interacting, both with the environment and with other users, accordingly.

For this work we use SimFleet [15], an agent-based fleet simulator. This software combines the possibilities offered by simulators with the flexibility of multi-agent systems, offering an ideal framework for the development and testing of solutions for improving urban traffic. However, SimFleet has a lot of room for improvement, as it is still in an early stage of development. Motivated by this, we propose different modules that improve SimFleet and, from a general perspective, the urban fleet simulation.

This paper approaches the enhancement of urban fleet simulations from two perspectives. On the one hand, the creation of more realistic simulation scenarios by basing element allocation and movement on real-world data. On the other hand, a more accurate reproduction of human behavior by modeling with rational self-interested agents. These two techniques enrich the simulations set in urban traffic domains and allow them to report more informed and interesting data and measurements.

The structure of the rest of the paper is as follows. Section 2 discusses about the strengths of multi-agent simulation and the use of self-interested agents. Section 3 describes SimFleet, the simulator used in our work. Section 4 presents the proposed infrastructure, describing the motivation and each of its modules. Finally, Sect. 5 assesses our work and comments on future extensions.

2 Related Work

In recent times, agent-based simulation has been crucial for creating more practical simulations with high scalability. There are several works that aim to conduct simulations in the urban mobility context to research issues such as traffic, citizen activity, crowds, emergency conditions, or the best placement of diverse facilities. For instance, in [14] authors research the optimal size of a carsharing system so as to maximize client satisfaction, evaluating different configurations through a simulator. Other works such as [7] compare various transportation services employing their own agent-based system.

Different methods have been developed to promote the modelling and creation of these simulations, enabling the implementation of experiments for the study of mobility within and between cities. The work in [2] includes a study of agent-based simulation methods for traffic and shipment.

Focusing on the urban traffic domain, there are many notable simulation tools that have aided in research activities. One of them is SUMO [3], an open-source traffic simulator which can be used to explore route choice, agent communication with different infrastructure, traffic management and even autonomous driving. SUMO uses a an origin/destination matrix to assign movement between zones of the city. Such a movement is described in terms of number of vehicles per time. In this line we also find MATSim [17], a framework for the implementation of demand-modeling and traffic flow simulations. Another example would be SIMmobility [1] which focuses on mobility demand impact prediction for smart shipment services. Finally, there exist also commercial tools such as VISSIM [5] which offers an array of technologies that can be combined to address multiple mobility and transportation problems.

In our work we make use of SimFleet [15], which is also an agent-based simulator, focused mainly on the development of strategies for the diverse vehicles of urban fleets. SimFleet allows complex simulations over cities with a large number of agents that can interact both among them as well as with certain city infrastructure, such as charging stations. We used SimFleet in the proposal of our infrastructure because of its flexibility, which allowed us to communicate it with several external modules. A more detailed description of SimFleet is presented in Sect. 3.

Self-interested agent interaction is a deeply researched topic in the field of game theory. From a theoretical perspective, authors in [6] study how cooperation emerges in different situations with self-interested individuals. There are other works with an approach similar to ours, like [11] in which authors consider multi-agent simulation to explore the effects of self-interested drivers on traffic when they act in a completely selfish manner.

With regard to self-interested agent coordination in urban traffic scenarios, techniques such as the ones in [8] and [10] could be applied. Those works present a two-game approach in which agents' possible plans are listed to obtain Nash equilibria that guarantee Pareto optimality and fairness by avoiding conflicts (which assume $-\infty$ utility). Another less computationally complex approach that obtains an equilibrium is the so-called best-response dynamics, presented in [13], and used to inspire the work in [9]. In this work we will present an agent coordination module that deals with the use of many self-interested agents in a single simulation scenario and sorts their actions through the techniques introduced in the two aforementioned papers.

3 SimFleet

SimFleet [15] is an agent-based urban fleet simulator built on the SPADE platform [4]. This fleet simulator was built to allow complex simulations over cities where a large number of agents interact to perform fleet coordination and management algorithms. SimFleet uses the multi-agent systems paradigm to allow the user to get access to autonomous, pro-active, intelligent, communicative entities. SPADE is a multi-agent systems platform that provides these features using a simple but powerful interface, which is why it was chosen for the development of SimFleet.

SPADE allows the creation of autonomous agents that communicate using the open XMPP instant messaging protocol [16]. This protocol is the standard protocol of the IETF and W3C for instant messaging communication and it (or some variant of it) is used in such important communication platforms as WhatsApp, Facebook or Google Talk. SPADE agents have also a web-based interface to create custom app frontends for agents, which is also used by SimFleet to show how every agent is moving through the city in a map that represents all routes made by agents. For the movement of such agents, SimFleet makes use of OSRM[2], a routing engine for finding shortest paths in road networks. Querying an OSRM server specifying the service *route* and passing origin and destination points returns the shortest route between those two points.

Finally, SimFleet is based on the Strategy design pattern, which allows the user to introduce new behaviors to the SimFleet agents without the need of modifying the simulator. This design pattern is used to introduce new algorithms that follow a common interface. In this case, introducing new coordination algorithms to an agent is as simple as building a *StrategyBehaviour* and loading it at SimFleet startup.

[2] http://project-osrm.org/.

SimFleet also provides some common agent classes that can be used (or inherit from them) to create a simulation. These agents represent the entities involved in a fleet simulator: fleet managers, transports, customers, and service directory. Next, we shortly describe these agent classes and how they interact during the simulation.

Fleet Manager Agents. SimFleet simulates an environment where there can be different kinds of fleets that provide services in a city. Each fleet has a fleet manager who takes care of the fleet, allows transports to be accepted in the fleet and puts customers and carriers in touch with each other to provide a service. An example of a fleet manager is a taxi company call center or a goods transport operator with trucks.

Transport Agents. These agents represent the vehicles that are moving through the city providing services. SimFleet supports any kind of city transport such as cars, trucks, taxis, electric cars, skateboards or even drones. However, the user can customize the kind of transport for its simulations. Transport may or may not belong to a fleet, but belonging to a fleet brings them some benefits like being found more easily and having a coordinator for its operations. Transport agents receive transport requests from customers and, if free, they will pick the customer up (or the package) and drive to its destination. However, before attending a request, a transport agent will make sure it has enough autonomy to do the whole trip. If not, the agent drops the request and goes to recharge its batteries or refuel to the nearest station. After serving one request, the agent awaits for more requests until the simulation is finished.

Customer Agents. Customers are the entities that want to perform an operation: calling a taxi to move from one place to another, send a package to a destination, etc. This entity is represented by the customer agent. In SimFleet, customers do not have the ability to move. They call a transport service which goes to the customer's position, picks up the package (or customer in case of a taxi, a bus, etc.), and transports the goods to a destination. Customer agents depend completely on the transport agents. To get a transport service the customer looks for an appropriate fleet in the directory and contacts to its fleet manager to get a transport service for the customer. The fleet manager broadcasts the requests to some or all of their registered transports (depending on its strategy) and any transport interested in attending it will send a proposal to the customer, who has to accept or refuse it (depending on the customer's strategy too). The customer waits until the transport agent picks it up and, once they arrive at the destination, it stops its execution.

4 Proposed Infrastructure

Our work revolves around SimFleet and its potential to aid in the improvement of urban traffic, providing accurate simulations of urban fleets which can be used both for research and testing. However, the current version of SimFleet has some limitations which we encountered while working on it. Therefore, we propose the enhancement of urban fleet simulations in SimFleet by introducing two main improvements: realistic simulation data generation, and self-interested agents for modeling the users of the urban traffic system.

The first improvement refers to the simulation data; the data that is feed to the simulator to characterize the simulation. In the context of urban simulation, examples of such data would be agent type, the amount of agents in a determined date and time, the areas of the city in which agents spawn and move around, the movement they perform, etc. All of these variables can be defined before the simulation execution so as to make it more realistic. Basing the values of these parameters on real-world data ensures that the scenario we are creating is a better representation of the system we want to analyze or experiment on. In addition, the real-world data can be replicated through different techniques, which would ensure the availability of new data to feed the simulator.

The second improvement is the introduction of self-interested agents in the simulations. These agents are selfish and take decisions based on their own private goals. Hence, they offer a better representation of some of the users of the urban traffic system such as drivers of private vehicles. When dealing with self-interested agents, coordination mechanisms involve game theory. The agents will propose their desired actions to every other agent in the simulation. At the same time, each agent will adapt their desired actions to the proposals of every other agent, aiming to avoid conflicts. This process obtains an equilibrium, a solution from which no agent will deviate, as all of them would be doing their best actions with respect to other agents' best actions.

Finally, we propose feeding both the generated data and the agent equilibrium to the simulator. The simulator illustrates the experiment by providing motion to the agents. With it, we can study and collect data about agent interaction, both with other agents as well as the elements of the scenario.

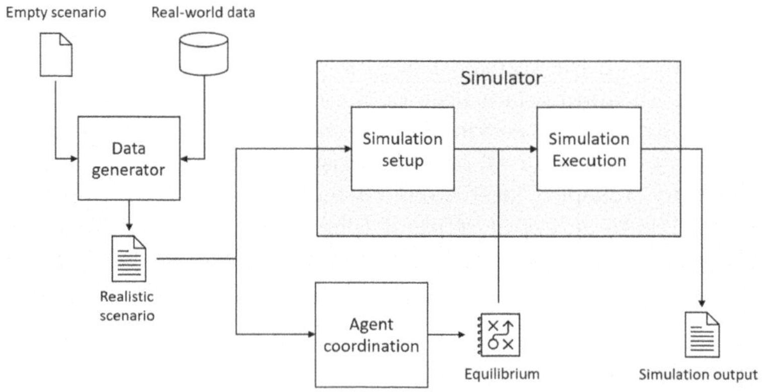

Fig. 1. Architecture of the proposed system

The proposed infrastructure can be seen in Fig. 1. The data generator module would take charge of the simulation data generation. It receives an empty simulation scenario and automatically fills it with the corresponding parameters. In addition, by using databases of real-world data, the obtained simulation scenarios can be more complex and realistic. Such scenario is then passed to the simulator for its setup. If self-interested agents want to be used, it would also be passed to the agent coordination module. This module outputs an equilibrium that coordinates the agents actions. Finally the equilibrium is read by the simulator to replicate the corresponding agent movements. All in all, our infrastructure pretends to ease the creation and execution of more lifelike simulations which, in turn, will report better and more interesting results from the experimentation. Following, we explain in detail each one of the modules of the infrastructure.

4.1 Simulation Data Generators

During our research, we identified the need of a system for the automatic generation of simulation scenarios. This means to provide the simulator users with a tool to test various distributions of elements and agents over the simulation area. In the context of urban simulations, we could use such a system to test different distributions of charging stations on a city. In addition, it could be used to locate vehicles and people with particular goals and create their associated traffic flow. With it, new opportunities arise to simulate different types of configurations, which can be very useful for the research community and even public organisms like city halls that want to test, for instance, the efficacy of charging stations in their towns.

However, one of SimFleet's main disadvantage arises when defining a simulation. A simulation is described in SimFleet by a JSON configuration file, which has to be manually written, including all agents and their attributes. This becomes specially troubling when trying to define big simulations, with a great number of agents.

To solve such issue, we propose two so-called generators [12], programs which, given a series of parameters, fill or create a new simulation scenario, leaving it ready for execution. Besides that, the generators would obtain more realistic simulations and agent distributions, being able to use cadastral, traffic, and Twitter information to obtain more accurate scenarios.

Firstly, we developed a *charging stations generator*, which locates the specified number of charging stations in the urban area where the simulation takes place. With it, various distributions of charging stations can be tested, seeing which ones achieve better traffic flow in the city.

Secondly, we introduced a *load generator* of movements in the city. It locates customer and/or transport agents in the urban area and creates routes for them; i.e.: defines their spawning and destination points. The movement of agents it creates can be random or informed by real-city data (extracted from open data portals) which make the routes of customer agents more realistic by choosing origin and destination points according to factors such as *population density*, *traffic information* and *Twitter activity*. The influence of each of these parameters is pondered by weights that are assigned to them.

4.2 Self-interested Agents Coordination

Multi-agent systems, as any system that is composed of many units, require coordination in order to correctly function. It is usual that such coordination is performed by an omniscient entity or centralized algorithm which informs each agent about what it has to do, thus removing the agents' own free will. As this approach can be fitting for elements of the urban traffic such as traffic lights or ambulance fleets, it is not realistic for modeling the majority of users. Private vehicle owners, pedestrians and many more are autonomous and take their decisions according to their own interests. Taking this into account, we propose the use of self-interested agents for modeling users of the urban traffic system.

Rational, self-interested agents have their own objectives and make their decisions accordingly to complete them. This "selfish" behavior can be a more realistic approach for representing certain types of users of an urban traffic system, such as taxis. Taxis (or other types of chauffeur-driven rental vehicles) are interested in serving the maximum number of customers possible, as they report a certain benefit. These vehicles usually belong to a fleet which, in turn, may belong to a company. However, generally, taxi drivers will give more importance to their own benefit rather than the overall benefit of their fleet, thus adopting a self-interested behavior. In order to preserve their free will, these agents are not coordinated by any centralized entity but rather by adapting their actions to the ones of every other agent in the system.

A rational entity will always prefer to obtain a reduced benefit than (to obtain) none because of a conflict. Although completely motivated by their private interests, autonomous agents are aware that the interests of another agent may be in conflict with their own. Therefore, the most rational behavior is to modify and adapt the desired actions, so as to still obtain the maximum possible benefit while taking into account the actions of others. This leads to avoid any arising conflict and, eventually, to an equilibrium.

In game theory, an equilibrium is a solution or set of agent actions from which no agent is incentivized to deviate. In other words, every agent has decided on a set of actions which reports them the maximum possible benefit with respect to the actions of every other participant agent. Applying this concept to our urban traffic simulator, with an equilibrium we obtain a coordinated simulation, as no conflict will originate during the agents' execution. As we commented on Sect. 2, the approach of [8] and [10] could be used to obtain a Pareto-optimal equilibrium. However, this approach is computationally expensive since listing the possible strategies of the agents is only possible for a reduced number of agents.

To obtain an equilibrium with lower computational cost for a considerable set of agents, we could approach it by means of best-response dynamics [13] and inspired by the work in [9]. This would require the definition of the coordination task as a multi-agent planning task. Therefore, the actions that each agent intends to perform during the simulation would be encoded in an agent plan. During the best-response process the agents propose their best plan (the one that reports them more benefits) in turns. After a whole round, the agents reevaluate the plan they proposed taking into account the plan of every other agent. If the actions of another agent are in conflict with theirs, the agent will propose a different plan which (1) avoids any conflict, and (2) is its current best plan. This process repeats iteratively until after a whole round no agent has modified its plan. This means that each agent is proposing their best plan with respect to every other agent's plan. Consequently, as no agent will obtain a benefit from switching plans, an equilibrium has been reached. The agents are coordinated, no conflicts will arise from the execution of their plans and their private interests have been preserved.

As can be seen in Fig. 1, the simulation scenario is inputted to the agent coordination module, as the agent actions are completely related to the elements of the simulation as well as their location, spawning time and other attributes. The module works by loading the simulation scenario, defining the agents, and running the coordination process. Once the equilibrium is obtained, it will be passed to the main simulator, which will make use of it to recreate conflictless agent movement.

4.3 Simulation Execution

The simulation task of our proposal would be performed by SimFleet. Nonetheless, any other simulator could be used as long as the integration between the different modules presented in this work is implemented.

SimFleet, as we have commented in Sect. 3, is oriented to the design and testing of agent strategies. As a complete simulator, it is perfectly capable of executing a simulation from the simulation scenario file, coordinating the actions and movements of the different agents. However, with our proposal, the coordination task falls on the agent coordination module and all that is left in the simulator is the visualization of the simulation as well as the capture of system interactions for the subsequent data analysis.

For this to be possible, it is necessary to integrate the different modules with SimFleet. The integration with the simulation data generators is trivial, as this module generates realistic simulation scenarios in a format that SimFleet is able to receive as input. To connect the agent coordination module, SimFleet must be able to receive the file describing the equilibrium that has been reached. By reading such file, SimFleet can assign each agent its actions and have them executed. As we have already mentioned, no conflicts should appear during execution since the best-response process avoids them.

Thus, with this integration, we complete the proposed infrastructure. An infrastructure which provides simulations in more realistic scenarios, based on real-world data, and preserves the private interests of the agents during the simulation, representing more accurately the urban traffic.

5 Conclusions

In this work we have proposed a different approach to urban fleet simulation which, integrated with SimFleet simulator, enhances its properties.

On the one hand, with the introduction of the simulation data generators, we have provided a tool for SimFleet users that allows them to easily define complex simulation configurations, providing a means to create different distributions of agents in the scenario. In addition, the use of the generators can make the simulations more realistic by giving the agents origin and destination points as well as movement based on real-world data. Nevertheless, we would like to improve the generators by studying better ways of distributing the agents and taking into account more parameters to take more advantage of the real-world data.

On the other hand, we have researched the use of rational, self-interested agents ultimately presenting an infrastructure which, together with SimFleet, solves urban simulations using transport agents that follow their private objectives. This involved the proposal of an agent coordination module that ensures self-interested agents both avoid conflicts and maximize their utility. For a correct completion of the coordination an equilibrium must be reached, that is, a stable solution from which no agent had incentive to deviate. For that, we proposed the use of the best-response dynamics, following the work in [9]. The equilibrium obtained by this process can easily be integrated with SimFleet or other simulators. In this way, we achieved the inclusion of self-interested agents in SimFleet simulations which accomplishes the goal of making the simulation more realistic.

In future works we want to create an instantiated version of the infrastructure presented here. Our interest is focused, most of all, on the agent coordination module, as we think self-interested agents are a feature worth exploring when it comes to urban traffic simulations. As we introduced in this work, for the coordination of self-interested agents game theory techniques must be developed and implemented. In addition, we would like to introduce planning so that every agent could decide on which actions are better for itself in every situation.

Acknowledgments. This work was partially supported by MINECO/FEDER RTI2018-095390-B-C31 project of the Spanish government.

References

1. Adnan, M., et al.: Simmobility: a multi-scale integrated agent-based simulation platform. In: 95th Annual Meeting of the Transportation Research Board Forthcoming in Transportation Research Record (2016)
2. Bazzan, A.L., Klügl, F.: A review on agent-based technology for traffic and transportation. Knowl. Eng. Rev. **29**(3), 375 (2014)
3. Behrisch, M., Bieker, L., Erdmann, J., Krajzewicz, D.: Sumo-simulation of urban mobility: an overview. In: Proceedings of SIMUL 2011, The Third International Conference on Advances in System Simulation. ThinkMind (2011)

4. Escrivà, M., Palanca, J., Aranda, G.: A jabber-based multi-agent system platform. In: Proceedings of the Fifth International Joint Conference on Autonomous Agents and Multiagent Systems, AAMAS 2006, pp. 1282–1284. Association for Computing Machinery, New York (2006). DOI https://doi.org/10.1145/1160633.1160866
5. Fellendorf, M.: Vissim: A microscopic simulation tool to evaluate actuated signal control including bus priority. In: 64th Institute of Transportation Engineers Annual Meeting, vol. 32, pp. 1–9. Springer, Cham (1994)
6. Hofmann, L.M., Chakraborty, N., Sycara, K.: The evolution of cooperation in self-interested agent societies: a critical study. In: The 10th International Conference on Autonomous Agents and Multiagent Systems, vol. 2, pp. 685–692 (2011)
7. Inturri, G., et al.: Taxi vs. demand responsive shared transport systems: an agent-based simulation approach. Transp. Policy **103**, 116–126 (2021). https://doi.org/10.1016/j.tranpol.2021.01.002
8. Jordán, J., Onaindía, E.: Game-theoretic approach for non-cooperative planning. In: Proceedings of the 29th AAAI Conference on Artificial Intelligence (AAAI), pp. 1357–1363 (2015)
9. Jordán, J., Torreño, A., de Weerdt, M., Onaindia, E.: A better-response strategy for self-interested planning agents. Appl. Intell. **48**(4), 1020–1040 (2017). https://doi.org/10.1007/s10489-017-1046-5
10. Jordán, J., Torreño, A., De Weerdt, M., Onaindia, E.: A non-cooperative game-theoretic approach for conflict resolution in multi-agent planning. Group Decis. Negot. (2020). https://doi.org/10.1007/s10726-020-09703-0
11. Lin, R., Kraus, S., Shavitt, Y.: On the benefits of cheating by self-interested agents in vehicular networks. In: Proceedings of the 6th international joint conference on Autonomous agents and multiagent systems, pp. 1–8 (2007)
12. Martí, P., Jordán, J., Palanca, J., Julian, V.: Load generators for automatic simulation of urban fleets. In: De La Prieta, F., et al. (eds.) PAAMS 2020. CCIS, vol. 1233, pp. 394–405. Springer, Cham (2020). https://doi.org/10.1007/978-3-030-51999-5_33
13. Monderer, D., Shapley, L.S.: Potential games. Games Econ Behav. **14**(1), 124–143 (1996)
14. Monteiro, C.M., et al.: Optimization of carsharing fleet size to maximize the number of clients served. Comput. Environ. Urban Syst. **87**, 101,623 (2021). https://doi.org/10.1016/j.compenvurbsys.2021.101623
15. Palanca, J., Terrasa, A., Carrascosa, C., Julián, V.: SimFleet: a new transport fleet simulator based on MAS. In: De La Prieta, F., et al. (eds.) PAAMS 2019. CCIS, vol. 1047, pp. 257–264. Springer, Cham (2019). https://doi.org/10.1007/978-3-030-24299-2_22
16. Saint-Andre, P.: Extensible messaging and presence protocol (xmpp): Core. RFC 6120, RFC Editor (2011). http://www.rfc-editor.org/rfc/rfc6120.txt
17. W Axhausen, K., Horni, A., Nagel, K.: The Multi-Agent Transport Simulation MATSim. Ubiquity Press, London (2016)

Modern Integrated Development Environment (IDEs)

Zakieh Alizadehsani[1](✉), Enrique Goyenechea Gomez[1], Hadi Ghaemi[2], Sara Rodríguez González[1], Jaume Jordan[3], Alberto Fernández[4], and Belén Pérez-Lancho[5]

[1] BISITE Research Group, University of Salamanca, Salamanca, Spain
{zakieh,egoyene,srg}@usal.es
[2] Computer Engineering Department, Ferdowsi University of Mashhad, Mashhad, Iran
hadi.qaemi@um.ac.ir
[3] Valencian Research Institute for Artificial Intelligence (VRAIN), Universitat Politècnica de València, Valencia, Spain
jjordan@dsic.upv.es
[4] Universidad Rey Juan Carlos, Madrid, Spain
alberto.fernandez@urjc.es
[5] Department of Computer Science and Automation, University of Salamanca, Salamanca, Spain
lancho@usal.es

Abstract. One of the important objectives of smart cities is to provide electronic services to citizens, however, this requires the building of related software which is a time-consuming process. In this regard, smart city infrastructures require development tools that can help accelerate and facilitate software development (mobile, IoT, and web applications). Integrated Development Environments (IDEs) are well-known tools that have brought together the features of various tools within one package. Modern IDEs include the advantages of Artificial Intelligence (AI) and Cloud Computing. These technologies can help the developer overcome the complexities associated with multi-platform software products. This paper has explored AI techniques that are applied in IDEs. To this end, the Eclipse Theia (cloud-based IDE) and its AI-based extensions are explored as a case study. The findings show that recommender system models, language modeling, deep learning models, code mining, and attention mechanisms are used frequently to facilitate programming. Furthermore, some researches have used NLP techniques and AI-based virtual assistance to promote the interaction between developers and projects.

Keywords: Integrated Development Environment (IDE) · Online IDEs · Software development · Artificial intelligence (AI) · Theia

1 Introduction

Today, cities have a strong desire to make their infrastructure smarter, which requires electronic infrastructure and developing the related software.

J. M. Corchado and S. Trabelsi (Eds.): SSCTIC 2021, LNNS 253, pp. 274–288, 2022.
https://doi.org/10.1007/978-3-030-78901-5_24

Unfortunately, software development is a time-consuming process. To enable cities to create electronic services faster, several tools have emerged which accelerate and facilitate the software development life cycle. Integrated Development Environments (IDEs) are well-known tools that have brought together the features of various tools within one package. The main goal of IDEs is to increase development speed, reduce errors, and increase the accuracy of the programming process [1,2]. According to the literature, there are many functionalities (features) such as debugging, autocomplete, etc. These features can fall into two categories, 1) Features that are usually running continuously in the background most of which can verify source code state such as live syntax checker and facilitate developments, including code autocomplete during program writing time. 2) Features that users can apply arbitrarily such as control versioning, code search, etc. However, IDEs could have more automation and intelligence to help developers. These features can be obtained by using Artificial Intelligence (AI) and Machine Learning techniques. Most IDEs include several tools to cover most aspects of software development like analyzing, designing, implementing, testing, documenting, and maintaining [3]. To increase the intelligence, these IDEs have embedded training models into the modern versions. This task can fall into two methods:

- Improving current functionalities (features)
- Adding new functionalities.

In this regard, Eclipse Theia [4] which is a cloud-based IDE, includes several AI-based extensions which have been explored in the case study. The findings of the present study can help researchers use this paper to identify popular IDE functionalities and related research areas, for example, the auto-complete functionality benefits from language modeling [5] and NLP techniques. Moreover, popular desktop and popular cloud IDEs have been investigated which can give knowledge about practical aspects.

The structure of the paper is as follows. Section 2 gives an overview of the available IDE functionalities. Section 3 reviews the related works. Section 4 conducts AI-based IDE functionalities. Section 5 describes the case study conducted with Theia Cloud IDE, and finally, Sect. 6 concludes the paper.

2 Background: IDE Functionality

Although programmers can develop without the IDEs and use simple text editors, IDEs are a set of tools that help programmers significantly. Syntax highlighting, debugging and editing features, which run in background, are the most common and basic features of IDEs [6]. Also, popular IDEs have provided advanced functionalities such as version controlling, terminal console, program element suggestion. These functionalities can be significantly improved by AI algorithms. Moreover, with emerging cloud-based services and the requirement of multiple development environments for different applications, IDEs are moving from desktop-based to cloud-based which are accessible through a web browser such as AWS Cloud9 [7], Codeanywhere [8] Eclipse Che [9].

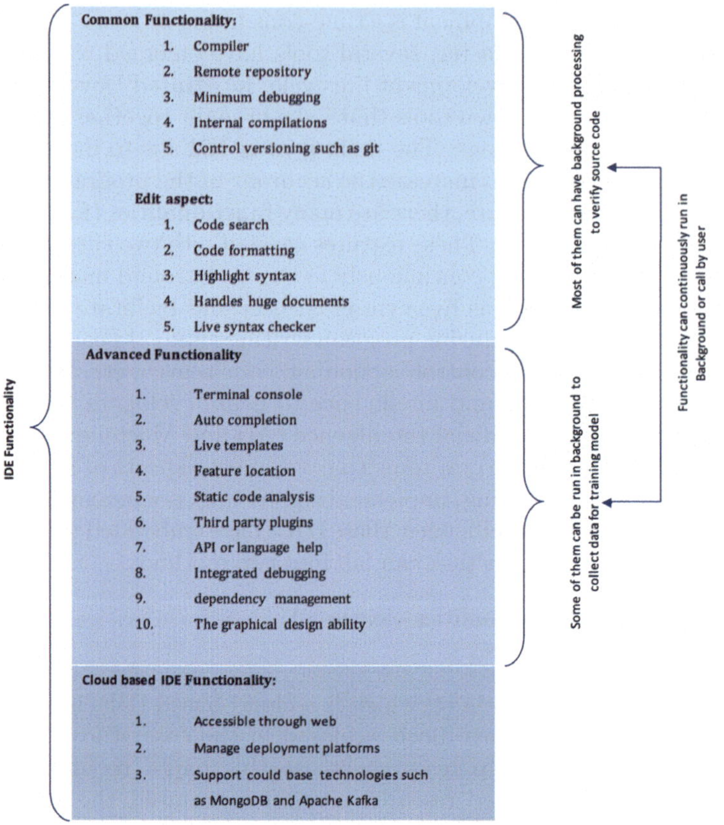

Fig. 1. Classified popular IDE features

Regarding related functionality topics, in the current paper, IDE functionalities are categorized within the three classifications 1) basic and common 2) advanced 3) cloud based. The details are illustrated in Fig. 1.

3 Related Works

Software development is a broad research area with different application domains in mobile, web, multimedia, IoT, etc. Thus, many studies have been conducted in the different fields of software development, which are mainly related to the stages of designing (identify requirements and dependencies), developing (implement, compile, run, test, debug), and optimizing (code reviews, integration).

IDEs combine tools that facilitate the development process. Several studies have investigated IDE technologies and their future [6,10,11]. Although secondary studies on IDE tools and identifying all related topics have received little attention, [12] has presented a valuable study that has investigated the role and applications of AI in classical software engineering. Their survey can serve as an informative guideline for researchers to build intelligent IDE tools [13].

Moreover, given the notable role of cloud technologies in modern IDEs, [14] has researched cloud-based IDEs. Also, [15] is a valuable work on Cloud IDE creation, which includes the required environment and reviewing the challenges posed to building cloud IDEs. From a more general perspective, [16] have investigated the effect of AI in software development and the role of learning from available codes. These codes can be collected from online code repositories or local code projects for IDEs. The methods and learning process from code have been investigated in depth in [17], which can be of help to researchers when designing intelligent IDE tools.

Besides, recent studies in the field of IDEs have been conducted to create a more appropriate and efficient interaction between the programmer and the programming environment, making the software production process more rapid. For example, some works used virtual assistants. [18] have trained the model which identified speech action types in developer question/answer conversations while bug repair.

The current study tries to give a big picture of popular AI approaches in desktop and online IDEs which have considered more practical aspects.

4 AI-Based IDE Functionality

Merging artificial intelligence with existing IDE functionality can bring new opportunities in most involved area in software development tools. This improvement can fall into two methods: 1) Improving current features 2) Adding new features.

4.1 Improve Current Functionalities

Existing features, such as code suggestion or code search, usually resort to recommender systems so as to provide more accurate results to developers [16].

Compiler: Compilers turn a programming language into low-level machine languages which can hide complexity from the developer and also help execute written code on different platforms [19]. This feature is essential and must be included as a basic IDE feature. However, it can be added as a third-party extension as well. Instruction selection, translation validation, and code optimization are some implementation issues in compiler construction. Recently, some efforts applied machine learning [20] or deepening [21] for code optimization.

Code Completion: In traditional IDEs, code completion can suggest lists of programming language function during program writing. Improving this functionality through AI can be divided into two groups 1) API/function/Class/variable suggestion improved by recommender systems 2) Automatic programming based on language modeling [5]. The first category can improve the accuracy of suggestions and provide an efficient list of API/function/variable suggestions [22].

The second group is for the code completion approach in which AI can be used, such as NLP techniques for language modeling [23]. These suggestions can consider local code features in the current IDE file. In other words, when developers are writing the code, IDEs automatically collect the data in the background and provide the data needed to train the models. These approaches usually use Deep Neural Networks (DNN) that need to learn from large amounts of code. In practical works, some efforts applied this solution in one-line code generation. For example, TabNine [24] has sped up programming by offering APIs. TabNine is a tool for code completion suggestions, trained on millions of open-source Java applications. This plugin works on the basis of the local code on which the developer is working.

Debugging: Debugging software is a process for the detection, location and fixing of the bugs during software testing which takes more than 50% of a programmer's time [25]. According to software development literature, automation has been used to speed up the software development process. Fully automated debugging is still being investigated. Semi-automatic debugging, which requires human participation, has also been investigated by researchers. Existing methods can be divided into two general categories: 1) Generate-and-validate approaches, which first produce a set of candidate patches, and are then tested and validated by the test suite. 2) Synthesis-based approaches, collect information by running a test suite while using this information to create a problem solver [26].

Recently, one of the most widely used methods of debugging is the use of automatic translation methods. For example, in [27,28], the patch created for debugging is based on the use of neural machine translation. The method applied in these studies is based on sequence-to-sequence learning.

Code Search: In the code search area, a search query can be written in natural languages or structured text (E.g. code). Therefore, search code AI models need both search queries and related codes. Most of the studies apply NLP techniques to search code. The main problem with these methods is that they do not take into account the difference between text and code which is structured code. In this regard, [29] introduced the DNN model which has used code snippets and description. Moreover, using AI approaches in code search requires a large code dataset, which might lead to an increase in training and test time [30]. Therefore, some efforts have employed embedded neural networks to reduce the dimension of huge code snippets and to make similar codes close together [29].

Automated Testing: Testing is an important part of quality assurance in which to verify that software is bug-free. In software engineering, testing can be applied at different levels and with different techniques [31]. Ideally, the goal of an automated data generation system for testing is to generate experimental data in a way that enables all branches of the program with the least possible computational effort. Machine learning methods are one of the most widely used methods in this field. For example, in [32] a productive statistical learning approach for files is introduced.

Although, IDE is used as an environment which facilitates code writing, some of these IDEs can provide automatic testing as well. However, some tests are usually presented in separate tools due to the high complexity. For example, to find security vulnerabilities, different inputs are repeatedly tested and the inputs are modified, which is called the fuzzing process. There are three general methods for fuzzing. Blackbox random fuzzing [33], Whitebox constraint-based fuzzing [34], grammar-based fuzzing [35]. Also, abnormally detection have been used for this purpose, as well.

Graphical Editor: Interactive graphical editors allow even non-expert users to create software [36]. MDA presents levels of abstraction which enable Visual Programming Language (VPL) [37] in graphical editors. These visual items can be modules, services, etc. When the user defines the sequence of visual items, the code generator can transfer them into executable code. Recently some works have used computer vision, to improve visual programming. The main idea behind these approaches is generating text from an image. However, in the automatic code generation field, the text is code (structured text) instead of natural language text [38]. Convolutional neural network (CNN) and recurrent neural network (RNN) have been used in this type of studies which are well-known deep neural networks [38–40].

Live Templates: Static IDs' templates can include class, function template, expression (loop, switch cases). Predefined placeholders in templates, usually filled by developers. However, the prediction ability of AI has led to the emergence of live templates in modern IDEs. These models can predict variables based on collecting information from local code projects [41].

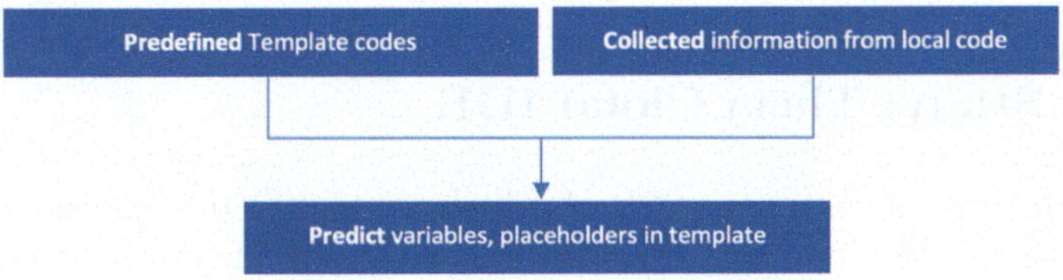

Fig. 2. Live template data

4.2 Adding New Functionalities

With the advent of online code repositories and improved data collection, it has become possible to add more new intelligent functionalities.

IDE Functionality Recommendations: [42] have found that users would rarely use all the functionalities of IDEs (on average they use just 42 commands out of the 1100 available commands). Therefore, some of the efforts [43,44] have presented new feature for offering IDE functionalities. These employ recommendation system models to make users aware of the existing functionalities. Some of theses functionalities have been outlined in this section.

Code Summarizing: The process of summarizing source code is to generate short descriptions in the natural language of written code. Code summarization can be considered as the basis of software documentation. The main idea of giving a short explanation to a programmer is to help them understand the code quickly and easily without having to read the code [45]. In general, source code memorization methods can be divided into two categories. 1) Pattern-based methods, 2) Artificial intelligence-based methods. For example, [46] has used pattern-based methods to create code explanations. The use of deep learning techniques, such as language modeling, has attracted the researchers' attention to this area as well [45,47].

AI Virtual Assistant: The AI-based virtual assistant is a program that can perform duties on the basis of commands or requests [48]. It identifies speech action types in developer question/answer conversations during bug repair. Google Assistant and Apple's Siri are popular assistants in the real-world. However, some works such as [18,49,50] have used virtual assistants in IDEs as well.

Comment Generation: Although comments are very useful, developers are not utilizing them enough in their coding, even if they add comments, they are not in the same style. Therefore, some research efforts have focused on generating descriptive comments for source code blocks [51]. Most of the academic works have focused on generating text description based on the source code functionality [52,53].

5 Case Study: Theia Cloud IDE

IDEs include several features, each of which is related to different disciplines, such as software development, user interaction, etc. Therefore, creating IDEs from scratch is troublesome. Under these circumstances, reusing other high-level frameworks can save time. These frameworks already provided the basic features, such as editing text (highlighting syntax, search, undo, find and replace), debugging, autocomplete for functions, block comment/uncomment [54]. As a result, using ready-made frameworks will facilitate the process for the developers, making it possible for them to invest more time in their custom IDEs. Table 1 shows the comparison between Eclipse Theia and Other IDEs.

There are different terms involved in Theia platform architectures. For example, the term "Theia" means Theia platform or Theia editor. Therefore, it is necessary to first define the terms used to refer to this technology.

Table 1. Comparison between Eclipse Theia and other IDEs

Title	Open source	Desktop/online	Building IDE product
VS Code	Only desktop version	+/+	−
Intelligent IDE	Freemium	+/+	−
Codeanywhere	Freemium	+/+	−
AWS Cloud9	Freemium	+/+	−
JSFiddle	Free	−/+	−
Eclipse Theia	Free	+/+	+

- **Workspace** is a specific machine, acting as a container that holds the project files, package managers, and an IDE-Interface. The interesting thing about workplaces is that developers can deploy a project and its dependencies in a workplace, then provide an image of it. This new workplace can be used as a basis for other projects [15]. However, it should be regarded that as the number of workplaces is increasing, using an orchestration tool for managing workplaces is essential. This is where that workplace server plays an important role.
- **Workspace server** supports the management of developer workplaces. It can give a dashboard for creating, ending, starting and sharing workspaces. This dashboard is similar to the management of a virtual machine on cloud technologies [55].
- **Theia platform**, Theia is a platform and some developers utilize products based on Theia (Che Theia). However, Che Theia is an editor for Eclipse Che based on the Theia platform and it is called Theia as well [56].
- **Eclipse Che**, is a developer workspace server for the Eclipse Theia platform. [57].

After understanding the basic concepts, it is helpful to learn about the structure of the Theia platform. Theia is available as a desktop and web-based application. To handle both architectures with a single source, Theia uses the client-server architecture. Moreover, Theia can also reuse other high-level frameworks, technologies, protocols, etc. Theia benefits from several technologies, including:

- **Keycloak**: In fact, Che Theia is an editor for Eclipse Che. Therefore, it is not responsible for multi-developer management on the system level, via containers or OS user rights. These services are related to Eclipse Che, which manages developer workspaces. Eclipse Che utilizes Keycloak to authenticate developers [57].
- **Monaco**: Not only Eclipse Theia has followed most of VS code objects but it uses VS code abilities such as Monaco code editor (the editor of VS Code), as well. This reuse allows developers to use VS Code extensions in Theia [58].
- **TypeScript**: Considering that JavaScript cannot support complex applications; Typescript is used to enhance maintainability. In this regard, Theia UI is fully implemented through Typescript [59].

- **InversifyJS**: Theia uses the dependency injection framework Inversify.js to compose and configure the frontend and backend applications. InversifyJS is a lightweight inversion of control (IoC) container for TypeScript and JavaScript apps. This feature contributes to extensibility and the future growth of products [60].
- **Language server protocol (LSP)**: Theia has a distributed architecture that needs to communicate between the client and the server. LSP is designed for communicating between a tool (the client) and a language smartness provider (the server). This protocol allows to implement the autocomplete feature of IDEs in client-server structure [61,62].

All of the mentioned technologies convert Theia into a platform for building IDEs and developing tools. Theia not only has most of the basic code editor features - highlight syntax, debugging, etc., but also, has significant expendability. Theia has core and other extensions that help to build extensions. This is where deep learning models can be extended as extensions for desktop or cloud IDE products. Also, researchers can use the benefits of VS code extensions in the VS code marketplace which contains several AI-based extensions [63]. This study has conducted some traditional and AI-based extensions that can be used as developer assistants (Table 2). In general, practical tools has focused on providing features. Therefore, finding theoretical background which is used in their development is difficult. However, there are some works on IDE extensions that have provided valuable information:

kite [64,65] is an auto-complete tool that supports many back-end and front-end languages such as Python, Java, JavaScript, etc. The unique feature of this plugin is the completion of multi-line codes. Kite uses GPT-2 (utilizing deep learning) which is a prepared general-purpose model called strong AI. This kind of learner can be used for different tasks, such as text translation, answering questions, summarizing passages, etc. This model used 25 million open-source code files that can be run locally.

Table 2. Traditional and AI-based extensions example

Title	Extension title	AI techniques	Ref
Kite	Auto-complete	DNN/GPT-2	[65]
Flutter	Debugger	–	[66]
DeepL	Auto-complete	DNN	[67]
VSearch	Code search	–	[68]
Voice-enabled programming	Virtual assistant	NLP techniques	[49]
Virtual assistant and skill templates	Virtual Assistant	–	[69]

DeepL, [67]. This API is a JSON-RPC API for direct translation of text from Visual Studio code. This plugin is a language translation service based on neural networks and deep learning algorithms, and is currently supported in German, English, French, Spanish, Italian, and Polish.

Voice-Enabled Programming Extension. [49] have introduced a VS Code extension that provides Voice-Enabled Programming. This model used the Language Understanding (LUIS) concept in the Microsoft Azure LUIS app. In general, extensions send query text to Azure and receive the results as JSON responses. Typical client extensions for LUIS are virtual assistants offering online chat communication via text or text-to-speech (chatbots).

In [15], the authors addressed the Eclipse Che structure. Figure 2 demonstrates the structure of AI-based extensions which can add to cloud based IDEs and bring the new structure (Fig. 3).

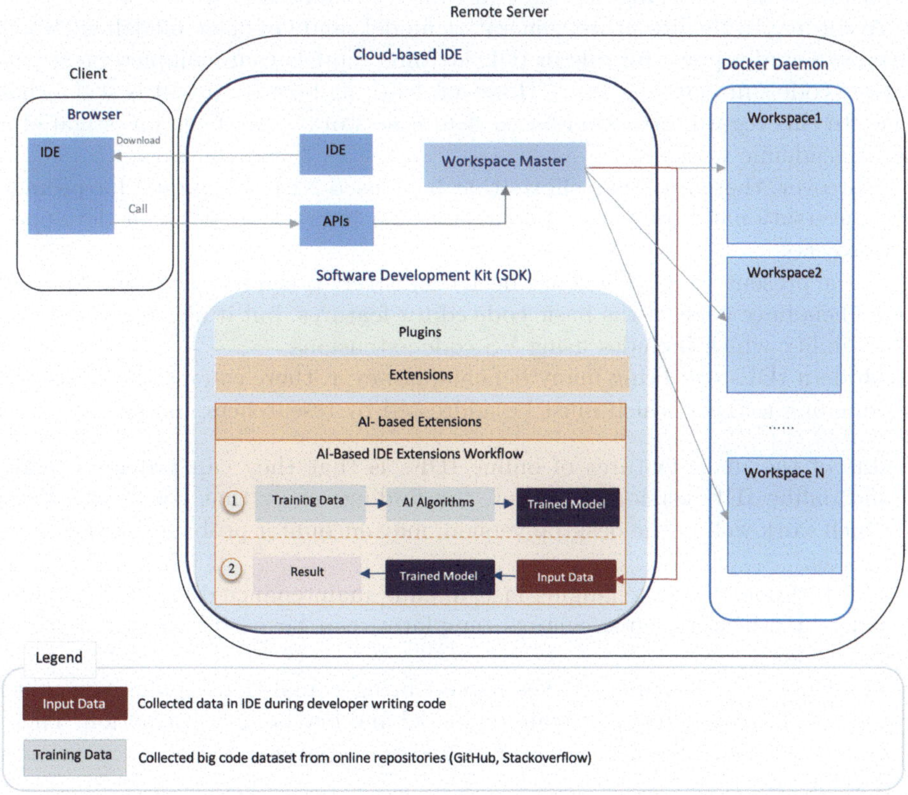

Fig. 3. Structure of AI-based extension in a Cloud IDE

6 Conclusion

There are some processes in software development such as implementing, debugging, bug detection, testing, etc. which can take the developers time. In this regard, IDEs include several tools that can be used to speed up these processes and provide automation. Recently, AI models have been utilized in IDEs to increase the automation. This growth can fall into two methods: 1) Improving current functionalities (features) 2) Adding new functionalities.

The first category includes improving current functionalities such as code autocomplete, code search, live templates, etc. These functionalities have been improved extensively by AI algorithms, especially through recommendation system models. The second category involves new IDE features such as AI virtual assistance, code summarization, etc. Some fields within AI have been more attended to so as to provide IDE features. Language modeling based on the sequence to sequence models, recommender systems, learning from existing code, and source code analyzing are some instances to name a few.

According to the literature, generative models and language modeling (RNN) have presented a powerful role in IDE learning models (autocomplete, test generation, code summarizing, etc.). However, learning long-term text is still a challenge. In this regard, attention-based neural networks have been investigated in recent academic works.

Moreover, there are some efforts that have used NLP techniques for promoting conversational experiences (Bots) and the interaction between developers and projects.

In the present study, Theia has been exclusively taken into account. Not only does Theia have most of the basic code editor features, but it also has significant extendibility which includes using VS code extensions.

Modern IDEs can bring many benefits, however, there can also be challenges in using online IDEs which must be addressed by researchers and engineers:

- One of the main features of online IDEs is that they can provide desktop and online IDE version. However, the findings show that some extensions, which work well in the desktop version, may encounter problems in the online version.
- Web technologies are changing fast, making maintenance difficult. Therefore, using high-level designs, such as templates, can be useful when developing extensions.
- The embedded learning models should bring a real-time application which needs to have a lightweight training model and low-latency prediction. Therefore, this feature should be considered when engineers choose an extension or develop it.
- When developing a Cloud IDE, it is important that it provide basic IDE features and support different languages, versioning tools, databases (SQL, NoSQL), cross-platform and multimedia development, online debugging, etc. Although most of the theses features can be found in extension market places, the serious projects need more guaranty. This is because maintenance, versioning, and extending these extensions may be challenging. Therefore, if a new IDE product is designed for the use of several extensions, engineers must carefully choose the extensions to minimize the risk that other extensions from other IDE functionalities have other requirements. Moreover, extensions should be investigated in terms of maintenance.

Acknowledgements. Supported by the project "Intelligent and sustainable mobility supported by multi-agent systems and edge computing (InEDGEMobility): Towards Sustainable Intelligent Mobility: Blockchain-based framework for IoT Security", Reference: RTI2018–095390-B-C32, financed by the Spanish Ministry of Science, Innovation and Universities (MCIU), the State Research Agency (AEI) and the European Regional Development Fund (FEDER).

References

1. Yigitcanlar, T., Butler, L., Windle, E., Desouza, K.C., Mehmood, R., Corchado, J.M.: Can building "artificially intelligent cities" safeguard humanity from natural disasters, pandemics, and other catastrophes? An urban scholar's perspective. Sensors **20**(10), 2988 (2020)
2. Chamoso, P., González-Briones, A., Prieta, F.D.L., Venyagamoorthy, G.K., Corchado, J.M.: Smart city as a distributed platform: toward a system for citizen-oriented management. Comput. Commun. **152**, 323–332 (2020)
3. Gasparic, M., Murphy, G.C., Ricci, F.: A context model for IDE-based recommendation systems. J. Syst. Softw. **128**, 200–219 (2017)
4. Theia, E.: Platform to develop Cloud & Desktop (2019). https://theia-ide.org/. Accessed 2020
5. Bengio, Y., Ducharme, R., Vincent, P., Jauvin, C.: A neural probabilistic language model. J. Mach. Learn. Res. **3**, 1137–1155 (2003)
6. Rustan, K., Leino, M., Wüstholz, V.: The Dafny integrated development environment. arxiv Preprint arxiv:1404.6602 (2014)
7. Cloud9, Cloud IDE. https://aws.amazon.com/cloud9/. Accessed 2021
8. Codeanywhere, Cloud IDE. https://codeanywhere.com/. Accessed 2021
9. Eclipse Che, Eclipse next-generation IDE. https://www.eclipse.org/che/. Accessed 2021
10. Omori, T., Hayashi, S., Maruyama, K.: A survey on methods of recording fine-grained operations on integrated development environments and their applications. Comput. Softw. **32**(1), 60–80 (2015)
11. Aho, T., et al.: Designing ide as a service. Commun. Cloud Softw. **1**(1) (2011)
12. Barenkamp, M., Rebstadt, J., Thomas, O.: Applications of AI in classical software engineering. AI Perspect. **2**(1), 1–15 (2020)
13. Corchado, J.M., et al.: Deepint.net: a rapid deployment platform for smart territories. Sensors **21**(1), 236 (2021)
14. Arora, P., Dixit, A.: Analysis of cloud IDEs for software development. Int. J. Eng. Res. General Sci. **4**(4) (2016)
15. Applis, L.: Theoretical evaluation of the potential advantages of cloud ides for research and didactics. In: SKILL 2019-Studierendenkonferenz Informatik (2019)
16. Lin, Z.-Q., et al.: Intelligent development environment and software knowledge graph. J. Comput. Sci. Technol. **32**(2), 242–249 (2017)
17. Allamanis, M., Barr, E.T., Devanbu, P., Sutton, C.: A survey of machine learning for big code and naturalness. ACM Comput. Surv. (CSUR) **51**(4), 1–37 (2018)
18. Wood, A., Rodeghero, P., Armaly, A., McMillan, C.: Detecting speech act types in developer question/answer conversations during bug repair. In: Proceedings of the 2018 26th ACM Joint Meeting on European Software Engineering Conference and Symposium on the Foundations of Software Engineering, pp. 491–502 (2018)
19. Cooper, K., Torczon, L.: Engineering a Compiler. Elsevier, Amsterdam (2011)

20. Wang, Z., O'Boyle, M.: Machine learning in compiler optimization. Proc. IEEE **106**(11), 1879–1901 (2018)
21. Chen, T., et al.: {TVM}: an automated end-to-end optimizing compiler for deep learning. In: 13th {USENIX} Symposium on Operating Systems Design and Implementation ({OSDI} 2018), pp. 578–594 (2018)
22. Nguyen, A.T., et al.: API code recommendation using statistical learning from fine-grained changes. In: Proceedings of the 2016 24th ACM SIGSOFT International Symposium on Foundations of Software Engineering, pp. 511–522 (2016)
23. Loaiza, F.L., Wheeler, D.A., Birdwell, J.D.: A partial survey on AI technologies applicable to automated source code generation. Technical report, Institute for Defense Analyses Alexandria United States (2019)
24. TabNine, Autocompletion with deep learning 2019. https://www.kite.com/. Accessed 2020
25. Gazzola, L., Micucci, D., Mariani, L.: Automatic software repair: a survey. IEEE Trans. Software Eng. **45**(1), 34–67 (2017)
26. Martinez, M., Monperrus, M.: Astor: exploring the design space of generate-and-validate program repair beyond GenProg. J. Syst. Softw. **151**, 65–80 (2019)
27. Hata, H., Shihab, E., Neubig, G.: Learning to generate corrective patches using neural machine translation. arXiv preprint arXiv:1812.07170 (2018)
28. Chen, Z., Kommrusch, S.J., Tufano, M., Pouchet, L.-N., Poshyvanyk, D., Monperrus, M.: SEQUENCER: sequence-to-sequence learning for end-to-end program repair. IEEE Trans. Softw. Eng. (2019)
29. Gu, X., Zhang, H., Kim, S.: Deep code search. In: 2018 IEEE/ACM 40th International Conference on Software Engineering (ICSE), pp. 933–944. IEEE (2018)
30. Cambronero, J., Li, H., Kim, S., Sen, K., Chandra, S.: When deep learning met code search. In: Proceedings of the 2019 27th ACM Joint Meeting on European Software Engineering Conference and Symposium on the Foundations of Software Engineering, pp. 964–974 (2019)
31. Portolan, M.: Automated testing flow: the present and the future. IEEE Trans. Comput. Aided Des. Integr. Circuits Syst. **39**(10), 2952–2963 (2019)
32. Godefroid, P., Singh, R., Peleg, H.: Machine learning for input fuzzing. US Patent App. 15/638,938, 4 October 2018
33. Sutton, M., Greene, A., Amini, P.: Fuzzing: Brute Force Vulnerability Discovery. Pearson Education (2007)
34. Godefroid, P., Levin, M.Y., Molnar, D.: Automated whitebox fuzz testing. In: Proceedings of NDSS (2008)
35. Gupta, R., Pal, S., Kanade, A., Shevade, S.: DeepFix: fixing common C language errors by deep learning. In: Proceedings of the AAAI Conference on Artificial Intelligence, vol. 31 (2017)
36. Casado-Vara, R., Rey, A.M.-d., Affes, S., Prieto, J., Corchado, J.M. : IoT network slicing on virtual layers of homogeneous data for improved algorithm operation in smart buildings. Future Gener. Comput. Syst. **102**, 965–977 (2020)
37. Coronado, E., Mastrogiovanni, F., Indurkhya, B., Venture, G.: Visual programming environments for end-user development of intelligent and social robots, a systematic review. J. Comput. Lang. **58**, 100970 (2020)
38. Vinyals, O., Toshev, A., Bengio, S., Erhan, D.: Show and tell: a neural image caption generator. In: Proceedings of the IEEE Conference on Computer Vision and Pattern Recognition, pp. 3156–3164 (2015)
39. Beltramelli, T.: pix2code: generating code from a graphical user interface screenshot. In: Proceedings of the ACM SIGCHI Symposium on Engineering Interactive Computing Systems, pp. 1–6 (2018)

40. Pang, X., Zhou, Y., Li, P., Lin, W., Wu, W., Wang, J.Z.: A novel syntax-aware automatic graphics code generation with attention-based deep neural network. J. Netw. Comput. Appl. **161**, 102636 (2020)
41. JetBrains, High-speed coding with Custom Live Templates. https://www.jetbrains.com/help/idea/using-live-templates.html. Accessed 2020
42. Murphy-Hill, E.: Continuous social screencasting to facilitate software tool discovery. In: 2012 34th International Conference on Software Engineering (ICSE), pp. 1317–1320. IEEE (2012)
43. Gasparic, M., Janes, A., Ricci, F., Murphy, G.C., Gurbanov, T.: A graphical user interface for presenting integrated development environment command recommendations: design, evaluation, and implementation. Inf. Softw. Technol. **92**, 236–255 (2017)
44. Gasparic, M., Gurbanov, T., Ricci, F.: Improving integrated development environment commands knowledge with recommender systems. In: Proceedings of the 40th International Conference on Software Engineering: Software Engineering Education and Training, pp. 88–97 (2018)
45. LeClair, A., Haque, S., Wu, L., McMillan, C.: Improved code summarization via a graph neural network. In: Proceedings of the 28th International Conference on Program Comprehension, ICPC 2020, pp. 184–195. Association for Computing Machinery, New York (2020)
46. Oda, Y., et al.: Learning to generate pseudo-code from source code using statistical machine translation (t). In: 2015 30th IEEE/ACM International Conference on Automated Software Engineering (ASE), pp. 574–584. IEEE (2015)
47. Iyer, S., Konstas, I., Cheung, A., Zettlemoyer, L.: Summarizing source code using a neural attention model. In: Proceedings of the 54th Annual Meeting of the Association for Computational Linguistics (Volume 1: Long Papers), pp. 2073–2083 (2016)
48. Bedia, M.G., Rodríguez, J.M.C., et al.: A planning strategy based on variational calculus for deliberative agents (2002)
49. Joshi, P., Bein, D.: Audible code, a voice-enabled programming extension of visual studio code. In: Latifi, S. (eds.) 17th International Conference on Information Technology-New Generations (ITNG 2020). AISC, vol. 1134, pp. 335–341. Springer, Cham (2020). https://doi.org/10.1007/978-3-030-43020-7_44
50. Virtual Assistant and Skill Templates. https://marketplace.visualstudio.com/items?itemName=BotBuilder.VirtualAssistantTemplate. Accessed 2020
51. Xu, F.F., Vasilescu, B., Neubig, G.: In-ide code generation from natural language: promise and challenges. arXiv preprint arXiv:2101.11149 (2021)
52. Wong, E., Yang, J., Tan, L.: Autocomment: mining question and answer sites for automatic comment generation. In: 2013 28th IEEE/ACM International Conference on Automated Software Engineering (ASE), pp. 562–567. IEEE (2013)
53. Xing, H., Li, G., Xia, X., Lo, D., Jin, Z.: Deep code comment generation with hybrid lexical and syntactical information. Empir. Softw. Eng. **25**(3), 2179–2217 (2020)
54. Sidhanth, N., Sanjeev, S., Swettha, S., Srividya, R.: A next generation ide through multi tenant approach. Int. J. Inf. Electron. Eng. **4**(1), 27 (2014)
55. Shi, S., Li, Q., Le, W., Xue, W., Zhang, Y., Cai, Y.: Intelligent workspace. US Patent 9,026,921, 5 May 2015
56. Eclipse Foundation (2020). https://ecdtools.eclipse.org/. Accessed 2021

57. Saini, R., Bali, S., Mussbacher, G.: Towards web collaborative modelling for the user requirements notation using Eclipse Che and Theia IDE. In: 2019 IEEE/ACM 11th International Workshop on Modelling in Software Engineering (MiSE), pp. 15–18. IEEE (2019)
58. Kahlert, T., Giza, K.: Visual studio code tips & tricks, vol. 1. Microsoft Deutschland GmbH (2016)
59. Bierman, G., Abadi, M., Torgersen, M.: Understanding typescript. In: Jones, R. (eds.) ECOOP 2014. LNCS, vol. 8586, pp. 257–281. Springer, Heidelberg (2014). https://doi.org/10.1007/978-3-662-44202-9_11
60. Inversify, lightweight inversion of control (IoC) container for TypeScript and JavaScript apps (2018). https://github.com/inversify/InversifyJS. Accessed 2021
61. langserver, Language Server protocol. https://langserver.org/. Accessed 2020
62. Bünder, H.: Decoupling language and editor-the impact of the language server protocol on textual domain-specific languages. In: MODELSWARD, pp. 129–140 (2019)
63. Microsoft. VS Marketplace, Extensions for the Visual Studio products. https://marketplace.visualstudio.com/. Accessed 2021
64. Kite, AI powered code completions (2019). https://www.kite.com/. Accessed 2020
65. Kite visualstudio. https://marketplace.visualstudio.com. Accessed 2021
66. Flutter, Dart-Code. https://marketplace.visualstudio.com/items?itemName=Dart-Code.flutter. Accessed 2021
67. deepl, AI powered code completions (2019). https://www.deepl.com/en/docs-api/. Accessed 2020
68. VSearch code. https://marketplace.visualstudio.com/items?itemName=mario-0.VSearch102. Accessed 2021
69. Virtual Assistant and Skill Templates. https://marketplace.visualstudio.com/items?itemName=BotBuilder.VirtualAssistantTemplate. Accessed 2021

Smart Cyber Victimization Discovery on Twitter

Niloufar Shoeibi[1(✉)], Nastaran Shoeibi[2], Vicente Julian[3], Sascha Ossowski[4], Angelica González Arrieta[5], and Pablo Chamoso[1]

[1] BISITE Research Group, University of Salamanca, Salamanca, Spain
Niloufar.Shoeibi@usal.es
[2] Babol Noshirvani University of Technology, Babol, Iran
[3] Department of Computer Systems and Computation, Universitat Politècnica de València, València, Spain
[4] Universidad Rey Juan Carlos, Madrid, Spain
[5] Department of Computer and Automation, University of Salamanca, Salamanca, Spain
https://bisite.usal.es/

Abstract. The advancement of technologies, the promotion of smartphones, and social networking have led to a high tendency among users to spend more time online interacting with each other via the available technologies. This is because they help overcome physical limitations and save time and energy by doing everything online. The rapid growth in this tendency has created the need for extra protection, by creating new rules and policies. However, sometimes users interrupt these rules and policies through unethical behavior. For example, bullying on social media platforms is a type of cyber victimization that can cause serious harm to individuals, leading to suicide. A firm step towards protecting the cyber society from victimization is to detect the topics that trigger the feeling of being a victim. In this paper, the focus is on Twitter, but it can be expanded to other platforms. The proposed method discovers cyber victimization by detecting the type of behavior leading to them being a victim. It consists of a text classification model, that is trained with a collected dataset of the official news since 2000, about suicide, self-harm, and cyberbullying. Results show that LinearSVC performs slightly better with an accuracy of 96%.

Keywords: Twitter · Cyberbullying · Suicide and self-harm · Cyber victim · Text classification · Text feature extraction

1 Introduction

Technological advancements, the popularity of online social networking sites, and having internet access, all contribute greatly to the quality of life but also have some ill effects, such as cyberattacks, cybercrimes, and cyberbullying. Therefore,

J. M. Corchado and S. Trabelsi (Eds.): SSCTIC 2021, LNNS 253, pp. 289–299, 2022.
https://doi.org/10.1007/978-3-030-78901-5_25

cybersecurity is a crucial matter for researchers, and detecting cyberbullies will lead to improving people's mental health and to making social networking sites safer [1].

Bullying is a repetitive, aggressive behavior that includes physical, verbal and social intimidation. Cyberbullying appeared as a new way of bullying and aggression with the use of digital technologies and can take place on social media and messaging platforms [2]. It appears as harassment, cyberstalking, cyberthreats, happy slapping, impersonation and denigration etc., and can lead to various health issues including mental, emotional and physical problems alongside face-to-face bullying. It has serious effects on both the victim and the aggressor. People who are bullied tend to be more insecure, can't concentrate, have depression, anxiety, self-harm, and even suicidal thinking and attempts. People who bully are more likely to abuse and harm others, do drugs and have behavioral issues [3].

Anyone can be a victim of cyberbullying, so it is important to identify it and report it to stop the cyber victimization. Social network platforms are trying their best to detect cyberbullying by improving their features and privacy policies. Even though there are lots of difficulties in implementing cyberbullying detection tools because of the human behavior is stochastic, and arbitrary, there are a lot of factors affecting the behavior of a person, the lack of datasets [4].

The proposed method focuses on discovering cyberbullying and prevent the future consequences and serious issues, such as self-harm and suicide attempts, in order to guarantee a peaceful and safe cybersociety. There are challenges in detecting cyberbullying; manual detection is time consuming, requires human involvement and is frustrating. There are few datasets available for this purpose, so most of them are labeled manually and because of the limited length of the tweets on twitter, only 168 characters can be used therefore it makes detection more difficult. Due to the lack of datasets and the need for a more complete dataset, a dataset has been created on the basis of the official news using official Google News API [5]since 2000 (New York Times news, etc.). The proposed architecture consists of two modules. One downloads the tweets from the Twitter platform and the other one is a text classification model which detects if the input text (tweet) is related to cyberbully, self-harm, and suicide with the accuracy of 96%.

This paper has been organized as follows: In Sect. 2, the related work is reviewed. Then, in Sect. 3, the overview and the architecture of the proposed method are presented. Finally, the results, conclusion and future work are discussed in Sect. 4.

2 Review of the State of the Art

In the field of user behavior mining on social media platforms, many studies have been carried out [6–9] and still, many doors are open to researchers in this area, to discover greater knowledge about human behavior in many different situations. In this paper, the focus is on cyber victimization detection and prevention, especially on Twitter, which is the most news-friendly social media platform and is the main target for investigating cyberbullying and the related psychological issues.

V. Balakrishnan et al. in [10] proposed a detection method to reduce cyberbullying in the basis of Twitter users' psychological characteristics like feelings and personalities. Users' personalities defined with Big Five and Dark Triad models, then they used machine learning classifiers like Naïve Bayes, Random Forest, and J48 for classifying tweets into four sections: bully, aggressor, spammer, and normal tweets. Results show that analyzing traits like extraversion, agreeableness, neuroticism, and psychopathy has a great impact on identifying online bullies.

In [11], researchers proposed techniques for the detection of cyberbullying and presented a comparative analysis, classifying multiple methods for cyberbullying detection. Many of them use the SVM classifier and have illimitable results, and one method used the unsupervised approach, and it is more complicated. The identification of cyber aggression is an essential factor in predicting cyberbullying, and user profile legitimacy detection plays a significant role in it.

In future smart cities [12–15] as well as the current physical world, issues such as bullying, harassment, and hate speech must be counteracted. Kumari et al. used the contents of social media to identify the cyberbullies in texts and images. It explains the single-layer convolutional Neural Network has better results with a unified representation of both text and image. Using text as an image is a more suitable model for data encoding. They applied three layers of text and three layers of a color image to interpret the input that presents a recall of 74% of the bullying class with one layer of Convolutional Neural Network [16].

Many studies concentrate on improving the cyberbullying detection performance of machine learning algorithms, as proposed models cause and strengthen unintended social biases. O. Gencoglu et al. in [17] introduced a model training method that uses fairness constraints and operates with different datasets. The result shows that varieties of unintended biases can be successfully mitigated without reducing the model's quality.

Muneer et al. in [18] applied seven different machine learning classifiers, namely, Logistic Regression (LR), Light Gradient Boosting Machine (LGBM), Stochastic Gradient Descent (SGD), Random Forest (RF), AdaBoost (ADB), Naive Bayes (NB), and Support Vector Machine (SVM) on a global dataset of 37,373 tweets from Twitter to detect cyberbullying without affecting the victims. These algorithms use accuracy, precision, recall, and F1 score as performance factors to conclude classifiers' recognition rate applied to the global dataset. Results indicate LR has a median accuracy of around 90.57%. Logistic regression obtained the best F1 score (0.928), SGD obtained the best precision (0.968), and SVM has the best recall (1.00).

Most of the existing cyberbullying detection techniques are supervised by a human and take time. However, Cheng et al. in [19] introduced an unsupervised cyberbullying detection model that has better performance than supervised models. This model includes two components: (1) a representation learning network for encoding social media using multi-modal features and (2) a multi-task learning network that identifies the bullies with a Gaussian Mixture Model. Their proposed model optimizes the parameters of both components for getting the liabilities of decoupled training.

Z. Abbass et al. in [20] proposed a three module framework: data preprocessing, classifying model builder, and prediction. For data classification, Multinomial Naïve Bayes (MNB), K-Nearest Neighbors (KNN), and Support Vector Machine (SVM) are used to build the prediction model. These algorithms achieve the precision, recall, and F-measure above 0.9. Also, the support vector machine performed better. This system has better accuracy than the existing network-based feature selection approach.

Balakrishnan et al. in [21]showed the relationship between personality traits and cyberbullying and introduced a way to detect cyberbullying by defining Big five and Dark Triad features. For cyberbullying classification, they used the Random Forest algorithm combined with a baseline algorithm including some Twitter features (i.e. amount of mentions, amount of followers and following, reputation, favorite count, status count, and the number of hashtags). Big Five and Dark Triad are notable in finding bullies, obtaining up to 96% (precision) and 95% (recall).

Automatic cyberbullying detection may help stop harassment and bullies on social media, using manually engineered features. Sadiq et al. in [22] applied multilayer perceptron and analyzed the state-of-the-art combination of CNNLSTM and CNN-BiLSTM in the deep neural network. This model identifies cyber harassments with 92% accuracy.

In Table 1, the summary of the selected papers related to social media user behavior mining focusing on cyberbully detection is presented, including the method proposed in each paper.

Table 1. The review of the state of the art on social media user behavior mining by focusing on the sub area of cyberbully detection.

Paper title	Area	Sub-area	Methodology
Improving cyberbullying detection using Twitter users' psychological features and machine learning[10]	Social Media Behavior Mining	Cyberbully Classification	Naïve Bayes, Random Forest, and J48
Taxonomy of Cyberbullying Detection and Prediction Techniques in Online Social Networks [11]	Social Media Behavior Mining	Cyberbully Detection	SVM
Towards Cyberbullying-free social media in smart cities: a unified multi-modal approach [16]	Social Media Behavior Mining	Cyberbully Detection	CNN
Cyberbullying Detection with Fairness Constraints [17]	Social Media Behavior Mining	Cyberbully Detection	Fairness constraints
A Comparative Analysis of Machine Learning Techniques for Cyberbullying Detection on Twitter [18]	Social Media Behavior Mining	Cyberbully Detection	LR, LGBM, SGD, RF, ADB, NB, and SVM
Unsupervised cyberbullying detection via time-informed gaussian mixture model	Social Media Behavior Mining [19]	Cyberbully Detection	Gaussian Mixture Model
A Framework to Predict Social Crime through Twitter Tweets By Using Machine Learning [20]	Social Media Behavior Mining	Cyberbully Detection	MNB,KNN, and SVM
Cyberbullying detection on twitter using Big Five and Dark Triad features [21]	Social Media Behavior Mining	Cyberbully Detection	Random Forest
Aggression detection through deep neural model on Twitter [22]	Social Media Behavior Mining	Cyberbully Detection	Combination of CNN-LSTM and CNN-BiLSTM

3 The Proposed Architecture for Cyber Victimization Detection

As has been discussed in the previous sections, finding cAs has been discussed in the previous sections, finding cyberbullying victims is crucial to stop self-harm and suicide attempts. It can help public organizations guarantee a safe cyber society by discovering the victims. Building a trustable dataset to solve this problem, has a significant value. For this reason, the official news released since 2000 has been taken into account. Figure 1 represents the distribution of the data by different Media channels.

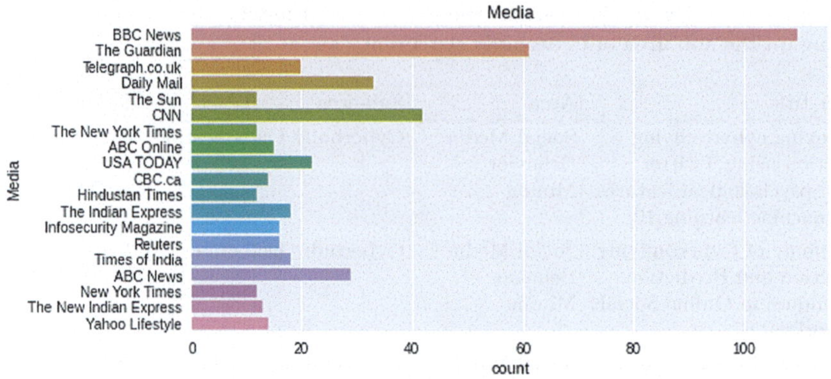

Fig. 1. Distribution of the data provided by official News Media.

Figure 2, represents the distribution of the data in the two different classes, also, it is understandable that the dataset is balanced, within the total of 1624 unique news articles.

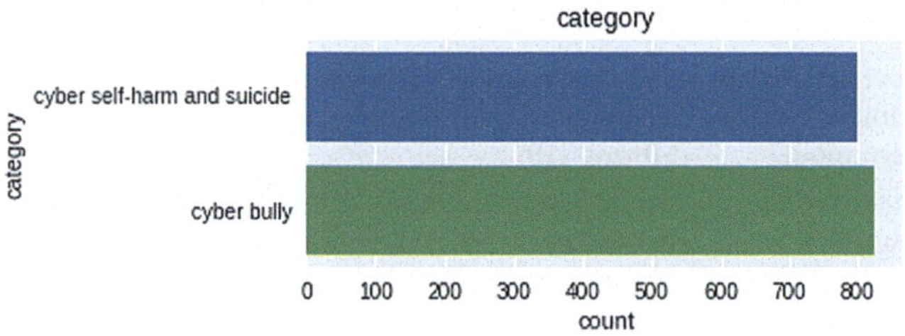

Fig. 2. Distribution of the data in the two categories.

The architecture proposed for cyber victimization detection has been presented in Fig. 3. This model consists of different stages, as discussed below.

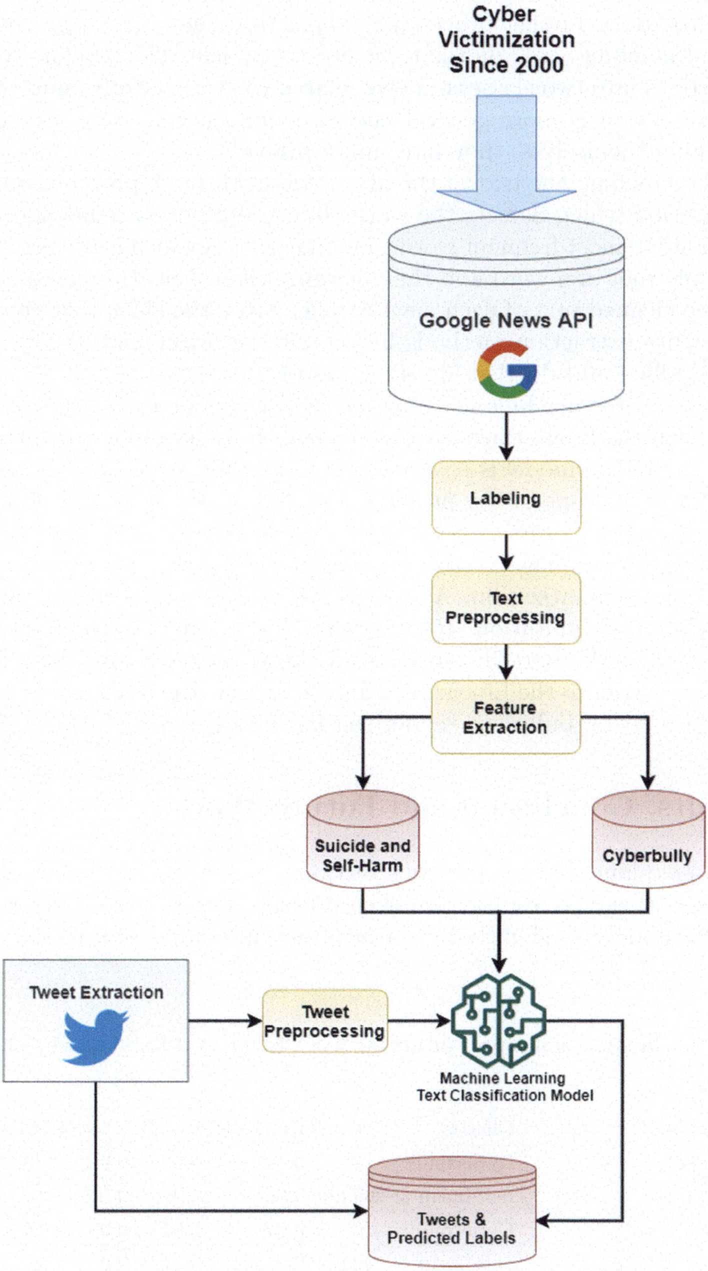

Fig. 3. The Architecture of Cyber Victimization Discovery.

First, a list of queries related to cyber victimization has been created, and using Google News official API, the news items related to each query are identified and stored for further processing. Then, these news items go through the procedure of labeling based on the topic of each article. The labeling component divides the data into two classes of news related to "CyberBully" and "Self-harm and Suicide" which is more general and also more urgent. If a user is posting about harming themselves, they have more priority.

After the labeling, the text of the articles goes through preprocessing including, tokenization which detects the words in the sentences, removing stopwords in English and 10 most frequent words, lemmatization which is the act of extracting the simple root of a word and then merging the tokens (preprocessed words) to create the cleaned text of each news article. Next, the clean text goes through the text feature extraction methods like count vectorizer and Tf-idf. Then, the dataset is shuffled and divided into train and test datasets.

In the end, three machine learning models are trained with this dataset and the model with the highest accuracy is selected to be used for the further steps.

As the aim of the model is to detect cyber victimization on Twitter, a query is done within the scope of the problem and the tweets are saved in a database ready to be processed. First, each tweet is preprocessed including tokenization, translation if the language of the tweet is not English, spell check, removing stopwords, and lemmatization. A spell check is necessary because on Twitter, due to the character limitation (168 characters), users tend to compact the words to be able to include more information in the tweet. After all these steps, the cleaned text is given to the Linear SVC model so that the label can be predicted. The labels are "cyberbully" or "cyber self-harm and suicide."

4 Results, Conclusion and Future Work

As has been explained in the previous section, three different classification models have been tried, the results have been presented in Table 2. It shows that the Linear SVC model has slightly better accuracy in comparison to the other two models.

Table 2. The classification report of Linear SVC, SGD, and Random Forest Classifier.

Model name	Classes	Precision	Recall	F1-score	Accuracy
LinearSVC	Cyberbully	0.96	0.95	0.96	0.96
	Self-harm & suicide	0.96	0.97	0.96	
SGD	Cyberbully	0.97	0.94	0.95	0.9569
	Self-harm & suicide	0.95	0.97	0.96	
Random Forest Classifier	Cyberbully	0.97	0.93	0.95	0.9507
	Self-harm & suicide	0.94	0.97	0.95	

Here are some samples of the tweets, completely anonymized that have been detected with the bully and self-harm and suicide related content presented in Fig. 4.

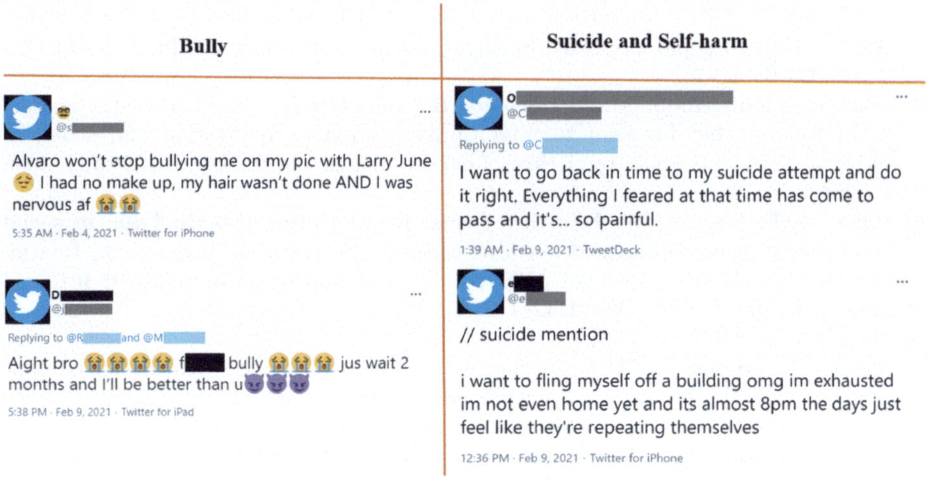

Fig. 4. The labeled tweet samples.

In this paper, a model has been presented that helps public organizations discover cyber victimization. The motivation behind building this model is to help the victims of cyberbullying and the victims who mention harming themselves or committing suicide. The architecture includes a text classification model whose accuracy is 96% (Linear SVC model) that has been trained with the official news published since 2000 within the scope of the cyber victimization by using count vectorizer and Tf-idf. Then, a query on Twitter is done by utilizing the official Twitter APIs. Later, tweets are preprocessed and after cleaning the tweets, this model is used to predict the label of the tweets related to this subject.

In the future, the plan is to expand the boundaries by identifying the aggressors as well as the victims, by reviewing the timeline and their activities to help them and motivate them to maintain a healthier lifestyle. Moreover, in the future the proposal will be implemented on the deepint.net platform which supports all types of data and contains a full suite of artificial intelligence techniques for data analysis, including data classification, clustering, prediction, optimization, and visualization techniques [23]. These abilities make it a perfect choice for implementing the proposal.

Acknowledgments. This research has been supported by the project "Intelligent and sustainable mobility supported by multi-agent systems and edge computing (InEDGE-Mobility): Towards Sustainable Intelligent Mobility: Blockchain-based framework for IoT Security", Reference: RTI2018-095390-B-C31/32/33, financed by the Spanish Ministry of Science, Innovation and Universities (MCIU), the State Research Agency (AEI) and the European Regional Development Fund (FEDER).

References

1. Bussey, K., Luo, A., Fitzpatrick, S., Allison, K.: Defending victims of cyberbullying: the role of self-efficacy and moral disengagement. J. School Psychol. **78**, 1–12 (2020)
2. Smith, P.K.: Research on cyberbullying: strengths and limitations. In: Vandebosch, H., Green, L. (eds.) Narratives in Research and Interventions on Cyberbullying Among Young People, pp. 9–27. Springer, Cham (2019). https://doi.org/10.1007/978-3-030-04960-7_2
3. Martínez-Monteagudo, M.C., Delgado, B., Díaz-Herrero, Á., García-Fernández, J.M.: Relationship between suicidal thinking, anxiety, depression and stress in university students who are victims of cyberbullying. Psychiatry Res. **286**, 112856 (2020)
4. Raza, M.O., Memon, M., Bhatti, S., Bux, R.: Detecting cyberbullying in social commentary using supervised machine learning. In: Arai, K., Kapoor, S., Bhatia, R. (eds.) FICC 2020. AISC, vol. 1130, pp. 621–630, Springer, Cham (2020). https://doi.org/10.1007/978-3-030-39442-4_45
5. Google news API
6. Shoeibi, N., Mateos, A.M., Camacho, A.R., Corchado, J.M.: A feature based approach on behavior analysis of the users on Twitter: a case study of AusOpen tennis championship. In: Dong, Y., Herrera-Viedma, E., Matsui, K., Omatsu, S., González Briones, A., Rodríguez González, S. (eds.) DCAI 2020. AISC, vol. 1237, pp. 284–294, Springer, Cham (2020). https://doi.org/10.1007/978-3-030-53036-5_31
7. Su, Y.-S., Wu, S.-Y.: Applying data mining techniques to explore user behaviors and watching video patterns in converged it environments. J. Ambient Intell. Humaniz. Comput. 1–8 (2021). https://doi.org/10.1007/s12652-020-02712-6
8. Shoeibi, N.: Analysis of self-presentation and self-verification of the users on Twitter. In: Rodráguez González S., et al. (eds.) DCAI 2020. AISC, vol. 1242, pp. 221–226, Springer, Cham (2020). https://doi.org/10.1007/978-3-030-53829-3_25
9. Marmo, R.: Social media mining. In: Encyclopedia of Organizational Knowledge, Administration, and Technology, pp. 2153–2165. IGI Global (2021)
10. Balakrishnan, V., Khan, S., Arabnia, H.R.: Improving cyberbullying detection using Twitter users' psychological features and machine learning. Comput. Secur. **90**, 101710 (2020)
11. Vyawahare, M., Chatterjee, M.: Taxonomy of cyberbullying detection and prediction techniques in online social networks. In: Jain, L.C., Tsihrintzis, G.A., Balas, V.E., Sharma, D.K. (eds.) Data Communication and Networks. AISC, vol. 1049, pp. 21–37. Springer, Singapore (2020). https://doi.org/10.1007/978-981-15-0132-6_3
12. Chamoso, P., González-Briones, A., De La Prieta, F., Venyagamoorthy, G.K., Corchado, J.M.: Smart city as a distributed platform: toward a system for citizen-oriented management. Comput. Commun. **152**, 323–332 (2020)
13. Yigitcanlar, T., Butler, L., Windle, E., Desouza, K.C., Mehmood, R., Corchado, J.M.: Can building "artificially intelligent cities" safeguard humanity from natural disasters, pandemics, and other catastrophes? An urban scholar's perspective. Sensors **20**(10), 2988 (2020)
14. Casado-Vara, R., Rey, A.M.-d., Affes, S., Prieto, J., Corchado, J.M.: IoT network slicing on virtual layers of homogeneous data for improved algorithm operation in smart buildings. Future Gener. Comput. Syst. **102**, 965–977 (2020)
15. González Bedia, M., Corchado Rodríguez, J.M., et al.: A planning strategy based on variational calculus for deliberative agents (2002)

16. Kumari, K., Singh, J.P., Dwivedi, Y.K., Rana, N.P.: Towards cyberbullying-free social media in smart cities: a unified multi-modal approach. Soft Comput. **24**(15), 11059–11070 (2020)
17. Gencoglu, O.: Cyberbullying detection with fairness constraints. IEEE Internet Comput. **25**, 20–29 (2020)
18. Muneer, A., Fati, S.M.: A comparative analysis of machine learning techniques for cyberbullying detection on Twitter. Future Internet **12**(11), 187 (2020)
19. Cheng, L., Shu, K., Wu, S., Silva, Y.N., Hall, D.L., Liu, H.: Unsupervised cyberbullying detection via time-informed Gaussian mixture model. arXiv preprint arXiv:2008.02642 (2020)
20. Abbass, Z., Ali, Z., Ali, M., Akbar, B., Saleem, A.: A framework to predict social crime through Twitter tweets by using machine learning. In: 2020 IEEE 14th International Conference on Semantic Computing (ICSC), pp. 363–368 (2020)
21. Balakrishnan, V., Khan, S., Fernandez, T., Arabnia, H.R.: Cyberbullying detection on Twitter using big five and dark triad features. Pers. Individ. Differ. **141**, 252–257 (2019)
22. Sadiq, S., Mehmood, A., Ullah, S., Ahmad, M., Choi, G.S., On, B.-W.: Aggression detection through deep neural model on twitter. Future Gener. Comput. Syst. **114**, 120–129 (2021)
23. Corchado, J.M., Chamoso, P., Hernández, G., Gutierrez, A.S.R., Camacho, A.R., González-Briones, A., Pinto-Santos, F., Goyenechea, E., Garcia-Retuerta, D., Alonso-Miguel, M., et al.: Deepint. net: A rapid deployment platform for smart territories. Sensors **21**(1), 236 (2021). Multidisciplinary Digital Publishing Institute

Reducing Emissions Prioritising Transport Utility

Holger Billhardt[1], Alberto Fernández[1(✉)], Sandra Gómez-Gálvez[1], Pasqual Martí[2], Javier Prieto Tejedor[3], and Sascha Ossowski[1]

[1] CETINIA, Universidad Rey Juan Carlos, Madrid, Spain
{holger.billhardt,alberto.fernandez,sandra.gomez.galvez, sascha.ossowski}@urjc.es

[2] Valencian Research Institute for Artificial Intelligence (VRAIN), Universitat Politècnica de València, Valencia, Spain
pasmargi@vrain.upv.es

[3] BISITE Research Group, University of Salamanca, Edificio Multiusos I+D+i, 37007 Salamanca, Spain
javierp@usal.es

Abstract. The regulation of mobility and traffic for the transportation of goods and the movement of people is one of the key issues local authorities are faced with, especially in large urban areas. The aim is to provide efficient mobility services that allow people the freedom to move within their cities as well as to facilitate the distribution of goods. However, the provisioning of transportation services should go in line with other general objectives, like reducing emissions and having more healthy living environments. In this context, we argue that one way to achieve objectives is to limit the use of transportation infrastructure elements and to assign the corresponding resources dynamically and in a prioritised manner to the traffic activities that have a higher utility from the point of view of the society, that is, activities that i) produce less pollution and ii) provide more value to society. Different mechanisms that restrict and control the access to an urban area based on pollution levels in that area are already in use in cities such as Madrid or London, but their level of dynamicity and adaptiveness is limited. In this paper we go beyond these approaches, and propose a prioritised access control approach that is highly dynamic, specific to individual vehicles, and that considers social utility or transportation efficiency. We provide a general model for our approach and instantiate it on a use case for last-mile delivery. We accomplish several experiments using the SUMO traffic simulation tool, to evaluate our proposal.

Keywords: Traffic management · Last-mile delivery · Prioritized resource allocation · Agreement technologies

1 Introduction

The organization of urban mobility and transportation is a field that has received tremendous changes, as well as a remarkable interest in the last years. Not only the desire of

J. M. Corchado and S. Trabelsi (Eds.): SSCTIC 2021, LNNS 253, pp. 300–311, 2022.
https://doi.org/10.1007/978-3-030-78901-5_26

people to move freely within the cities, but also the transformation of customer habits towards a more and more online acquisition of goods and the subsequent logistic requirements, tend to increase the traffic in big cities. At the same time the consciousness and sensibility has grown regarding environmental pollution and its effects on public health and the quality of life of citizens. In this context, authorities of big cities all over the world are faced with the problem of providing efficient transportation solutions reducing at the same time traffic-related problems like traffic jams or environmental pollution. In parallel to this trend, both, research and industry, have proposed and provided new innovative solutions for more environmental-friendly means of transportations, like new types of vehicles (e.g., electric cars, scooters, bikes and so on) or new transportation services that are based on the concept of "collaborative economy" or "collaborative consumption" [1] and aim at a more efficient usage of available resources, e.g., the sharing of transportation means for several transportation tasks.

We argue that in this new context also the management of the usage of transportation infrastructures can help to control environmental pollution while facilitating at a same time the transportation of goods and people in big cities. Here we understand as infrastructures all those facilitating elements or resources of a transport system that are used in a shared manner by different users and at different times, such as traffic lights, roads, tracks and lanes, parking spaces, etc. The availability of such resources is usually limited, and we believe that new usage schemas should be set up that prioritize vehicles or transportation services that are more environmental-friendly and have a higher transportation efficiency or are more important from a social point of view. A simple, already existing example of this idea is the ability of ambulances to cross any crossroad when they carry patients in life-threatening situations.

In this paper, we present a novel approach towards smart infrastructures that adaptively limits the access of vehicles to certain parts of a city based on the measured pollution. The idea is to specify regulation devices at all entry points of a specific sensible area of a city that tracks the current pollution and dynamically grants or restricts the access of vehicles based on admissible pollution values. In this process, the devices prioritize the access of vehicles that have a higher importance from a social point of view.

Section 2 presents some related work and Sect. 3 gives a general formalisation for our prioritized access model. In Sect. 4 we instantiate the model with a use case of a parcel delivery service. Here, the importance of a delivery task is measured in terms of the number of parcels a vehicle is transporting. We present different experiments using the traffic simulation tool SUMO [10]. Finally, Sect. 5 concludes the paper.

2 Related Work

A lot of works can be found in the Intelligent Transportation Systems literature that aim at smart solutions to traffic control in big cities. The expansion of smart road infrastructures, supported by vehicle-to-vehicle and vehicle-to-infrastructure communications, opened a field for experimenting with a variety of methods and approaches to address different challenges.

The prioritised use of road traffic infrastructures has been commonly regulated by means of traffic lights or smart intersections [2]. While traffic light actions (phases) make no distinction of the vehicles demanding access, intersection management can deal with individual agents. This is the case of the reservation-based control system proposed by [3], which allows autonomous vehicles to negotiate with intersection managers time and space slots to cross the intersection. That system has been extended by different authors, for example by market-based mechanisms to prioritise the access to networks of intersections [4].

Traffic control systems are mostly focused on regulating traffic flow to avoid congestion, thus indirectly reducing pollution. Recently, approaches that specifically account for pollution emissions are gaining interest. [5] analysed two control actions to reduce air pollution in urban areas caused by traffic, namely reducing speed and environmental restricted zone. The latter consists in restricting access to the most contaminant vehicles, which was based on their classification according to the European Emission Standards. The implementation was static, i.e. vehicles were classified in four categories and the two less polluting categories were permitted access. [6] evaluated different intersection control algorithms, showing that platoon-based algorithms obtain less pollutant emissions (higher throughput) but lower fairness than FIFO. [7] studied the effectiveness of traffic signal control and variable message signs for reducing traffic congestion and pollutant emissions. [8] proposed a Pareto-optimal Max Flow algorithm, which obtains multiple distinct possible paths with maximum flow between a pair of points. Thus, these solutions can be used to distribute traffic and pollution more evenly through a city. These works focus on simulating and assessing the performance of specific static actions. A dynamic traffic light control system based on traffic and air pollution was presented in [9].

Our objective is proposing a prioritised access control approach that is highly dynamic, specific to individual vehicles and that considers social utility or transportation efficiency.

3 Prioritised Access to Transport Infrastructures

In this section we propose a general model for a prioritized allocation of transport infrastructure resources. As pointed out earlier, we understand by transport infrastructure all elements that are provided to the general public facilitating mobility. Such elements may be static, like for instance, streets, lanes of a street, crossroads, parking spaces, etc. Other elements may be mobile, like vehicles of sharing systems or more classical public transportation facilities, like buses, trains or subways, and so on. Infrastructure elements may be used by any person, and their usage is usually regulated through specific norms or conventions. For example, the usage of a lane of a particular street is regulated through the corresponding traffic norms. Also, many elements may be used without charge (usually this holds for most static elements) and others may have some cost (e.g., public transportation).

Transportation infrastructure elements are intrinsically limited and typical traffic problems like traffic jams or excessive delays in movements arise when the usage demand of certain elements exceeds the available resources or capacities. Such mismatches between demands and available resources usually arise in big cities where the population density is very high. In addition, in many big cities, there might also be an interest in putting additional limitations on the use of certain infrastructure elements. For instance, in many European cities, traffic is restricted in some ways in the centre with the aim of having more human-friendly environments or reducing the pollution.

Typically, the aforementioned situations lead to a problem of assigning limited resources to an excessive demand and the decision who should be allowed to access a given resource. In the traffic domain, such decisions are typically not taken in a goal-directed manner. Rather the rule of "who comes earlier wins" applies. In contrast, we believe that from the point of view of improving social welfare, limited infrastructure capacity should be preferably assigned to users or tasks that are more "important" or less harmful with respect to some global, social parameters. That is, the access to or use of limited transport infrastructures should be prioritized.

Our model is based on the notion of a *trip*. A trip refers to a movement activity carried out by an individual vehicle for accomplishing a certain transportation task (e.g., transporting one or more persons, parcels, etc.). Trips are accomplished by *vehicles*. Vehicles have different characteristics, like size, type, emissions, etc. During a trip, a vehicle will use elements of the transport infrastructure.

The accomplishment of a trip has a certain utility for the issuer of the trip (the user) that refers to the preferences of fulfilling the underlying task in an optimal manner. In general, the aim of a mobility infrastructure is to provide a set of services that contributes to the welfare of society as a whole. From this perspective, each trip can be conceived as providing some level of global utility to society.

Given a trip t, we define its utility as a function of two factors i) the global transportation utility of t ($UT(t)$), and ii) the cost of the trip $C(t)$:

$$U(t) = g(C(t), UT(t))$$

UT(t) can be considered as the importance of the trip (e.g., of transporting the associated elements to a destination place) from the point of view of the society. For example, an ambulance movement will be more important than a simple movement of a person in her private car. Thus, the transportation utility will be higher. Regarding $C(t)$, we assume that any mobility activity, and the subsequent use of the transportation infrastructure, has an associated cost. This cost depends on the vehicle that carries out the trip and will include direct costs (e.g. due to the usage of infrastructure elements) as well as externalities as, for instance, the emissions produced. Usually, the global utility of a trip is correlated with $UT(t)$ and is inversely correlated with $C(t)$.

It should be noted that the global utility may be aligned with the individual utility of a user, but this must not be the case in all situations. For example, in the case of a patient who must be transported urgently to a hospital, the global utility will be very high (as for the patient herself, of course). In contrast, if a person moves in order to go shopping, the importance for the society will be usually much higher from that person's perspective than from the point of view of the society.

A trip requires the use of elements in the transport infrastructure. If such elements are scarce, e.g., the demand exceeds the available resources, then the use or assignment of infrastructure elements should be prioritized in order to optimize global utility. This could be done in the following way. Let I be an infrastructure element with capacity *cap*(I) during some time interval $\Delta time$. Furthermore, let $T = \{t_1, t_2, \ldots, t_m\}$ be the trips that claim some of I's capacity during $\Delta time$, where $capR(I, t_i)$ denotes the portion of I that is requested by the trip t_i. The resources or capacities of I during the interval $\Delta time$ should be assigned by some *control strategy*. Here we define the objective of the control strategy to assign all available resources of I to the set of trips with highest utility. That is to a set $T' \subseteq T$, with:

$$\sum_{t_i \in T'} capR(I, t_i) = cap(I)$$

$$\forall t_i \in T', t_j \notin T' : U(t_i) \geq U(t_j)$$

We implicitly assume that a vehicle requests the use of element I in order to fulfil the underlying transportation task in the best possible manner. Thus, if the requested capacity is denied, the vehicle needs to find an alternative solution which may lead to a worse task completion.

In general, we consider that different elements of the transportation infrastructure should be regulated with different control strategies. The general objective here is to increase the global utility of the whole transportation system in a city in terms of the aggregation of the utilities of all transport trips and this can be obtained by prioritizing the trips with higher utility. In addition, giving privileges to more efficient trips (trips with less cost and higher importance) will promote such trips and may encourage users to invest in vehicles with less social costs or to optimize the loads of their trips.

The different parameters of the general model described in this section may be difficult to calculate. Usually, parameters like utilities and costs have to be estimated and will depend on the concrete application case. In the next section, we instantiate the described model for a use case of last-mile delivery and the regulation of the access to a certain area in a city in order to limit the pollution in that area.

4 Use Case. Last-Mile Delivery

4.1 Scenario

We consider a Last-Mile Delivery scenario where different vehicles deliver parcels in a city. In addition, there is an area in the city centre (the *control zone*) with dynamic access restrictions in order to keep the environmental pollution in this area below some threshold. In this context and with respect to the general model described in Sect. 3, the infrastructure element whose use is regulated through a control strategy is the access to the control area. Here, the proposed control system does not only have to assign limited resources to trips (e.g., permitting or denying the access to the control zone). It also controls dynamically the capacity of the control zone, that is, which vehicles are allowed to enter the control zone and which vehicles are rejected and need to take a detour. In particular, our aim is to dynamically increase or decrease the area access restrictions depending on the pollution in a given moment.

In the sequel we instantiate the model proposed in Sect. 3 to the use case under consideration. In this article, our aim is to analyse the general validity of our proposal. In order to keep the clarity of our description we apply a set of simplifying assumptions.

We assume that each trip includes one or more parcels, and we consider that all parcels in a vehicle trip have the same origin and destination. With this, we define the transportation utility of a trip by:

$$UT(t) = \frac{n(t)}{d(t)+1}$$

where $n(t)$ is the number of parcels included in the trip t and $d(t)$ denotes the delay of the arrival time of the trip if it cannot access the control zone.

With respect to the cost of a trip t, our objective is to reduce the emissions in the control area and thus, we define:

$$C(t) = e(t)$$

where $e(t)$ is a measure representing the average emission the vehicle carrying out the trip t would emit in the control zone. We consider that vehicles belong to different emission types, which are known a priori and which are used to estimate $e(t)$.

4.2 Control Strategies

We analyse different control strategies. Each strategy determines the access restrictions to be applied at each moment and decides which vehicles can enter the control area.

We consider that the system implementing a control strategy works as follows. A vehicle that wants to enter the restricted area requests access at an entry point, and the control strategy either grants or denies this access.

The idea of the strategy is to restrict access to the control zone in such a way that the measured pollution in the area p_t is kept below a certain maximum at any time t.

For this we apply the following idea. We calculate an *access permission level* k_t to the control zone as follows:

$$k_t = \begin{cases} 1 & \text{if } p_t \leq \theta_L \text{ (no restrictions)} \\ 0 & \text{if } p_t \geq \theta_H \text{ (no vehicles allowed)} \\ \frac{(\theta_H - p_t)}{(\theta_H - \theta_L)} & \textit{otherwise} \end{cases}$$

where θ_L and θ_H are two threshold values. θ_H represents the maximum allowed pollution, a value that should not be exceeded. θ_L is a lower bound on p_t that is used as a control point from which access restrictions are applied.

Given the access level k_t, we define the following different control strategies:

Baseline (B).

At each moment t, the value of k_t determines the ratio of trips that are allowed to enter the control zone. This strategy does not prioritize the access for different trips. That means, implicitly it employs a definition of $U(t_i)=c$, where c is a constant.

The strategy is implemented by randomly granting access with probability k to each trip that request access to the restricted area. The strategy is used as a baseline and represents the ad hoc idea of reducing the emissions by a proportional reduction of vehicles in the control zone.

Vehicle Emission (VE).
As in the baseline strategy, at each moment t, the value of k_t determines the ratio of trips that are allowed to enter the area. However, this strategy prioritizes trips having lower emissions. That is, the strategy employs a utility function:

$$U(t_i) = \frac{1}{C(t_i)} = \frac{1}{(1 + e(t_i))}$$

That is, access is allowed for the $(k_t \cdot 100)\%$ of vehicles with lowest emissions.

Vehicle Emission per Package (VEP).
Like the previous two strategies, at each moment t, the value of k_t determines the ratio of trips that are allowed to enter the area. The access is prioritized with respect to the emissions of a trip and the importance of a trip (number of parcels carried), using the following definition of global utility:

$$U(t_i) = \frac{UT(t_i)}{C(t_i)} = \frac{n}{(d(t_i) + 1) * (1 + e(t_i))} = \frac{n}{(1 + e(t_i))}$$

Note that $d(t_i) = 0$, if the trip is granted access to the control zone. The strategy allows the access for the $(k_t \cdot 100)\%$ of vehicles with the lowest emissions per package.

Ratio Reduction Emission (RRE).
In this case, the access value k_t is not applied to the ratio of cars that can enter the restricted area. Instead, k_t represents the ratio of emissions that are allowed to be generated with respect to the normally generated emissions in the same moment or time frame. Given k_t, we calculate the ratio k_t' of vehicles with lowest emissions (with respect to the normal demand) that together produce the $(k_t * 100)\%$ of the emissions normally generated in the same moment or time frame. It holds that $k_t' \geq k_t$. Afterwards, the strategy applies the same prioritization schema as *VE*.

Ratio Reduction Emission per Package (RREP).
As the RRE strategy, here k_t is translated to a ratio of vehicles k_t'. Then, the same prioritization schema as in VEP is employed with the new ratio.

In the strategies VE and VEP we use historical data to estimate de threshold value for emissions (emissions per package) to determine whether a vehicle belongs to the k_t vehicles with lowest emissions and access is granted to all these vehicles. In a similar way, in the strategies RRE and RREP, we use historical data to estimate the ratio k_t' and to decide whether or not a vehicle can access the area.

4.3 Experiments

We carried out experiments using SUMO [10], a popular open-source microscopic traffic simulator. SUMO includes several emission models. We used HBFA3, which is based on the database HBEFA[1] version 3.1. The model simulates several vehicle emission pollutants, and we chose NO_x as the reference in our experiments.

[1] http://www.hbefa.net/.

SUMO implements several vehicle emission classes including heavy duty, passenger and light delivery emission classes, combined with different EU emission standards (levels 0–6). In the experiments, we chose 7 different types of vehicles with different emission classes: eVehicle ("Zero/default"), gasolineEuroSix ("HBEFA3/LDV_G_EU6"), dieselEuroSix ("HBEFA3/LDV_D_EU6"), hovDieselEuroSix ("HBEFA3/PC_D_EU6"), normalVehicle ("HBEFA3/PC_G_EU4"), highEmissions ("HBEFA3/PC_G_EU3"), and truck ("HBEFA3/HDV_D_EU4").

While SUMO is able to provide information about pollutants emitted by vehicles, it does not include a model of how those pollutants evolve in the air. These values depend on many different factors, not only different pollutant emission sources but also on weather conditions (wind, rain, temperature changes, etc.).

Typical access policies to city centres are taken based on pollution data measured by atmospheric stations, and not on the direct emissions of vehicles. This is the approach followed by the strategies proposed in Sect. 4.2, which are based on the pollutants in the air at a certain point in time t (denoted by p_t).

In our experiments, we have simulated air quality data through a simple model as follows. The basic idea is that pollution at time t is the sum of the previous pollution plus the amount emitted by vehicles during the last time period minus a quantity that is removed by atmospheric effects: $p_t = p_{t-1} + e_t - \lambda_t \cdot 45000$, where p_t is the pollution in the air at time t, e_t is the pollution emitted by vehicles between the time interval from t-1 to t, $\lambda_t \in [0.8, +1.2]$ is a uniformly randomly generated factor that represents a ratio of pollutants removed from the air in time t (we set the constant 45000 empirically). While we recognise that this is a simplification of the real world, this measure allows us to analyse and compare the different control strategies.

For the experiments we designed a virtual city as shown in Fig. 1. It includes a 5 × 5 km square control zone with eight access points. The network is made up of road segments connected by intersections. All segments are bidirectional. The strategies control the access to the *control zone* (marked in red).

There is a set of vehicles carrying packages from an origin to a destination location. Each vehicle is characterised by origin, destination, number of packages and average pollutant emissions (mg/s). All trips start and end at one of the external points and, in principle, would pass through the control zone if access was granted. Upon reaching the border of the control zone, vehicles ask for permission to enter. If access is granted, they cross the control zone. If not, they have to find a route that bypasses the city centre, which will usually take longer.

We are only interested in trips whose planned routes cross the control zone. The rest does not provide any value for our evaluation. For this reason, trips were generated randomly connecting North-South and East-West, in both directions.

We run a simulation of 2.5 h. During the first hour, trips are generated at a rate of about 1000 per hour; during the next 30 min, at a rate of 2000; and during the last hour, with a rate of 1000. Origin and destination, number of packages (between 1 and 20), and vehicle type are chosen randomly. All vehicle types have the same probability, except *truck* and *highEmissions*, which have half the rate of the others.

We carried out simulations without control zone restrictions to set some initial parameters. Namely we set $p_o = 10^6$, $\theta_L = 0.8 \cdot 10^6$ and $\theta_H = 10^6$. The time interval to update k is set to 60 s.

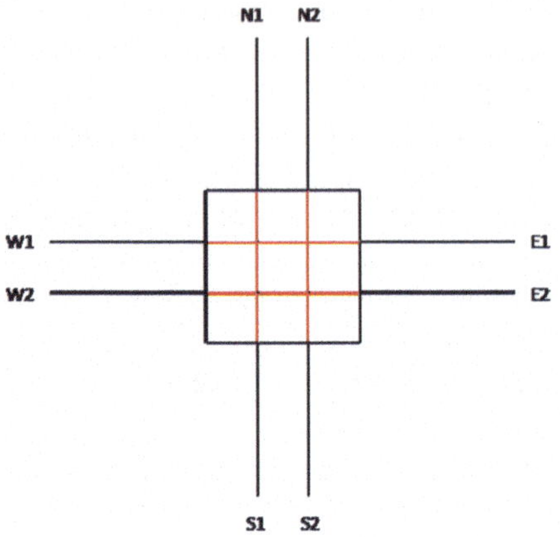

Fig. 1. Road network used in the experiments. Red lines represent the control zone.

4.4 Results

Table 1 shows a summary of the results of the experiments. For each strategy we present the number of vehicles (out of a total 3108 vehicles) that did not enter the control zone due to access limitations, the total amount of NO_x (mg) emitted by vehicles in the control zone, the average transportation time per parcel and the average time per trip. We also included the results without access limitations.

We confirmed that applying control strategies the amount of pollution emitted by vehicles in the control zone was reduced about 33%. All strategies obtain similar results following different approaches as described in Sect. 4.2. This is due to the fact that they all use the same low and high thresholds (θ_L, θ_H) of allowed pollution. The different

Table 1. Simulation results.

Strategy	#vehicles no access control zone	NOx emitted control zone	Time per parcel (s)	Avg. trip time (s)
No control	15/0.5%	12439311	1359	1359
Baseline	1077/34.7%	8061291	1501	1501
VE	627/20.2%	8002710	1455	1457
VEP	553/17.8%	7959436	1436	1445
RRE	301/9.7%	8100536	1389	1391
RREP	242/7.8%	8088261	1374	1379

strategies achieve this goal by limiting access to more or less vehicles. Besides the access limitations, vehicles may select to bypass the city because of traffic congestions. This is the case for example for the 15 cars with the "no control" strategy that did not enter the control zone.

As expected, the *baseline* strategy restricts the access to more vehicles than the others, since it randomly chooses which ones are granted access. *VE* and *VEP*, obtain similar results granting access to much more vehicles than *baseline.* These strategies' approach is to reduce the same percentage of vehicles as *baseline* but they select the less contaminant vehicles. The effect is that less contaminants are released, thus pollution is reduced and, consequently, more vehicles are allowed to enter the control zone. *RRE* and *RREP* are the strategies that allow more vehicles to enter the control zone. Their approach is to allow access to less contaminant vehicles that jointly add up a percentage of expected emissions. That is, they restrict access to a few but high contaminant vehicles (e.g. trucks represent 80% of total emissions in our experiments).

Average trip times are in line with the rate of vehicles that could/could not access the control zone. Strategies that took into account the number of packages benefited of slightly lower travel times per package as compared to just considering emissions. However, the differences are rather small.

Figure 2 shows the evolution of NO_x pollution (p) in the control zone over time. We present the results after a "warm up" period of about 1000 s. The curve for the no control strategy escapes the graph at time 4400 s reaching a maximum value higher than $5{\cdot}10^6$, so we opt for not showing that part and keep the other curves visible.

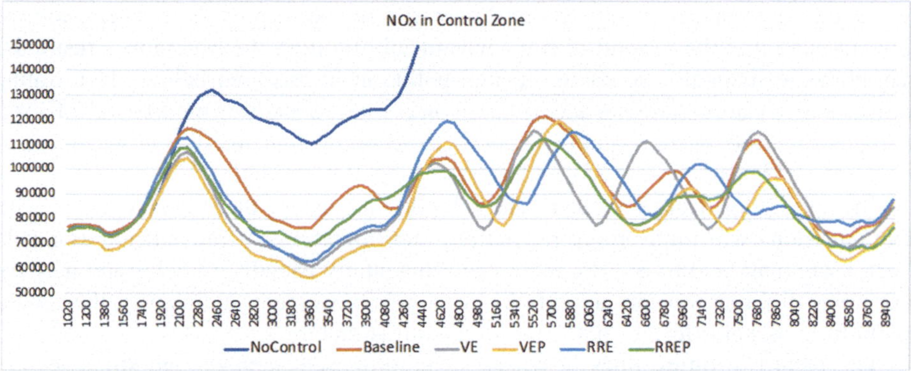

Fig. 2. Evolution of pollution (NO_X) in control zone through the simulation time (seconds)

As it is shown in Fig. 2, all control strategies help to limit the level of NO_X emissions and there is no clear difference among the different strategies in that sense. The strategies start limiting the access when the threshold value of 800.000 is reached and close the access at 1.000.000. As it can be noted, triggering limitations does not have an immediate effect on reducing p and the limitations have to be set for more time in order to reduce p under the lower threshold of 800.000. In fact, even if the control zone is closed at all, p may still increase because of the vehicles that are already in the control zone. The effect is that p, may pass the upper threshold (here 1.000.000) on several occasions. This also produces the effect of p to oscillate roughly between 800.000 and 1.200.000. The oscillation will only occur if the demand is high and would produce too many emissions if no control strategy is used. This happens in the experiment from about 4000 s.

5 Conclusions

The demand for using the transportation infrastructure has increased tremendously in the last years, especially in highly dense urban areas. In this paper, we have argued that the use of infrastructure elements should be regulated and prioritized with the aim of fostering global utility. In particular, we present a general assignment model that assigns limited traffic resources to the traffic activities (trips) that have a higher utility from the point of view of the society. Here utility is considered to have two components: i) cost (e.g. emissions), and ii) social "importance" of the transportation activity.

We have instantiated the model for the case of last-mile delivery in a city with an access restricted area. We present a control system that dynamically determines the access limitation level for the control zone based on the current measures of environmental pollution. Furthermore, the system employs a prioritization strategy that determines which vehicles (trips) can enter the area and which ones need to bypass it. We have proposed different strategies: i) all trips have the same priority, ii) low-emission cars have higher priority, and iii) priority depends on the "importance" of the trips and on the emissions of the cars. In the latter case, "importance" is measured in terms of packages a vehicle is delivering.

We have carried out several experiments with the traffic simulation tool SUMO to analyse the performance of our proposal with different strategies. As a conclusion we can determine that the general idea of dynamically limiting the access to a restricted area allows to maintain the environmental pollution in this area below given limits. Furthermore, a prioritization of access based on emissions and/or "importance" of a trip improves the utility of the system and allows to accomplish in an efficient way more of the important transportation tasks under the given pollution limits.

Given the prioritization methods, less important tasks or the use of vehicles with more emissions will imply more restrictions and limitations of movement. As a side effect, users may tend to acquire more environmentally friendly vehicles and may try to combine different transportation tasks in a single movement. In this way, they could benefit from higher priorities in the use of the infrastructure.

With regard to future lines of research, the proposed methods and strategies are still at an early stage of research. We plan to specify the general model for prioritized access in more detail. Furthermore, with regard to the access limitation for an area, we want to analyse more sophisticated methods for specifying the access levels using gradient minimization models. Another line or research is the definition of "importance" of different types of trips, probably using semantic technologies.

Acknowledgments. This work has been partially supported by the Spanish Ministry of Science, Innovation and Universities, co-funded by EU FEDER Funds, through project grants InEDGEMobility RTI2018–095390-B-C31/32/33 (MCIU/AEI/FEDER, UE) and by the Regional Government of Madrid (grant PEJD-2019-PRE/TIC-16575), cofunded by EU ESF Funds.

References

1. Botsman, R.: Defining the sharing economy: what is collaborative consumption-and what isn't? fastcoexist.com. available at http://www.fastcoexist.com/3046119/defining-the-sharing-economywhat-is-collaborative-consumption-and-what-isnt (2015)
2. Namazi, E., Li, J., Lu, C.: Intelligent intersection management systems considering autonomous vehicles: a systematic literature review. IEEE Access **7**, 91946–91965 (2019)
3. Dresner, K., Stone, P.A.: Multiagent approach to autonomous intersection management. J. Artif. Intell. Res. **31**, 591–656 (2008)
4. Vasirani, M., Ossowski, S.: A market-inspired approach for intersection management in urban road traffic networks. J. Artif. Intell. Res. **43**, 621–659 (2012)
5. Vergés, J.T.: Analysis and Simulation of Traffic Management Actions for Traffic Emission Reduction. TU, Berlin (2013)
6. Lemos, L.L., Pasin, M.: Intersection control in transportation networks: opportunities to minimize air pollution emissions. In: IEEE 19th International Conference on Intelligent Transportation Systems (ITSC) (2016)
7. Mascia, M., et al.: Impact of traffic management on black carbon emissions: a microsimulation study. Netw. Spat. Econ. **17**, 269–291 (2017)
8. Kamishetty, S., Vadlamannati, S., Paruchuri, P.: Towards a better management of urban traffic pollution using a Pareto max flow approach. Transportation Research Part D: Transport and Environment, **79**, 102194 (2020)
9. Artuñedo, A., del Toro, R.M., Haber, R.E.: Consensus-based cooperative control based on pollution sensing and traffic information for urban traffic networks. Sensors **17**(5), 953 (2017)
10. Alvarez Lopez, P., et al.: Microscopic traffic simulation using SUMO. In: 21st IEEE International Conference on Intelligent Transportation Systems, Maui, USA, pp. 2575–2582 (2018)

Doctoral Consortium

Gamification Proposal of an Improved Energy Saving System for Smart Homes

David García-Retuerta[1(✉)] and Juan M. Corchado[1,2,3]

[1] University of Salamanca, Patio de Escuelas Menores, 37008 Salamanca, Spain
{dvid,corchado}@usal.es

[2] Air Institute, IoT Digital Innovation Hub, 37188 Salamanca, Spain

[3] Department of Electronics, Information and Communication, Faculty of Engineering, Osaka Institute of Technology, Osaka 535-8585, Japan
https://bisite.usal.es/en/group/team/David

Abstract. The residential sector accounts for 17% of the final energy consumption in the world, and its energy-related improvements are characterised by expensive remodelling, destruction and reconstruction. However, the increasing digitalisation taking place all around the world has paved the way to new, revolutionary methods such as gamification-based approaches. The goal of a meaningful gamification is to provide the user with an gameful and engaging experience designed to create medium-term and long-term habits in the users. This proposal suggest that IoT monitoring systems in Smart Homes can make use of gamification to raise energy awareness among its users and reinforce energy efficient habits.

Keywords: Gamification · Energy saving · IoT · Smart homes

1 Introduction

Gamification seeks takes the concept of energy systems in smart homes to a new level. New tools and technologies have proven their usefulness in improving digitisation and environmental awareness. However, an effort is needed to provide a engaging, dynamic and personalised application to all the interconnected house IoT sensors and to make full use of their potential. This new technology should adapt well to individual users, create a competitive environment for energy efficiency, respect their data privacy, raise self-awareness, and adjust itself in real time. Furthermore, it should have a medium-term focus on behaviour in order to reinforce good habits among its users. This project focuses on the application of energy optimisation techniques and gamification to IoT systems with real utility [1–5]. During the review of the state of the art, it was noted that machine learning has not been used in this regard; although it has a great research potential [6].

In this proposal, the many vectors for promoting energy savings using gamification techniques are grouped into the following:

1. Ranking or leaderboard, by block of flats.

J. M. Corchado and S. Trabelsi (Eds.): SSCTIC 2021, LNNS 253, pp. 315–317, 2022.
https://doi.org/10.1007/978-3-030-78901-5_27

2. Definition of several energy efficiency levels.
3. Push notifications and badges for achievements. These are sub-divided in 2 types: medium-term and long-term rewards.
4. Community challenges, to be defined by the users.
5. Annual carbon footprint per floor. General recommendations to reduce the footprint, options to offset it and comparison against similar neighbours.
6. Real-time information about energy consumption.
7. Privacy. Clear and simple explanations of how the data is used and processed.

Wherever possible, there should be a "more information" button explaining to the user how the system works or what each recommendation, achievement, screen, etc.; is based on. The goal of this is to make it possible to have very-highly engaged users [7–11].

2 Conclusion

This work proposes a novel system for improving the energy efficiency of smart homes, with a focus on both medium-term and long-term habits. Habits are created via instant energy recommendations (via notifications) and weekly energy recommendations. The algorithm is going to be applied to a set of houses in Salamanca (Spain) in the future. The main goal of the algorithm is to adapt well to a vast variety of households, always allowing them to dig deeper into energy saving optimisation. Furthermore, the designed algorithm makes privacy key and the scoreboards do not reveal information about the user's neighbours' score, preventing them from disliking each other.

This paper provides a novel, rules-based and real-time system, which enables a correct recommendation system to work seamlessly in Smart Homes. We also address its different applications, as the guideline here described can be adapted to a vast variety of situations, cities and countries. In future research, we will extend the system to consider more inputs and will test which recommendations are more effective among the users.

Acknowledgements. This research has been supported by the project "Intelligent and sustainable mobility supported by multi-agent systems and edge computing (InEDGE-Mobility): Towards Sustainable Intelligent Mobility: Blockchain-based framework for IoT Security", Reference: RTI2018-095390-B-C32, financed by the Spanish Ministry of Science, Innovation and Universities (MCIU), the State Research Agency (AEI) and the European Regional Development Fund (FEDER).

References

1. Yigitcanlar, T., et al.: Artificial intelligence technologies and related urban planning and development concepts: how are they perceived and utilized in Australia? J. Open Innov. Technol. Market Complex. **6**(4), 187 (2020)
2. Malek, J.A., Lim, S.B., Yigitcanlar, T.: Social inclusion indicators for building citizen-centric smart cities: a systematic literature review. Sustainability **13**(1), 376 (2021)

3. Garcia-Retuerta, D., Casado-Vara, R., Calvo-Rolle, J.L., Quintián, H., Prieto, J.: Deep learning for house categorisation, a proposal towards automation in land registry. In: de la Cal, E.A., Villar Flecha, J.R., Quintián, H., Corchado, E. (eds.) Hybrid Artificial Intelligent Systems. HAIS 2020. LNCS, vol. 12344. Springer, Cham (2020). https://doi.org/10.1007/978-3-030-61705-9_58
4. Yigitcanlar, T., Butler, L., Windle, E., Desouza, K.C., Mehmood, R., Corchado, J.M.: Can building "artificially intelligent cities" safeguard humanity from natural disasters, pandemics, and other catastrophes? An urban scholar's perspective. Sensors **20**(10), 2988 (2020)
5. García-Retuerta, D., Casado-Vara, R., Martin-del Rey, A., De la Prieta, F., Prieto, J., Corchado, J.M.: Quaternion neural networks: state-of-the-art and research challenges. In: Analide, C., Novais, P., Camacho, D., Yin, H. (eds.) IDEAL 2020. LNCS, vol. 12490, pp. 456–467. Springer, Cham (2020). https://doi.org/10.1007/978-3-030-62365-4_43
6. Sepasgozar, S., et al.: A systematic content review of artificial intelligence and the Internet of things applications in smart home. Appl. Sci. **10**(9), 3074 (2020)
7. Corchado, J.M., et al.: Deepint.net: a rapid deployment platform for smart territories. Sensors **21**(1), 236 (2021)
8. García-Retuerta, D., Canal-Alonso, A., Casado-Vara, R., Rey, A.M., Panuccio, G., Corchado, J.M.: Bidirectional-pass algorithm for interictal event detection. In: Panuccio, G., Rocha, M., Fdez-Riverola, F., Mohamad, M.S., Casado-Vara, R. (eds.) PACBB 2020. AISC, vol. 1240, pp. 197–204. Springer, Cham (2021). https://doi.org/10.1007/978-3-030-54568-0_20
9. Chamoso, P., González-Briones, A., de la Prieta, F., Venyagamoorthy, K.G., Corchado, J.M.: Smart city as a distributed platform: toward a system for citizen-oriented management. Comput. Commun. **152**, 323–332 (2020)
10. Casado-Vara, R., Martín, Á., del Rey, S., Affes, J.P., Corchado, J.M.: IoT network slicing on virtual layers of homogeneous data for improved algorithm operation in smart buildings. Future Gener. Comput. Syst. **102**, 965–977 (2020)
11. González Bedia, M., Corchado, J.M.: A planning strategy based on variational calculus for deliberative agents. Comput. Inf. Syst. **9**(2002), 2–13 (2002)

A Decision Support System for Transit-Oriented Development

Aya Hasan AlKhereibi[1,2](✉)

[1] Mechanical and Industrial Engineering, College of Engineering, Qatar University, Doha, Qatar
aalkhereibi@qu.edu.qa

[2] Qatar Transportation and Traffic Safety Center, College of Engineering, Qatar University, Doha, Qatar

Abstract. This research is obtained for the exact purpose of developing a decision support system that addresses the built environment, transportation, travel behavior, and system dynamics aspects of Transit-Oriented Development (TOD). Mainly, the research will be composed of several stages. First, a comprehensive review for research trends and the project's operations of metro systems in similar regions, as well as transport, land use. The second stage consists of conducting an empirical built environment analysis of the integration between rail systems and adjacent land uses for an existing situation at Doha Downtown. Then a sophisticated models for metro ridership prediction concerning station precincts and corridors to be developed. The fourth stage includes investigate people's travel behavior for citizens given their travel diary, socio-demographic characteristics to define a citizens' travel pattern. The fifth stage is to develop a system dynamics model integrating infrastructure and economic aspects of TOD aiming to simulate the system behavior over time. The final stage of the dissertation is to develop a decision support system concerning the main pillars of Transit Oriented Development. This research is expected to contribute to the body of knowledge and Qatar-based transport and land use research capacity as well.

Keywords: Transit-oriented development · Decision support system · System dynamics · Travel behavior · Machine learning

1 Introduction

The planning decisions are tailored to the information, and it needs a collaboration of a large number of entities and stakeholders (Ibraeva et al. 2020), especially, the transportation-urban planning decision. It is crucial since it evolves several separate entities with dissimilar perspectives and aims that could result in a trade-off in the final decision.

In Transit-Oriented Development (TOD), the main purpose focuses on making the built environment in compliance with the transportation system planning and the encouragement of people to shift their travel mode choice (Ganning and Miller 2020).

Tamakloe and Hong (2020) stated that the concern of TOD is to acclimatize a transit-supportive environment that could accommodate walking, cycling, and public transit which could increase the number of transit users and still reserve the urban fabric.

J. M. Corchado and S. Trabelsi (Eds.): SSCTIC 2021, LNNS 253, pp. 318–323, 2022.
https://doi.org/10.1007/978-3-030-78901-5_28

1.1 Problem Statement

Research obtained in the field of TOD has particularly investigated the issue from one perspective of its application, yet the proposed dissertation intends to study the case as a complete system including several aspects. This research advocates its applicability by investigating a uniques case of the State of Qatar, where the majority of developed land use forces residents to drive to reach their destinations. This refers to several reasons, first; Qatar's urban fabric is known as low-dense with mostly single-use developments. Another reason is Qatar is known as a wealthy nation, in which its people are more willing to own their vehicles rather than using public transport, thus in general the private cars are the dominant transport choice. The climate conditions are also another point needed to be taken into consideration in this context since climate conditions limit the ability to walk especially in summer.

1.2 Contribution to the State of the Art

The research is significant and can substantially contribute to the state of the art for several reasons. First, it will provide a review of best practices in integrating transit and land use; and idealized futures for the Doha transit system (Doha Metro) and station precincts at various scales; 1) Improved built environment analysis and TOD index assessment for application in similar countries. 2) Metro ridership prediction model considering built environment by station; 3) A travel behavior analysis model that captures the pattern of citizens travel in Doha which could be applied to similar countries; 4) System dynamics behavior of the planning, operation, and economic aspect of Transit-oriented Development system; 5) A decision support system framework that could be in help of decision-making of Transit Oriented Development strategic planning.

2 Methods and Materials

Conceptually the dissertations' approach is inter-disciplinary; involving transportation/transit planning, urban design, urban land use planning, travel behavior studies, and system dynamics. Harnessing the methods and theory of these fields into one coherent research framework provides the dissertation with the necessary rigor and analytical power to meet the research objectives. The approach and methods will be delivered in five inter-connected parts. As shown in Fig. (1) the dissertation will cover five aspects. The first part will review and map TOD research trends, then the built environment, an empirical analysis for an existing district will be performed to examine the TOD characteristics and assess the TOD index. The third aspect is the operations of the transportation system, the research focus will be to predict rail ridership using machine learning methods. The fourth pillar is the travel behavior analysis, where a sample of citizens movement tracks data will be taken to define travel patterns using machine and deep learning methods. Then a system dynamics model will be built to incorporate the economic aspect of the system and simulate the TOD system behavior with respect to the built environment, travel behavior, and metro ridership.

The dissertation argues for a search for a Transit-Oriented system that may lead to people-oriented sustainable development, and improve livability. This research seeks sixteen months of investigation to support the development and implementation of the various research methods along the Doha rail corridors and to allow for sufficient attention to data analysis and dissemination of research findings.

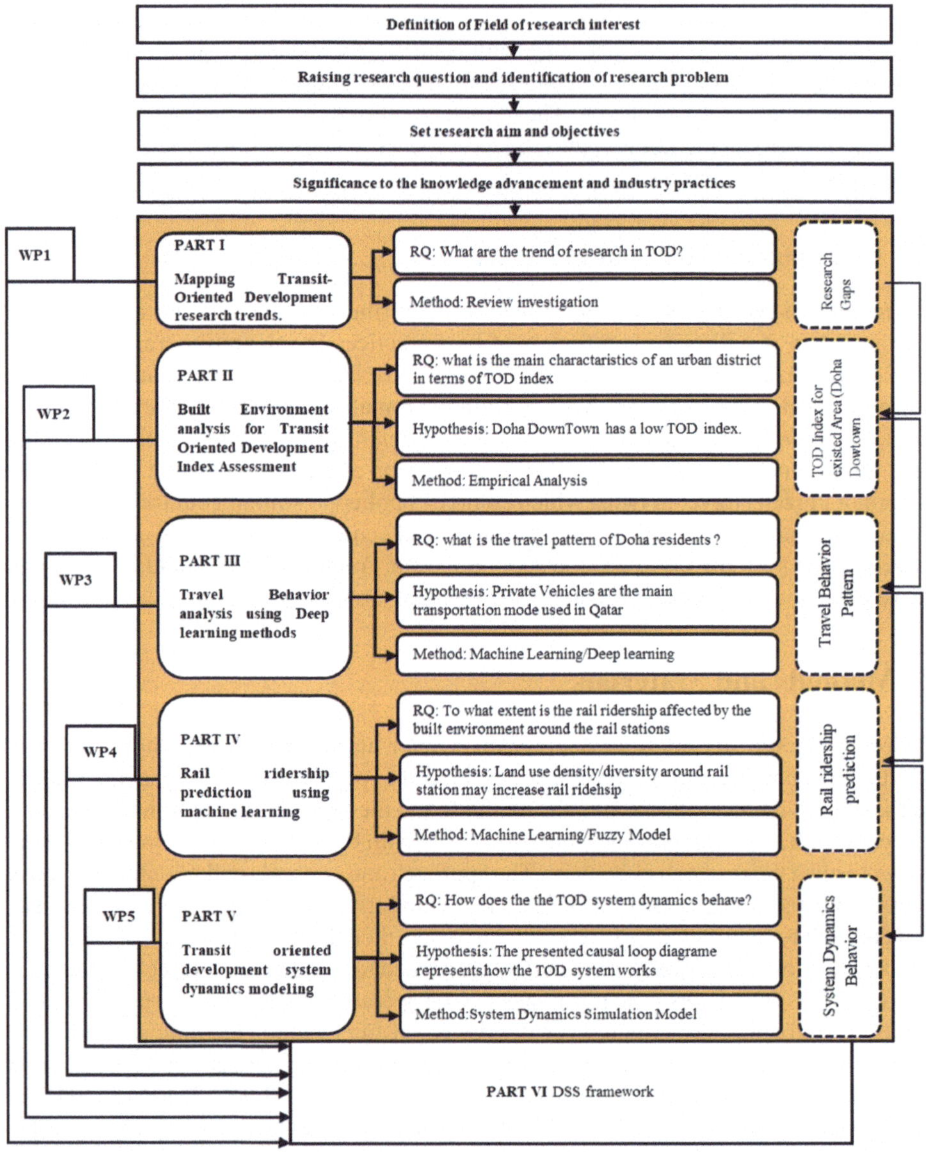

Fig. 1. Dissertation methodology

2.1 PART I Mapping Transit-Oriented Development Research Trend

The first chapter of the ongoing dissertation will focus on mapping the research trend of Transit-Oriented Development by going to the historical lens of the early concept initiates heading to the contemporary transport planning, urban planning, and urban design research findings to be in help of providing comprehensive solutions to the current problems. The chapter will mainly provide an up-to-date systematic review for TOD research trends, starting from presenting the concept and how it was farmed in the urban planning theories. Then, the vast research is dedicated to the research of the TOD effect, distinguishing its impacts on travel behavior, land use, economy, and transportation operation. The last part of the review evolves the planning aspect of TOD focusing on decision support tools and policy-making issues. The outcome of this part of the research will provide research gaps that may be filled in the coming parts of the dissertation.

2.2 PART II Built Environment Analysis for Transit-Oriented Development Index Assessment

In this chapter, an empirical analysis model will be developed and applied to the Doha Downtown district, which has three adjacent metro stations: Msheireb station, Souq Waqif station, and National Museum. The roads in this area offer a high degree of accessibility and connectivity to many focal points within the district. The data on current land use density is used to develop several scenarios for further development of the area. It is to note that as some of the areas in the district are being developed or redeveloped, smart growth planning of the area based on the availability and accessibility to transportation can provide better liveable conditions to the people. The outcomes of the analysis performed in this chapter expected that developing all the areas and diversity of land use can create a better environment. The chapter will focus on the TOD index; it might be worthwhile to consider land-use development for bike-sharing or an area-based free street tram system from one station to another within the study area.

2.3 PART III Travel Behavior Analysis in an Urban Planning Context

The third chapter attempts to define citizens travel patterns by clustering the movement data, raw data preprocessing and filtering will be undertaken in the first stage of the analysis. The second stage includes a combination of different machine learning techniques, using clustering and classification methods. Support Vector Machine will be employed to identify the number of clusters, then using the bagged clusters algorithm and the C-Means algorithm the results will be validated. The third stage is to examine the interdependencies between the resulted clusters and the socio-demographic attributes for the sample undertaken using crosstabs analysis. Then a classification fuzzy method will be employed to define the travel pattern for the sample under investigation. It is assumed that the application of this framework proves the strength of the Machine learning technique to identify the travel behavior pattern, as found by (Dumbliauskas and Grigonis 2020).

2.4 PART IV Rail Ridership Prediction Using Machine Learning

The fourth chapter aims to develop a Support Vector Machine (SVM) model to predict the metro station-based ridership utilizing the land use densities in the vicinity of metro stations. The ridership data are obtained from Qatar Rail, and the land use data are obtained from the Ministry of Municipality and Environment (MME), State of Qatar. The land use densities in the catchment area of 800 m around the metro stations will be considered in this study. This model will use 80% of the data for training and the remaining 20% for validation. Three indicators will be used to evaluate the model efficiency, namely MAE, RMSE, and R^2 (Wang et al. 2018). The developed model can be utilized by both Urban and Transport planners in their processes to plan the land use around metro stations and predict the transit demand from those plans while designing land use developments and planning related transit operations around the metro stations and achieve the optimal use of the transit system i.e. TOD.

2.5 PART V Transit-Oriented Development System Dynamics Modeling

Transit-oriented development is an urban planning approach that facilitates the achievement of Sustainable Development Goals from urban planning and transportation perspective (Liu et al. 2020). The primary concept of transit-oriented development is mainly to minimize private vehicle usage by enhancing using other movement modes such as; walking, cycling, and public transportation (Phani Kumar et al. 2020). This chapter hypothesizes that the proper application of Transit-oriented development is by measuring the typical causality characteristics of the Transit-oriented development index, not only by measuring the current state of the city urban fabric but also by investigating the dynamicity of the interacted measures over time. These assessments and modeling will crucially draw the lines for the policymaker to provide more effective policies. In the chapter, a conceptual system dynamics model for the Transit-oriented development index is developed. Typically, Transit-oriented development studies investigate the development around the transit node, but this research study specialized in the interaction between defined variables of land use, transit, and economy.

3 Conclusion

To sum up everything that has been stated so far, the last part of the dissertation will attempt to propose a decision support system that is informed by the previous parts covering built environment analysis, travel behavior, metro ridership, and system dynamics. The proposed system could be of help to propose appropriate strategies for application in support of decision-makers, urban planners, and transportation operators.

References

Dumbliauskas, V., Grigonis, V.: An empirical activity sequence approach for travel behavior analysis in Vilnius city. Sustain. Swit. **12**(2), 468 (2020). https://doi.org/10.3390/su12020468

Ganning, J., Miller, M.M.K.: Transit oriented development and retail: is variation in success explained by a gap between theory and practice? Transp. Res. Part D Transp. Environ. **85**(May), 102357 (2020). https://doi.org/10.1016/j.trd.2020.102357

Ibraeva, A., de Almeida Correia, G.H., Silva, C., Antunes, A.P.: Transit-oriented development: a review of research achievements and challenges. Transp. Res. Part A Policy Pract. **132**(2019), 110–130 (2020). https://doi.org/10.1016/j.tra.2019.10.018

Liu, L., Zhang, M., Xu, T.: A conceptual framework and implementation tool for land use planning for corridor transit oriented development. Cities **107**(2019), 102939 (2020). https://doi.org/10.1016/j.cities.2020.102939

Phani Kumar, P., Ravi Sekhar, C., Parida, M.: Identification of neighborhood typology for potential transit-oriented development. Transp. Res. Part D Transp. Environ. **78**(2019), 102186 (2020). https://doi.org/10.1016/j.trd.2019.11.015

Tamakloe, R., Hong, J.: Assessing the efficiency of integrated public transit stations based on the concept of transit-oriented development. Transportmetrica A Transp. Sci. **16**(3), 1459–1489 (2020). https://doi.org/10.1080/23249935.2020.1753849

Wang, X., Zhang, N., Zhang, Y., Shi, Z.: Forecasting of short-term metro ridership with support vector machine online model. J. Adv. Transp. **2018**, 13 (2018). https://doi.org/10.1155/2018/3189238

Smart-Heritage: An Intelligent Platform for the Monitoring of Cultural Heritage in Smart Cities

Marta Plaza-Hernández[1(✉)] and Juan Manuel Corchado Rodríguez[1,2,3]

[1] BISITE Research Group, University of Salamanca, Edificio Multiusos I+D+i, Calle Espejo 2, 37007 Salamanca, Spain
{Martaplaza,corchado}@usal.com
[2] AIR Institute, IoT Digital Innovation Hub, 37188 Salamanca, Spain
[3] Department of Electronics, Information and Communication, Faculty of Engineering, Osaka Institute of Technology, Osaka 535-8585, Japan

Abstract. Conservation of cultural heritage is crucial to safeguarding society's history and memory. This research proposal aims to develop a smart platform that manages cultural heritage in cities, contributing to the generation and transfer of knowledge, working as a supporting tool for decision-making.

Keywords: Cultural heritage · Internet of Things · Edge computing · Artificial intelligence · Smart cities

1 Introduction

Monitoring cultural heritage in cities is needed to protect tangible testimonies while ensuring its accessibility to future generations. Conservation solutions, from the evaluation and analysis of the heritage's status until the restoration activities, are required not only for in-situ preservation but also for utilising the available information for decision-making [1].

With the current data availability, the challenge is to identify smart and adaptive ways of combining the information to create relevant knowledge [2]. The Internet of Things (IoT) technology, which refers to the connection of multiple and heterogeneous objects with electronic devices to collect and provide data, has grown rapidly, finding applications in many sectors [3]. In cities, IoT systems improve institutions and companies' transparency and efficiency, safeguard citizens' safety and well-being, reduce infrastructure's risks, and costs, and minimise the environmental impact [4]. The use of sensors is one of the most cost-effective practices for heritage evaluation, facilitating the monitoring of environmental changes [5–8].

This research proposal aims to develop an Edge-IoT architecture that allows the incorporation of Artificial Intelligence (AI) algorithms and models to monitor cultural heritage. It will contribute to the generation and transfer of knowledge, working as a support tool for decision-making.

J. M. Corchado and S. Trabelsi (Eds.): SSCTIC 2021, LNNS 253, pp. 324–327, 2022.
https://doi.org/10.1007/978-3-030-78901-5_29

2 Smart-Heritage

A basic IoT-Edge platform includes three layers [9, 10] (Fig. 1). *Layer 1* comprises the IoT sensors and actuators responsible for the data collection. *Layer 2* includes the edge nodes responsible for data processing. *Layer 3* covers the cloud services responsible for data analysis and visualization. Solutions based on this architecture has been implemented in several scenarios [11–13]. The architecture presented here (Fig. 2) will follow this structure.

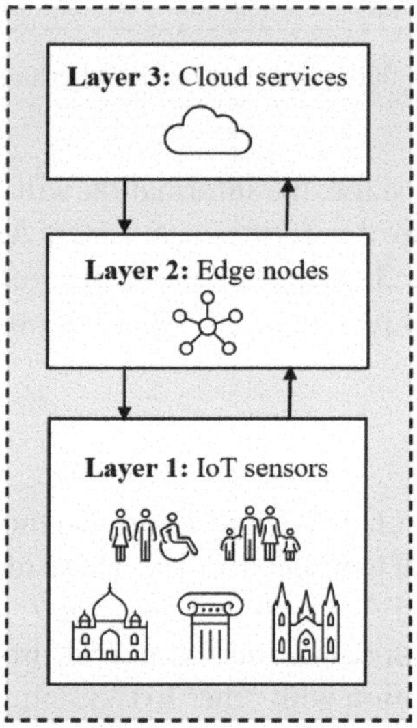

Fig. 1. IoT-edge basic architecture. Adapted from [9, 10].

Smart-heritage will have the ability to combine data stored in databases with information collected in real-time. The platform will have the flexibility to distribute intelligence to the network's Edge, so that data can be filtered and pre-processed before it is transmitted to the Cloud, reducing its volume and costs. The platform will focus on the following dimensions: *Heritage Management* and *Crowd Management*. These modules will communicate and exchange information between them. The platform will be able to operate independently or in collaboration with other available IoT systems.

The platform will present the current state of cultural heritage, so that conservation approaches and restoration activities can be employed in-situ. Additionally, it will predict the effect of degradation phenomena, especially due to climate change.

The information will be a combination of real-time data gathered by the IoT sensors and data from public databases. The sensors will measure key environmental factors that participate in the degradation (temperature, humidity, chemical pollution, etc.), in particular for heritage located outdoors. Real-time images will be collected to analyse degradation over time. Information from public databases will include physicochemical properties, age, previous restoration works, etc. The different stages of the data are displayed in the figure below (Fig. 2):

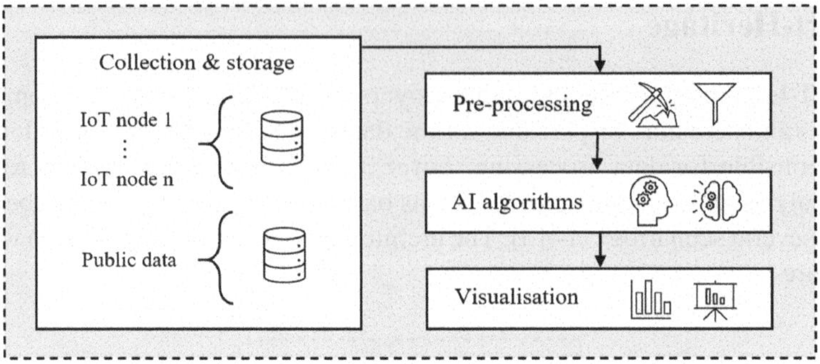

Fig. 2. Proposed data stages for cultural heritage management.

During the pre-processing stage, the information will be extracted and transformed into a comprehensible structure for further use. Later, AI algorithms will be selected depending on the type of data. Lastly, the knowledge generated will be transferred to the city council and public and private organisations for decision-making.

3 Conclusions

This PhD project aims to research on AI algorithms and models that support the monitoring and management of cultural heritage in cities. These techniques will be incorporated into a platform based on IoT-Edge technologies. *Smart-heritage* will combine information stored in databases with data collected in real-time. It will be autonomous and scalable, facilitating its integration with other IoT systems.

Acknowledgements. This work has been supported by the project "XAI - Sistemas Inteligentes Auto Explicativos creados con Módulos de Mezcla de Expertos", ID SA082P20, co-financed by Junta Castilla y León, Consejería de Educación, and FEDER funds.

References

1. UNESCO. Managing Cultural World Heritage, file:///C:/Users/Marta/Downloads/activity-827–1.pdf. Accessed 15 Jan 2021
2. Silva, B.-N., Khan, M., Han, K.: Towards sustainable smart cities: a review of trends, architectures, components, and open challenges in smart cities. Sustain. Cities Soc. **38**, 697–713 (2018)
3. CRS, The Internet of Things (IoT): An overview. https://crsreports.congress.gov/product/pdf/IF/IF11239. Accessed 7 Feb 2021
4. UNESCO and NETEXPLO: Smart cities: shaping the society of 2030. United Nations Educational, Scientific and Cultural Organization (UNESCO), Paris, France (2019). (ISBN 978–92–3–100317–2)
5. Chianese, A., Piccialli, F., Jung, J.-E.: The internet of cultural things: towards a smart cultural heritage. In: 2016 12th International Conference on Signal-Image Technology & Internet-Based Systems (SITIS), pp. 493–496 (2016)

6. Lerario, A.: The IoT as a key in the sensitive balance between development needs and sustainable conservation of cultural resources in Italian heritage cities. Sustainability **12**(17), 6952 (2020)
7. Lerario, A., Varasano, A.: An IoT smart infrastructure for S. Domenico church in matera's "Sassi": a multiscale perspective to built heritage conservation. Sustainability **12**(16), 6553 (2020)
8. Nisiotis, L., Alboul, L., Beer, M.: A prototype that fuses virtual reality, robots, and social networks to create a new cyber–physical–social eco-society system for cultural heritage. Sustainability **12**(2), 645 (2020)
9. Yu, W., et al.: A survey on the edge computing for the internet of things. IEEE Access **6**, 6900–6919 (2017)
10. Sittón-Candanedo, I., Alonso, R.-S., Corchado, J.-M., Rodríguez-González, S., Casado-Vara, R.: A review of edge computing reference architectures and a new global edge proposal. Future Gener. Comput. Syst. **99**, 278–294 (2019)
11. Sittón-Candanedo, I., Alonso, R.-S., García, O., Muñoz, L., Rodríguez-González, S.: Edge computing, IoT and social computing in smart energy scenarios. Sensors **19**(15), 3353 (2019)
12. Alonso, R.-S., Sittón-Candanedo, I., García, O., Prieto, J., Rodríguez-González, S.: An intelligent edge-IoT platform for monitoring livestock and crops in a dairy farming scenario. Ad Hoc Netw. **98**, 102047 (2020)
13. Alonso, R.-S., Sittón-Candanedo, I., Casado-Vara, R., Prieto, J., Corchado, J.-M.: Deep reinforcement learning for the management of software-defined networks and network function virtualization in an edge-IoT architecture. Sustainability **12**(14), 5706 (2020)

Correction to: Intrusion Detection System for MQTT Protocol Based on Intelligent One-Class Classifiers

Esteban Jove, Jose Aveleira-Mata, Héctor Alaiz-Moretón, José-Luis Casteleiro-Roca, David Yeregui Marcos del Blanco, Francisco Zayas-Gato, Héctor Quintián, and José Luis Calvo-Rolle

Correction to:
Chapter "Intrusion Detection System for MQTT Protocol Based on Intelligent One-Class Classifiers" in: J. M. Corchado and S. Trabelsi (Eds.): *Sustainable Smart Cities and Territories*, LNNS 253, https://doi.org/10.1007/978-3-030-78901-5_22

In the original version of the book, the following belated correction has been incorporated: The misspelled author name "Francico Zayas-Gato" has been corrected to "Francisco Zayas-Gato" in the Frontmatter, the Backmatter, and Chapter 22.

The updated version of this chapter can be found at
https://doi.org/10.1007/978-3-030-78901-5_22

J. M. Corchado and S. Trabelsi (Eds.): SSCTIC 2021, LNNS 253, p. C1, 2022.
https://doi.org/10.1007/978-3-030-78901-5_30

Retraction Note to: Sustainable Smart Cities and Territories

Juan M. Corchado and Saber Trabelsi

Retraction Note to: Chapter 4 in: J. M. Corchado and S. Trabelsi (Eds.): *Sustainable Smart Cities and Territories*, LNNS 253, https://doi.org/10.1007/978-3-030-78901-5_4

The Series Editor and the publisher have retracted this chapter. An investigation by the publisher found that a number of chapters, including this one, from multiple conference proceedings raise various concerns, including but not limited to compromised editorial handling, inappropriate or unusual citation behavior and undisclosed competing interests. Based on the findings of the investigation, the Series Editor and the publisher no longer have confidence in the results and conclusions of this chapter.

The authors have not responded to correspondence regarding this retraction.

Retraction Note to: Chapter 9 in: J. M. Corchado and S. Trabelsi (Eds.): *Sustainable Smart Cities and Territories*, LNNS 253, https://doi.org/10.1007/978-3-030-78901-5_9

The Series Editor and the publisher have retracted this chapter. An investigation by the publisher found that a number of chapters, including this one, from multiple conference proceedings raise various concerns, including but not limited to compromised editorial handling, inappropriate or unusual citation behavior and undisclosed competing interests. Based on the findings of the investigation, the Series Editor and the publisher no longer have confidence in the results and conclusions of this chapter.

The authors disagree with this retraction.

The retracted version of these chapters can be found at
https://doi.org/10.1007/978-3-030-78901-5_4
https://doi.org/10.1007/978-3-030-78901-5_9
https://doi.org/10.1007/978-3-030-78901-5_19

J. M. Corchado and S. Trabelsi (Eds.): SSCTIC 2021, LNNS 253, p. C3–C4, 2024.
https://doi.org/10.1007/978-3-030-78901-5_31

Retraction Note to: Chapter 19 in: J. M. Corchado and S. Trabelsi (Eds.): *Sustainable Smart Cities and Territories*, LNNS 253, https://doi.org/10.1007/978-3-030-78901-5_19

The Series Editor and the publisher have retracted this chapter. An investigation by the publisher found that a number of chapters, including this one, from multiple conference proceedings raise various concerns, including but not limited to compromised editorial handling, inappropriate or unusual citation behavior and undisclosed competing interests. Based on the findings of the investigation, the Series Editor and the publisher no longer have confidence in the results and conclusions of this chapter.

Author Juan M. Corchado has stated on behalf of all authors that they disagree with this retraction.

Author Index

J. M. Corchado and S. Trabelsi (Eds.): SSCTIC 2021, LNNS 253, pp. 329–330, 2022.
https://doi.org/10.1007/978-3-030-78901-5

The manufacturer's authorised representative in the EU is Springer Nature Customer Service Centre GmbH, Europaplatz 3, 69115 Heidelberg, Germany. If you have any concerns regarding our products, please contact ProductSafety@springernature.com

Printed and bound by CPI Group (UK) Ltd, Croydon, CR0 4YY
30/09/2026
02225448-0012